The Scientific Works of
Robert Grosseteste

The Scientific Works of Robert Grosseteste

*Mapping the Universe: Robert Grosseteste's
De sphera—On the Sphere*

Volume II

Series Editors
GILES E. M. GASPER
TOM C. B. McLEISH
HANNAH E. SMITHSON
SIGBJØRN OLSEN SØNNESYN

*A collaborative monograph with edition, translation, and analysis under
the aegis of the Ordered Universe Research Project*

UNIVERSITY PRESS

Great Clarendon Street, Oxford, OX2 6DP,
United Kingdom

Oxford University Press is a department of the University of Oxford.
It furthers the University's objective of excellence in research, scholarship,
and education by publishing worldwide. Oxford is a registered trade mark of
Oxford University Press in the UK and in certain other countries

© Giles E. M. Gasper, Tom C. B. McLeish, Hannah E. Smithson,
and Sigbjørn Olsen Sønnesyn 2023

The moral rights of the authors have been asserted

First Edition published in 2019

All rights reserved. No part of this publication may be reproduced, stored in
a retrieval system, or transmitted, in any form or by any means, without the
prior permission in writing of Oxford University Press, or as expressly permitted
by law, by licence or under terms agreed with the appropriate reprographics
rights organization. Enquiries concerning reproduction outside the scope of the
above should be sent to the Rights Department, Oxford University Press, at the
address above

You must not circulate this work in any other form
and you must impose this same condition on any acquirer

Published in the United States of America by Oxford University Press
198 Madison Avenue, New York, NY 10016, United States of America

British Library Cataloguing in Publication Data
Data available

Library of Congress Control Number: 2022918369

ISBN 978-0-19-880552-6

DOI: 10.1093/oso/9780198805526.001.0001

Printed and bound by
CPI Group (UK) Ltd, Croydon, CR0 4YY

Links to third party websites are provided by Oxford in good faith and
for information only. Oxford disclaims any responsibility for the materials
contained in any third party website referenced in this work.

This volume is dedicated to Tom McLeish, for his universal vision of academic collaboration and his total commitment to the values and practice of collaborative, interdisciplinary, working, and to whom we all owe so much.

'But where can wisdom be found? Where can we learn to understand?'
Job 28.12

Preface

The second volume of *The Scientific Works of Robert Grosseteste* takes as its focus the treatise *On the Sphere*. It is the longest, by far, of the shorter scientific works, and, by the evidence of over fifty-five surviving manuscript copies, Grosseteste's most popular work. *On the Sphere* is a complex and intricate treatise, more so than it might appear at first glance, or than might be implied by its designation by many commentators as a textbook. Its subject, astronomy, was accorded the highest honour in Grosseteste's earliest known work, *On the Liberal Arts* (edited, translated, and analysed in Volume I of this series). The current volume explores what Grosseteste's intentions were in writing *On the Sphere*, its contents, sources, argument, illustrations, and themes, alongside its wider intellectual setting and historical context.

Addressing these multifaceted themes is made easier, and more stimulating, through multi-disciplinary working. This is exactly the framework adopted and developed by the Ordered Universe Project, under the aegis of which the current volume, as the others in the series, is produced. *On the Sphere* is presented here as a collaboration between medievalists of various specialisms: history, art history, history of science, Arabic studies, philosophy, Latin and palaeography, physics, engineering, vision science, and psychology. The variety of perspectives brought to bear on the elucidation of the text makes for a richer interpretation of the thirteenth-century treatise. Analysis of the mathematical concepts with which Grosseteste operated are integrated with textual and historical investigation, to mutual benefit. The volume offers fresh insight not only on the purpose and scope of the treatise, but on its content, in particular on its individual treatment of trepidation and lunar motion, and Grosseteste's wider activities and intellectual interests in the same period, including the literature of pastoral care in an era famous for church reform.

viii PREFACE

The presentation of *On the Sphere* in this volume emerges from collaborative reading symposia held in Rome and Durham in 2016. The Ordered Universe methodology is founded upon the practice of reading and discussing together in a multi-disciplinary forum. All of Grosseteste's scientific texts to be presented in the series have been subjected to the same process, with draft Latin editions and English translations treated to repeated and in-depth scrutiny by participants representing a wide range of disciplines. Having a particular text as the singular point of focus gives a common end for the variety of insights and questions, and an anchor-point for the discussion. It is from these multi-layered conversations between contemporary disciplinary practitioners, and the resulting interpretation of a medieval thinker and his world, that the lines of approach for commentary and analysis take shape. From collaborative reading the process moves to collaborative writing which takes place over a longer arc.

Alongside the multi-disciplinary nature of its commentary on Grosseteste, the aim of the *Scientific Works* series to is provide modern editions of the shorter treatises on natural phenomena. The new editions will replace those made by Ludwig Baur in 1912. An important contribution in its period, Baur's editions nevertheless suffer from shortcomings, chiefly the limited range of manuscript witnesses identified. In the century since then many more have come to light. In some cases, however, more recent critical editions exist. This is the case for *On the Sphere* with a critical edition from 2001 compiled by Cecilia Panti, and it is this edition which is re-printed in this volume.

On the Sphere is the only one of Grosseteste's scientific treatises to carry manuscript illumination with the text. These extend from an image of the universe in hemisphere common to a high proportion of copies, to far more elaborate schema which appear in a more limited number. The fact that illuminations are present in the medieval tradition opens, in a more direct manner than is possible for other treatises, questions connected to diagrammatic representation and visualization. Geometric content is a regular theme within Grosseteste's works, scientific, pastoral, and theological, and is, unsurprisingly, a dominant feature of *On the Sphere*. The treatise asks the reader to imagine the constructions it describes, from the sphericity of the universe, to the mechanics of solar and lunar eclipses. To aid readers who, like a number within the research group, require a little more assistance with mental geometry, the commentary to the edition in Chapter 4 is accompanied by diagrams to elucidate the places in Grosseteste's text where he discusses geometric constructions. This volume is accompanied also by a virtual three-dimensional model of the phenomena described in the treatise which features as a web resource to accompany the written analysis <https://ordered-universe.com/de-sphera-visualisation/>. An account of the principles behind the model, and a users' guide, are included within this volume. The model was created in 2016 by Jack Smith, then an undergraduate engineering student at Pembroke College, University of Oxford, to assist the interdisciplinary work on the treatise, and has since been developed and refined for publication. It is an interesting and rewarding exercise to read *On the Sphere* at the same time as being able to see the structures and working of the medieval universe visualized through modern computing technology.

One of the more unusual collaborations to emerge in the course of Ordered Universe research has been, since 2015, that with The Projection Studio, UK-based sound and light projection specialists, and with Ross Ashton and Karen Monid in particular. To date the project has inspired seven sound and light shows which have played to audiences of over a million people across Europe, the UK, and the USA. *On the Sphere* holds a special place in this collaboration, with two shows designed around its contents. The first, *Horizon*, was co-commissioned by the Napa Lighted Art Festival, USA, and Light Up Poole in the UK, for January and February 2019 respectively, with Grosseteste's reflections contrasted to contemporary research on bio-luminescence from NASA's Jet Propulsion Laboratory. The second, *Zenith*, was a second commission for Light Up Poole, 2019. Quite apart from delight in the aesthetic qualities of the projections, this collaboration opened up new avenues for public dissemination of research, with various team members taking part in talks and interactive activities to support the projections at Napa and Poole. The engagement with Ashton and Monid goes further. Both attended Ordered Universe symposia, bringing their unique perspectives to the work of interpretation.

In a similar vein, *On the Sphere* featured as the centrepiece for a programme drawing on research by the Ordered Universe Project to encourage access to university from school students, especially from less-advantaged backgrounds. The programme was created under the aegis of OxNet, a national and international initiative to widen and diversify student participation at university run from Pembroke College, University of Oxford. The Ordered Universe's programme was associated especially with the North East of England, working with Southmoor Academy, Sunderland, and a network of schools across the region, growing in number from five to sixteen across the five years that it ran. *On the Sphere* was the set-text for the programme's Easter Schools in 2020 and 2021, with the cohort of students engaging in-depth with the treatise and some of the questions it raises. The quality of their questions and thoughts were such as to challenge their tutors, and it is right to acknowledge the impact of these students in the broader exploration of *On the Sphere* that follows.

Any academic enterprise incurs debts of gratitude, and for one of this scale these are multiple. In the first instance funding from the Arts and Humanities Research Council (UK) for the Ordered Universe Project (reference AH/N001222/1) must be acknowledged; without this support none of the underpinning research activities would have been possible. Thanks are due to the hosts of the symposia at which the primary collaborative readings for *On the Sphere* took place: Professor Cecilia Panti from the University of Rome 'Tor Vergata', and Professor Giles Gasper, Durham University, their respective local support teams, and the then project administrator Dr Rachael Matthews. Aspects of various of the chapters featured in a postgraduate conference involving students from McGill University, Durham University, and the University of Oxford, organized in 2018 as part of the Ordered Universe programme; in a presentation for the medieval seminar at University College London; and in public talks at the University of York and Durham University, and at Thinking 3D: Space & Time at the University of Oxford in 2019, in addition to the Light Festival activities described above. For schools outreach thanks are due to OxNet director, Dr Peter Claus, the overall programme co-ordinator Felix Slade, and the North-East hub co-ordinator Claire Ungley.

This volume includes a number of different visual media. The computer modelling of *On the Sphere* was made possible through a Pembroke College, Oxford, Rokos Awards internship for Jack Smith in 2016 at the

Department of Engineering Science, University of Oxford. This was supervised by Professors Clive Siviour and Hannah Smithson, with additional input from Professors Tom McLeish and Giles Gasper, and Dr Sigbjørn Sønnesyn. Drs Laura Cleaver, and Sarah Gilbert secured permissions for the reproduction of the images of the medieval illustrations of *On the Sphere* and Grosseteste's handwriting from: the Bibliothèque Mazarine, Paris; University Library, University of Cambridge; the British Library; Verdun, Bibliothèque municipale; the Bodleian Library, University of Oxford; Pembroke College Library, University of Cambridge; and the Parker Library, Corpus Christi College, University of Cambridge. A third element in the visual elements to this volume are the illustrations which accompany the translation of *On the Sphere*. These, designed to aid the reader's understanding of the text, were prepared by Professor Brian Tanner and Rosie Taylor. Rosie was also responsible for the map in the first chapter of the volume.

Planning for the volume took place at a workshop part-funded by the University of York and hosted by Professor Tom McLeish at that university's Centre for Medieval Studies, in July 2019. All of the authors within the volume are very grateful for the contributions of all who attended the collaborative reading symposia. They are deeply indebted as well to the continuing support from Oxford University Press, and the good offices and high professional standards of Tom Perridge, Karen Raith, Katie Bishop, Jo Spillane, and Joy Mellor. It is a pleasure in this respect to acknowledge another level of collaboration, in this case between international publishing houses. The series editors are grateful to colleagues at SISMEL for their kind permission to re-print the critical edition of Grosseteste's *On the Sphere* prepared by Cecilia Panti. In addition the authors would like to thank Laura Napran for preparing the index to the volume.

It is exciting to bring *On the Sphere* to a wider audience and as part of a continuing collaboration between arts and humanities disciplines and the natural sciences. It has been the consistent experience of the Ordered Universe Project team that such collaboration brings results which are surprising and unexpected but also always coherent with the task of presenting and analysing an 800-year-old treatise. It is important to stress that the participation of scientists in the production of this series, and in this volume in particular, in no way implies that Grosseteste is

xii PREFACE

presented as a modern scientist in this activity. Nor does the Ordered Universe's interdisciplinary engagement of scholars in the sciences together with those in the humanities imply that his science (his astronomy in this case) is evaluated by modern standards. Rather, the engagement of modern scientists adds more ways in which to sharpen the questions asked of Grosseteste's methods, assumptions, and insights.

That considerable value is added to the interpretation of the treatise though a multi-disciplinary team is shown in what follows. Tracing Grosseteste's steps is a complex task, demanding linguistic analysis, historical context, and the intellectual elucidation of his sources and methods. Knowledge of the physical phenomena that Grosseteste describes, is also important, and it is here that modern scientific perspectives play an integral role in the exegesis. Although the conceptual frameworks in which these phenomena are situated are quite different the phenomena themselves have not changed between the thirteenth century and now. This makes modern science, especially in psychology's focus on human processing of sensory information about such phenomena very useful to the interpretation of textual description of them. In the particular case of astronomy, the predictive (and retrodictive) capacity of the science of planetary positional astronomy has also proved helpful, allowing accurate reconstruction of the historical appearance of the sky (an example in the current treatise addresses Grosseteste's discussion of 'Toledan Trepidation'). In a different vein, as well as requiring humanities traditions of textual investigation, Grosseteste's thinking inspires scientific analysis of the principles on which his discussion of natural phenomena is based. Different perspectives cast different light onto his approach to natural philosophy and the considerable implications to be inferred from the internal logic of his treatments. These include the mathematical structure, and content, of his treatises which is often implied rather than explicit. A notable example in the current treatise is to be found in his discussion of the motion of the moon, which implies a more quantitative content than is at first apparent.[1]

---

[1] For a wider reflection on the interdisciplinary practice of the Ordered Universe, see Giles E. M. Gasper, Tom McLeish, and Hannah E. Smithson, 'Listening between the Lines: Medieval and Modern Science' *Palgrave Communications*, 2 (2016), 16062.

Collaboration across the disciplines contributes to another distinctive, and surprising, element of Ordered Universe research, namely the generation of new research questions for contemporary exploration of physical phenomena. Many of the resulting investigations have been reported in articles in leading scientific journals, all of which have been written collaboratively across the Ordered Universe team.[2] These support the wider research enterprise of the project, further hone the collaborative methodology, and test individual questions or treatises in a process which then informs the production of volumes in this series. A similar class of collaborative publications have been produced by team members for leading history of science journals, as well as contributions to the edited collections on particular themes perhaps more familiar to humanities publishing.[3] In this sense *The Scientific Works of Robert Grosseteste* sits as part of a wider and dynamic collaborative eco-system. All of this brings to the fore a richer and fuller appreciation of Grosseteste's achievement. While A. C. Crombie may have been overly enthusiastic in his claims for Grosseteste as the forerunner of the experimental science, an idea justly criticized by contemporary and subsequent historians of science, there is an important extent to which the connection between modern scientists and the deeper past of their fields should be emphasized, and which Grosseteste's work illustrates particularly well.[4] It is this connection that the current approach seeks to address.

On the Sphere is no simple introduction to astronomy, nor are its purposes merely pedagogical. While Grosseteste was master of his subject, the ultimate purpose for such knowledge is integrally bound to his Christian faith. Some fifteen or so years after the composition of the

---

[2] The most recent, of eight such articles, is J. S. Harvey et al., 'A Thirteenth-Century Theory of Speech', *Journal of the Acoustic Society of America*, 146 (2019), 937–47.

[3] Rebekah C. White et al., 'Magnifying Grains of Sand, Seeds, and Blades of Grass. Optical Effects in Robert Grosseteste's *De Iride* (*On the Rainbow*) (circa 1228–1230)', *Isis*, 112 (2021), 93–107; Giles E. M. Gasper, Brian K. Tanner, Sigbjørn O. Sønnesyn, and Nader El-Bizri, 'Travelling Optics: Robert Grosseteste and the Optics behind the Rainbow', in Christian Etheridge and Michele Campopiano (eds.), *Medieval Science in the North: Travelling Wisdom 1000–1500* (Turnhout: Brepols, 2021), 25–60; Giles E. M. Gasper and Brian K. Tanner, '"The Moon Quivered Like a Snake": A Medieval Chronicler, Lunar Explosions, and a Puzzle for Modern Interpretation', *Endeavour*, 44 (2020), 100, 750: doi:10.1016/j.endeavour.2021.100750.

[4] A. C. Crombie, *Robert Grosseteste and the Origins of Experimental Science 1100–1700* (Oxford: Oxford University Press, 1953).

xiv PREFACE

treatise Grosseteste, in his *Dictum* 89 reflected on the importance of order within the universe:

> For it is the desire for unity that forms the universe in being, preserves it and orders it. Indeed from this desire the elements of this world, and the heavens that contain all things, and the celestial bodies, namely the sun, moon and stars, which are formed into round and spherical figures, which of all figures is the most capacious and beautiful and the strongest against any eventuality, not having any protruding and piercing angle or any cavity that might collect filth, from this same desire they serve the same order of their gathering, so that the heavier and less luminous will be lower, and lighter and purer higher. And still from the same desire all the members of an animate body adhere to depend on a single principle, that is, the heart, or [a principle] proportionate to it.[5]

The geometry of celestial bodies and their intrinsic beauty outlined here is entirely consistent with the natural philosophical positions Grosseteste explores in *On the Sphere* and other related texts, for example, *On Light*. *Dictum* 89 uses the arrangement of astronomical unity as an analogy for pastoral care, and the church's mission to the world. As will be seen in what follows, Grosseteste's considerable expertise in the literature of pastoral care, the responsibility of the church to its congregations, developed conterminously with, but not independently from, his scientific works.

---

[5] Grosseteste, *Dictum* 89: 'Appetitus namque unitatis res universas format, in esse conservat, et ordinat. Ex hoc namque appetitu mundi istius elementa, et celum quod continet universa, corporaque celestia, videlicet sol et luna et sydera, que formata sunt in figuras rotundas et spericas, que est omnium figurarum capacissima et pulcherima, et contra omnes casus munitissima, non habens anguli prominenciam pungentem, nec depressionis cavitatem sordes colligentem, ex hoc eodem appetitu servant eundem collacionis sue ordinem, ut videlicet ponderosiora minusque lucencia sint inferius et leviora et puriora superius. Ex eodem quoque appetitu, omnia corporis animati menbra unico adherent innitunturque principio, cordi videlicet, aut ei proporcionato'.

Contents

List of Figures	xix
List of Boxes	xxiii
List of Tables	xxv
List of Plates	xxvii
Abbreviations and Short Titles	xxix
List of Contributors	xxxiii
Author Contributions	xxxv

Introduction	1
1. Dating *On the Sphere* and Locating Robert Grosseteste	**10**
1. Dating *On the Sphere*	10
2. Historical Context of the Period *c.*1210–*c.*1220	17
3. Where Was Grosseteste?	25
3.1 Archdeacon at Large and a Family Affair?	26
3.2 Chancellor of Oxford?	29
3.3 Grosseteste at Court: Legal Activities	35
3.3.1 Papal Judge-Delegate	35
3.3.2 A Brush with Royal Justice	41
3.3.3 Haughmond Abbey	45
2. Mathematics and Pastoral Care: Contextualizing Grosseteste's *On the Sphere*	**48**
1. MS Savile 21: Contents and Script	50
1.1 Grosseteste's Handwriting	51
1.2 Dating the Grosseteste Sections of MS Savile 21	55
1.3 Resources for Study	57
2. Pastoral Care	60
2.1 The Shaping of Pastoral Care	62
2.2 Grosseteste's *Pastoralia* to *c.*1221	65
2.2.1 *On the Way of Making Confession* (*De modo confitendi*)	66
2.2.2 *Meditations*	69
2.3 Audience and Intention	71
2.3.1 A Vignette from Gerald of Wales	73
2.4 *On the Temple of God* (*Templum Dei*)	76

xvi CONTENTS

3. *On the Sphere*: Manuscripts, Translation,
Historiography, and Synopsis — 83
  1. Manuscript Transmission of *On the Sphere* — 83
  2. Principles of This Translation — 88
    2.1 Particular Problems of the Present Translation — 89
      2.1.1 §1 and Passim: *Figura, Situs, Machina* — 89
      2.1.2 §5 Rationibus Naturalibus et Experimentis Astronomicis — 90
      2.1.3 §13 Cingulus Signorum — 90
      2.1.4 §20 Cenith Capitis — 90
      2.1.5 §31 Sphera Recta et Obliqua — 91
      2.1.6 §40 Aux/Oppositum Augis — 91
      2.1.7 §40 Circulus Egresse Cuspidis — 91
  3. Synopsis — 92

Latin Edition and English Translation of *De Sphera* (*On the Sphere*) — 95
Edited by Cecilia Panti *and translated by* Sigbjørn Olsen Sønnesyn

4. Commentary on the Model of the World Machine — 132

5. Intellectual Inheritances: Absence and Presence — 167
  1. Islamicate Astronomy — 169
  2. Saturn, Jupiter, Mars, Venus, and Mercury: Grosseteste
and the Toledan Tables — 173
  3. Tools for Astronomical Observation in Grosseteste's Lifetime — 185
  4. Latin Traditions — 191
    4.1 The World Soul and the Elements — 192
    4.2 A Tale of Two Spheres — 194
    4.3 Petrus Alfonsi, Sphericity, Eclipses, and the City of Arim — 199

6. Astronomy from Liberal Art to Aristotelian Science in
Grosseteste's Thought — 205
  1. Situating *On the Sphere* — 207
  2. Astronomy as Liberal Art — 208
  3. Astronomy as an Aristotelian Science: The *Posterior
Analytics* — 216
    3.1 ἐπιστήμη as Understanding — 217
    3.2 ἐπιστήμη as a Science — 222
    3.3 The Hierarchical Ordering of Sciences and the Concept of
Subalternation — 223
  4. Grosseteste on Aristotelian Scientific Understanding — 226
    4.1 Subalternation and the Science of Astronomy — 227
    4.2 Demonstration *Propter Quid* and Demonstration *Quia* — 230
    4.3 The Role of Observation — 237

CONTENTS xvii

5. Grosseteste's Account of the Discipline of Astronomy
as a Framework for *On the Sphere* 240

7. Structure, Scope, and Sources of *On the Sphere* 242
1. *On the Sphere* and the Definition of Sphericity: Astronomy
as Subalternated to Geometry 245
2. *On the Sphere* and the Demonstration of Sphericity:
Astronomy Subalternated to Natural Philosophy 250
3. *Situs:* The Basic Reference Points of the Spherical Universe 260
4. Intervals of Time 264
 4.1 The Sun and the Zodiac 265
 4.2 The Sun's Eccentricity 266
5. The Fixed Stars 267
6. The Moon 268
7. The Scope of the Treatise 274

8. Trepidation or Precession: The Turning Point in a
Tradition 279
1. The Precession of the Earth's Axis and the Alternative
of Trepidation 279
2. Historical Background 286
3. *On the Sphere* 293
4. Climate Change and Trepidation 294
5. A Comparison of Historical Observations with Calculations 300

9. Kinematic Descriptions, Epicycles, and Modelling
On the Sphere 305
1. The Ptolemaic System of the *Almagest* 307
 1.1 Ptolemaic Planetary Theory 308
 1.2 Ptolemaic Lunar Theory 310
2. Modern Planetary Kinematics 311
 2.1 Orbits 312
 2.2 Descriptions of Elliptical Orbits 314
 2.3 Epicycles for the Sun-Earth System 318
 2.4 Epicycles for the Earth-Moon System 324
 2.5 Implications 329

10. Illuminating *On the Sphere* 331
1. The World Machine Diagram 335
2. Other Diagrams 346
3. Diagrams and Instruments 352

Concluding Reflections 362

xviii CONTENTS

The Virtual Celestial Model of *On the Sphere*	368
1. Overview	369
Part 1: A Geometric Construction	369
Part 2: The Sun	369
Part 3: Season	370
Part 4: The Zodiac	371
Part 5: Eccentricity	371
Part 6: Trepidation	371
Parts 7 and 8: The Moon and Eclipses	372
2. Instructions for Using the Virtual Celestial Model	372
Appendix 1. *Commissio Cancellarii Universitatis Oxon'*—Concerning the Commission to the Chancellor of the University of Oxford	375
Appendix 2. Manuscripts of *On the Sphere* with Diagrams	379
Bibliography	383
Index	411

List of Figures

1.1. Grosseteste's activities in Herefordshire and environs. Created by Rosie Taylor — 40

4.1. (a) A semicircle drawn about point O on diameter AB. (b) The sphere generated by rotating the semicircle about the diameter AB — 133

4.2. (a) Second semicircle, about the same origin O, of smaller diameter DF. (b) Creation of a second sphere, contained within the first, by rotation of the semicircle around DF — 133

4.3. Cross-section of nested spheres containing the five elements — 134

4.4. (a) A model universe with a non-spherical body. (b) For circular motion of and in a non-spherical body, there must be unfilled space between bodies — 135

4.5. (a) The horizon is the same for persons on a flat earth. (b) The horizon differs, depending on the latitude of a person situated on the earth, if the earth is round in a north-south direction — 136

4.6. The variation in the times of a lunar eclipse demonstrates that the earth is round in an east-west direction — 137

4.7. Diurnal circular motion of the fixed stars about the Pole star — 138

4.8. The principal circles on the celestial sphere as described in the text — 139

4.9. A new *colure* drawn 'according to the sides of a square' — 140

4.10. Position of the ecliptic on the celestial sphere with respect to the equinoctial — 141

4.11. The diurnal motion of the sun, assuming it is in a fixed position with respect to the fixed stars for two possible positions on the ecliptic — 143

4.12. The combination of the sun's diurnal motion and annual motion on the ecliptic results in a tight spiral motion — 143

4.13. The horizon and the visible celestial hemisphere viewed from two different positions on the earth — 144

XX   LIST OF FIGURES

4.14. Horizon as seen by a person below the equinoctial circle. Circles centred on and perpendicular to the polar axis are at right angles to the horizon — 145

4.15. Shadow of a person on the equator when the sun is in the various zodiac signs — 146

4.16. Shadow of a person north of the equator, but south of the tropic of Cancer, points south at the solstice — 147

4.17. Coincidence of the planes of the horizon and ecliptic — 149

4.18. The formation of the 'Midnight Sun' for someone below the Arctic circle when the sun is in the sign of Cancer — 150

4.19. The right sphere of an observer situated at the equator. Here, the stars rise at right angles — 151

4.20. More vertical and more oblique rising — 152

4.21. Sketch of the projection on the heavens of some zodiac and near-ecliptic constellations. (a) The rising of an equinoctial point. (b) The rising of a tropic point — 153

4.22. The eccentric of the sun, its circular orbit not being centred on the earth — 155

4.23. The seven climes in the inhabited part of the earth and the two great oceans — 157

4.24. The trepidation model of Thābit ibn Qurra — 159

4.25. Eccentric of the moon with additional epicycles — 160

4.26. (a) Mean position of moon. (b) Position of coincidence of mean solar and lunar positions. (c) Displacement after one day, corresponding to an angular increment δ. (d) Position of opposition of sun and moon — 161

4.27. Inclination of the moon's orbit with respect to that of the sun — 163

4.28. The phases of the moon — 164

4.29. Apparent and real positions of the moon on the firmament — 165

4.30. Resolution of the diversity of aspects of the moon into latitudinal and longitudinal components — 166

8.1. (a) Precession of the Equinoxes. (b) Trepidation — 284

8.2. Predictions of precession and trepidation — 285

8.3. Comparison of the predictions of Toledan trepidation and modern precession — 301

LIST OF FIGURES xxi

8.B.1.	Elliptical orbit of the earth around the sun	282
8.B.2.	Spherical trigonometric construction	303
9.1.	The planetary scheme of the *Almagest*	309
9.2.	Illustration of Kepler's laws	313
9.3.	Key parameters used to describe the position of a planet relative to the sun	315
9.4.	Graphical representation of the first three epicyclic terms of the sun's orbit	323
9.5.	Comparison of the position of the sun according to different models	324
9.6.	Description of the moon's orbit	328
9.7.	Graphical representation of the first three epicyclic terms of the moon's orbit	329
9.B.1.	Increasingly accurate approximations to a square wave	317
9.B.2.	Fourier Series in two dimensions	317
9.B.3.	A simple epicyclic orbit	319
9.B.4.	Ellipse as the sum of two circles	320
9.B.5.	First five terms in the Fourier Series	322
9.B.6.	Comparison of approximations to an ellipse	322
9.B.7.	Two approximations to an elliptical orbit	327
9.B.8.	Geometric interpretation of equation for a body in orbit around a focus	327
10.1.	British Library, Harley MS 4350, f. 4 © The British Library Board	332
10.2.	British Library, Harley MS 3735, f. 74 © The British Library Board	339
10.3.	Cambridge University Library MS Ff.6.13, f. 18. Reproduced by kind permission of the Syndics of Cambridge University Library	340
10.4.	Oxford, Bodleian Library MS Laud Misc. 644, f. 143 © The Bodleian Libraries, The University of Oxford	344
10.5.	Verdun, Bibliothèque municipal MS 25, f. 31. Reproduced by kind permission of Michaël George	345

xxii LIST OF FIGURES

10.6. Paris, Bibliothèque Mazarine MS 3642 f. 89 © Bibliothèque Mazarine   350

10.7. British Library, Harley MS 3735, f. 23 © The British Library Board   355

10.8. British Library, Harley MS 3735, f. 78v © The British Library Board   357

List of Boxes

8.1.	Milankovitch Cycles	281
8.2.	Notes on Calculation of Right Ascension and Declination under Precession and Trepidation	303
9.1.	Kepler's Laws	314
9.2.	Epicycles and the Fourier Series, Part 1	316
9.3.	Epicycles and the Fourier Series, Part 2	319
9.4.	Epicycles and the Fourier Series, Part 3	320
9.5.	Refinement for Orbital Motion	325

List of Tables

8.1.	Historical Longitude Observations of α Leonis	300
9.1.	Key Parameters of the Earth's Orbit	318
9.2.	A Simplified Lunar Orbit	328

List of Plates

1. Clockwise from top left: Oxford, Bodleian Library MS Savile 21, f. 156v. Cambridge, University Library MS Ff.1.24, f. 42v. Cambridge, Corpus Christi College MS 480, f. 9v. Cambridge, University Library MS Ff.1.24, f. 25r.
2. Cambridge, Pembroke College MS 7, f. ii v.
3. Detail from Cambridge, University Library MS Ff.1.24, f. 107v (above). Cambridge, University Library MS Ff.1.24, f. 229v (below).
4. Oxford, Bodleian Library MS Savile 21, f. 157r.
5. A selection of Grosseteste's Hindu-Arabic numerals in Oxford, Bodleian Library, MS Savile 21 (above) and Cambridge, University Library MS Ff.1.24 (below).

Abbreviations and Short Titles

AL, *Physica*
: *Aristoteles Latinus* editions: *Physica*, ed. Fernard Bossier and J. Brams, *AL* VII.1, fasc. 2 (Turnhout: Brepols, 1990), 7–340; text from the *Aristoteles Latinus Database*.

Aristotle, *De caelo*
Aristotle, *De generatione et corruption*
Aristotle, *Physica*
Aristotle, *Analytica Posterioria*
Aristotle, *Meteorologica*
: References to Aristotle's work are given in the standard form, referring to: August Immanuel Bekker, *Aristotelis Opera edidit Academia Regia Borussica, ex Recognitione Immanuelis Bekkeri*, 5 vols (Berlin: Georgium Reimerum, 1831–70). English translations from Jonathan Barnes (ed.), *The Complete Works of Aristotle: The Revised Oxford Translation*, 2 vols. (Princeton: Princeton University Press, 1995).

Baur, *Die philosophischen Werke*
: *Die Philosophischen Werke des Robert Grosseteste, Bischofs von Lincoln*, ed. Ludwig Baur (Münster: Aschendorff, 1912).

CCCM
: Corpus Christianorum Continuatio Mediaevalis

CCSL
: Corpur Christianorum, series Latina

CSEL
: Corpus Scriptorum Ecclesiasticorum Latinorum

Grosseteste, *Comm. Post. An.*
: Robert Grosseteste, *Commentarius in Posteriorum Analyticorum libros*, ed. Pietro Rossi (Florence: Olschki, 1981).

Grosseteste, *Comm. Phys.*
: Robert Grosseteste, *Commentarius in VIII Libros Physicorum Aritotelis*, ed. Richard C. Dales (Boulder: University of Colorado Press, 1963).

XXX ABBREVIATIONS AND SHORT TITLES

Grosseteste, *Compotus*	Robert Grosseteste's *Compotus*, ed. and trans. Alfred Lohr and C. Philipp E. Nothaft (Oxford: Oxford University Press, 2019).
Grosseteste, *De artibus liberalibus*	Robert Grosseteste, *De artibus liberalibus*, ed. and trans. Sigbjørn O. Sønnesyn, in Giles E. M. Gasper et al., *Knowing and Speaking: Robert Grosseteste's De artibus liberalibus* ('On the Liberal Arts') and *De generatione sonorum* ('On the Generation of Sounds') (Oxford: Oxford University Press, 2019), 74–95.
Grosseteste, *De sphera* Grosseteste, *De cometis* Grosseteste, *De motu supercelestium*	Cecilia Panti, *Moti, virtù e motori celesti nella cosmologia di Roberto Grossatesta. Studio ed edizione dei trattati 'De sphera', 'De cometis', 'De motu supercelestium'* (Florence: SISMEL—Edizioni del Galluzzo, 2001).
Grosseteste, *Dicta/Dictum*	Robert Grosseteste, *Dicta*, transcription of Oxford, Bodleian Library MS Bodley 798 (SC 2656), ed. Joseph W. Goering and Edwin J. Westermann (2003), https://ordered-universe.com/dicta/
Harrison Thomson, *Writings*	S. Harrison Thomson, *The Writings of Robert Grosseteste, Bishop of Lincoln 1235–1253* (Cambridge: Cambridge University Press, 1940).
Knowing and Speaking	Giles E. M. Gasper et al., *Knowing and Speaking: Robert Grosseteste's De artibus liberalibus* ('On the Liberal Arts') and *De generatione sonorum* ('On the Generation of Sounds') (Oxford: Oxford University Press, 2019).
McEvoy, *Philosophy*	James McEvoy, *The Philosophy of Robert Grosseteste* (Oxford: Oxford University Press, 1982).
MGH	*Monumenta Germaniae Historica*

ABBREVIATIONS AND SHORT TITLES   xxxi

Panti, *Moti, virtù e motori celesti*

Cecilia Panti, *Moti, virtù e motori celesti nella cosmologia di Roberto Grossatesta. Studio ed edizione dei trattati 'De sphera', 'De cometis', 'De motu supercelestium'* (Florence: SISMEL—Edizioni del Galluzzo, 2001).

PL

Patrologiae cursus completus, series Latin, ed. J.-P. Migne, 221 vols. (Paris, 1844–65).

Ptolemy, *Almagest*

J. L. Heiberg, *Claudii Ptolemaei opera quae exstant omnia*, I: *Syntaxis mathematica*, 2 vols. (Leipzig: Teubner, 1898–1903), with English translation in *Ptolemy's Almagest*, trans. Gerald J. Toomer (London: Duckworth, 1984).

Sacrobosco, *De sphera*

John of Sacrobosco, *De sphera*, in Lynn Thorndike (ed. and trans.), *The Sphere of Sacrobosco and Its Commentators* (Chicago: University of Chicago Press, 1949), 76–143.

Thorndike, *The* Sphere *of Sacrobosco*

Lynn Thorndike, *The Sphere of Sacrobosco and Its Commentators* (Chicago: University of Chicago Press, 1949).

Southern, *Grosseteste*

Richard W. Southern, *Robert Grosseteste: The Growth of an English Mind in Medieval Europe*, 2nd ed. (Oxford: Oxford University Press, 1992).

List of Contributors

Laura Cleaver is Senior Lecturer in Manuscript Studies at the School of Advanced Study, University of London.

Jack P. Cunningham is Reader in Ecclesiastical History in the Department of Humanities at Bishop Grosseteste University, Lincoln.

Nader El-Bizri is Dean of the College of Arts, Humanities, and Social Sciences at the University of Sharjah.

Seb Falk is a Fellow of Girton College, University of Cambridge.

Giles E. M. Gasper is Professor of High Medieval History in the Department of History at Durham University.

Sarah Gilbert is a Project Cataloguer in the Department of Manuscripts at Cambridge University Library.

Sarah Griffin is Assistant Archivist, Lambeth Palace Library, London.

Anne Lawrence-Mathers is Professor of Medieval History in the Department of History at the University of Reading.

Tom C. B. McLeish FRS was Professor of Natural Philosophy Emeritus in the Department of Physics and at the Centre for Medieval Studies and Humanities Research Centre at the University of York.

C. Philipp E. Nothaft is a Research Fellow at Trinity College Dublin.

Clive R. Siviour is Professor of Engineering Science in the Department of Engineering Science and Tutorial Fellow at Pembroke College, University of Oxford.

Jack Smith is an Engineering Science Graduate from the University of Oxford.

Hannah E. Smithson is Professor of Experimental Psychology in the Department of Experimental Psychology and Tutorial Fellow at Pembroke College, University of Oxford.

Sigbjørn Olsen Sønnesyn is Lecturer in Medieval Christianity in the Department of Religion and Theology, University of Bristol.

Brian K. Tanner is Emeritus Professor of Physics in the Department of Physics at Durham University.

Rosie Taylor is an Artist.

David Thomson is Honorary Research Fellow in the Department of History at Durham University.

Author Contributions

Mapping the Universe takes the form, as do all volumes in the *The Scientific Works of Robert Grosseteste*, of a jointly-written monograph, with edition and English translation, in this case, of Robert Grosseteste's *On the Sphere*. As such the chapters are presented not as contributions to an edited volume of essays but as integral elements to the whole volume. This method stems from the collaborative reading symposia of the Ordered Universe Research Project and is extended to joint publication and collaborative writing. All individual contributions to the volume are recorded below, an important statement, especially for early career colleagues, of the time and expertise the authors have devoted to the volume, in its parts and as a whole. All would agree, nevertheless, that the experience of co-writing produces something much greater than the individual contribution represents. Although difficult to quantify, collaboration at this level is a powerful act of academic generosity and the courage and willingness of the authors to undertake joint publication in this way should be acknowledged. It is an unusual mode for arts and humanities, and though collaborative writing is standard practice in the natural sciences the scale of what is offered here is not. The management of the volume is no small enterprise either. In all chapters one of the authors took an editorial lead for clarity of purpose. General management and editorial oversight of the volume was provided by Gasper with assistance from Sønnesyn, Smithson, and McLeish, the other series editors. All authors contributed to review and scrutiny of chapters, some many times over; the bibliography was compiled by Thomson.

Preface and Introduction	authorship divided between Gasper, Sønnesyn, McLeish, and Smithson. Lead was Gasper.
Chapter 1	authored by Gasper, with artwork from Taylor.
Chapter 2	principal author was Gasper, with contribution from Sønnesyn to the sections on pastoral care; material on Savile 21 provided by Gilbert, with additional

xxxvi AUTHOR CONTRIBUTIONS

	material from Nothaft and Lawrence-Mathers. Chapter lead was Gasper.
Chapter 3	principal author was Sønnesyn for description of manuscripts and principles of translation; Gasper for historiographical notes; Thomson for synopsis. Chapter lead was Gasper.
Latin edition	prepared by Panti.
English translation	prepared by Sønnesyn.
Chapter 4	principal author was Tanner, with substantial contribution from McLeish and Falk, additional contribution from Sønnesyn, artwork from Taylor. Chapter lead was Tanner.
Chapter 5	authorship was divided between El-Bizri on Islamicate astronomy; Lawrence-Mathers on astrology; Nothaft on observation; Sønnesyn, McLeish, and Gasper on Latin traditions, with Cunningham on the *anima mundi*. Chapter lead was Gasper.
Chapter 6	authored by Sønnesyn.
Chapter 7	authored by Sønnesyn.
Chapter 8	principal authors were McLeish and Nothaft, with substantial contribution from Tanner. Chapter lead was McLeish.
Chapter 9	principal authors were Siviour and McLeish. Chapter lead was McLeish.
Chapter 10	principal authors were Cleaver and Griffin. Chapter was co-led.
Concluding Reflections	authorship divided between Gasper, McLeish, Sønnesyn, and Smithson. Co-led by Gasper, McLeish, Sønnesyn.
Virtual Celestial Model	principal authors were McLeish, Siviour, Smith, with contribution from Gasper. Chapter lead was McLeish.
Appendix 1	principal authors were Sønnesyn and Gasper.
Appendix 2	principal authors were Cleaver and Griffin.

Introduction

Robert Grosseteste's *On the Sphere* is a more complex treatise than it might at first appear. It has been unfavourably compared to Grosseteste's later works as well as to rival astronomical texts of a similar or earlier date. Yet in quantitative terms the treatise appears as Grosseteste's greatest literary success. It survives in more copies than any other of Grosseteste's works with over fifty identified to date (Ch. 3, §1), this in comparison to the *Compotus* which has thirty-eight and the *Commentary on Posterior Analytics* with thirty-two. By contrast the other shorter scientific works survive in far fewer numbers, five for *On the Liberal Arts*, seven for *On the Generation of Sounds*, and fourteen for *On Light*. The *Hexaemeron*, the most complete and finished work by Grosseteste alongside the *Commentary on Posterior Analytics*, survives in only seven manuscripts. The chronological range of the surviving manuscripts, as well as its presence in five sixteenth-century printed editions, also show that *On the Sphere* was copied and consulted throughout the later Middle Ages and into the Early Modern period. That Grosseteste's *On the Sphere* was popular can, therefore, be asserted with some confidence.

What is not as clear is the purpose for which Grosseteste wrote his treatise. The popularity of the work alone might suggest an authorial intention for a broader audience rather than more specialist readers. From this position it would be a not unreasonable inference that *On the Sphere* was intended as a useful and perhaps introductory text. How Grosseteste's treatise is understood is also affected by the interpretation of its relationship to a near-contemporary work of the same title by John of Sacrobosco. The relationship between the two works, and in particular their respective chronologies and the use that one made of the other (Grosseteste of Sacrobosco), is discussed in more detail later (Ch. 1, §1). For the present purpose it is worth noting that Sacrobosco's

2 THE SCIENTIFIC WORKS OF ROBERT GROSSETESTE

treatise was certainly intended as an elementary text as its structure and pedagogical techniques, for example mnemonic verses, indicate (Ch. 5, §4.2), to say nothing of the tradition of glossing and commentary that it generated subsequently, and its place in university curricula.[1] The manuscripts in which it survives are, as Lynn Thorndike put it, 'legion', and its early modern reception equally pervasive.[2] The connection between the two treatises *On the Sphere* extended far beyond the circumstances of composition. Within the far smaller manuscript corpus for Grosseteste's text it is common to find it placed alongside that by Sacrobosco, as well the works on computus by both authors (Ch. 3, §1). These circumstances might also lead to the assumption that contemporaries understood the treatises to be similar in design, and therefore that Grosseteste's *On the Sphere* had the same intention as that of Sacrobosco.

Many of these assumptions about Grosseteste's treatise are to be found in two influential studies, by Thorndike, and by Sir Richard Southern. For Thorndike the two treatises, although differently executed (Grosseteste's being shorter, rougher, and unfinished, Sacrabosco's longer, polished, and complete), exhibit the same broad plan and order.[3] While original elements to Grosseteste's work, such as trepidation, are noted, Thorndike placed greater emphasis on the fundamental similarities of the two works. Southern's description of the purpose of *On the Sphere* is unfortunate. He characterized the treatise as both a comprehensive guide to astronomy and as an elementary text.[4] Although Southern highlights Grosseteste's independent thought, in this case the manner in which he approached his sources, such as the focus on Thābit's correction of Ptolemy by 'experiment', further comments reinforce an interpretation of *On the Sphere* as unbalanced by idiosyncrasy. It is Grosseteste's 'strong preference for the visible and the

---

[1] Thorndike, *On the Sphere*, 18–41, 42–6.

[2] Thorndike, *On the Sphere*, 74. Matteo Valleriani (ed.), *De sphaera of Johannes de Sacrobosco in the Early Modern Period* (Cham: Springer, 2020); Kathleen Crowther, Ashley Nicole McCray, Leila McNeill, Amy Rodgers, and Blair Stein. 'The Book Everybody Read: Vernacular Translations of Sacrobosco's *Sphere* in the Sixteenth Century', *Journal of the History of Astronomy*, 46 (2015), 4–28; Corinna Ludwig, 'Die Karriere eines Bestsellers. Untersuchungen zur Entstehung und Rezeption der Sphaera des Johannes de Sacrobosco', *Concilium medii aevi*, 13 (2010), 153–85; Jürgen Hamel, *Studien zur 'Sphaera' des Johannes de Sacrobosco* (Leipzig: AVA, Akademische Verlagsanstalt, 2014).

[3] Thorndike, *On the Sphere*, 10–13.   [4] Southern, *Grosseteste*, 142.

INTRODUCTION 3

concrete' which inspires his regular appeal to visual images. The treatise also includes 'rash speculation' on the habitability of the earth, and finally, an overly long account of solar and lunar motion.[5] The latter was caused, Southern suggests, by Grosseteste's particular fascination with eclipses, not for reasons of astrology but to account for the shape of the moon and the behaviour of light. The unusual structure of *On the Sphere* is then used by Southern as a reason for the relative lack of success for the treatise as a schools text, with comparison to the very different reception of Sacrobosco's treatise.

There is a circularity in these arguments which stems from the assumptions that *On the Sphere* was intended as a general, and introductory, account of the astronomical system, and that it therefore serves as a poor relation to Sacrobosco's treatise.[6] However, rather than operating from the basis that Grosseteste's *On the Sphere* is a less successful 'textbook' than Sacrobosco's, or that it was designed to be one in the first place, it is perhaps more fruitful to think about what it is that Grosseteste actually sets up in his treatise. Bruce Eastwood pointed the way in a review of Southern's treatment, remarking that:

> The description of Grosseteste's *De sphaera* as a brief outline for beginners is incorrect. No medieval text attempting to explain trepidation can be called a primer. Nor is it helpful to say that Grosseteste found Thābit ibn Quarra to have disproved Ptolemy experimentally... Grosseteste presumably understood that Thābit had made a more sophisticated model of stellar motion to account for a supposed variation in the rate of precession, which Ptolemy considered a constant rate.[7]

Southern did accept that an interpretation of the intention of *On the Sphere* as a school textbook was mistaken, eventually instead suggesting it was meant for private study (which its reception somewhat belies); he consistently maintained that Grosseteste's treatise was 'quite unsuitable

---

[5] Southern, *Grosseteste*, 145–6.   [6] Southern, *Grosseteste*, 145–6.
[7] 'Robert Grosseteste: The Growth of an English Mind in Medieval Europe by R. W. Southern', review by Bruce Eastwood, *Speculum*, 63 (1988), 233–7.

4 THE SCIENTIFIC WORKS OF ROBERT GROSSETESTE

for the schools'.[8] The most extensive recent discussion of the treatise, that by Cecilia Panti, also notes the introductory nature of the treatise. She calls particular attention to the elementary and superficial qualities of both Sacrobosco's and Grosseteste's work, especially with respect to the knowledge shown of Ptolemy's *Almagest* in comparison to that shown in Arabic summaries also available in Latin translation, notably by Alfraganus, and in Geber's critique of the Ptolemy. While this is true, however, neither work states any particular aim to provide an introduction to the *Almagest*.[9]

What *On the Sphere* was designed to do remains an open question and one that is explored over the course of the present volume. Grosseteste sets out his own aim succinctly: 'Our purpose in this treatise is to describe the shape of the world machine and the [relative] position and shapes of its constituent bodies, and the movements of higher bodies and the shapes of their orbits (DS §1)'. In carrying out his intentions Grosseteste concentrated on particular aspects of astronomy, from spherical geometry and various proofs for the sphericity of the world, to the measurement of night and day. While all constituent bodies of the world machine are mentioned, the fixed and wandering stars (planets), his focus is on two in particular, the sun and the moon. The movements of both are explored in some detail and the mechanics of how eclipses occur. Throughout the treatise Grosseteste discussed the whole universe without restriction, that is to say he treats the super- and sub-lunary realms together and of a piece.

This approach to the whole universe is illustrated neatly in the case of one of the most original elements of *On the Sphere*, the discussion of the trepidation of the eighth sphere or 'sphere of fixed stars' (Ch. 8). Grosseteste does indeed appear to be one of the first Latin authors to use this theory in detail, which he encountered, almost certainly, in a treatise *On the Motion of the Eighth Sphere*, translated by Gerard of Cremona, and attributed, wrongly, to the ninth-century astronomer Thābit ibn Qurra. Amongst Islamicate astronomers trepidation was used to address an apparent variation in the precession of the stars, that is, in

---

[8] Southern, *Grosseteste*, lix, n. 47, and 146.
[9] Panti, *Moti, virtù e motori celesti*, 45, 67–132 at 88.

INTRODUCTION 5

their movement and displacement with respect to the equinoxes, that had arisen through historical observational errors. Grosseteste, however, does not merely treat trepidation as a model for explaining the observable movements of celestial bodies, but also for explaining the observable climatic effects of these celestial movements on the various regions of the earth (Ch. 8, §4).

Where Grosseteste encountered *On the Motion of the Eighth Sphere* is an intriguing question, and of a sort that is difficult to answer with currently available information. He was not, so far as can be ascertained, attached formally to any institution with a library, and he was a secular cleric as opposed to a monk. Presumably he acquired his own books and booklets as opportunity and circumstances allowed, perhaps in France, possibly Paris, during the interdict, perhaps elsewhere. This makes a manuscript now in the Bodleian Library, Oxford, MS Savile 21, of particular interest. S. Harrison Thomson identified Grosseteste's handwriting within the manuscript, a claim that later scholarship has accepted, albeit somewhat tentatively. The first section in Grosseteste's hand includes two horoscopes calculated for a date in 1216, which gives at least the possibility that he copied out the other works in this section at or around the same time. These works include *On the Motion of the Eighth Sphere* and mathematical treatises by Jordanus. All of this was reading matter well suited to the subject of *On the Sphere*. The case for Grosseteste's handwriting is reviewed thoroughly in what follows (Ch. 2, §1), allowing a firmer insight into the resources he had gathered by 1216, and providing additional evidence for dating *On the Sphere* to some point between 1216 and 1219 (Ch. 1, §1).

The date of the treatise also involves issues connected to Grosseteste's familiarity, or not, with other sources. The striking absence of *On the Motions of the Heavens* by al-Biṭrūjī (d. c.1204 CE) a work which Grosseteste clearly knew in the 1220s, and which had been translated into Latin in 1217 by Michael Scot, allows further speculation on the date of *On the Sphere* (Ch. 1, §1). The relationship between Grosseteste's treatise and that of the same name by Sacrobosco is also explored in some detail in what follows (Ch. 1, §1; Ch. 5, §4.2; and Chs. 6 and 7). *On the Sphere* shows Grosseteste's critical engagement with Ptolemy's *Almagest*, in whatever form it was known to him (Ch. 7, §1;

6 THE SCIENTIFIC WORKS OF ROBERT GROSSETESTE

Ch. 9, §1), and its legacy in Islamicate astronomy. Grosseteste makes use especially of al-Farghānī (d. 861 CE) and that independently of Sacrobosco (Ch. 7), as well as Thābit.

The two foundational thinkers for Grosseteste's *On the Sphere* are, however, Euclid and Aristotle. The treatise is characterized by its sustained application of Aristotelian notions of science and of Euclidean geometry. As set out in Chapters 6 and 7, Grosseteste deploys Aristotle's principles of demonstration alongside Euclid's geometrical axioms and propositions. This speaks to a longer acquaintance with Aristotle's natural philosophy, for example the *Posterior Analytics*, than has been emphasized by some in the scholarly tradition.[10] Latin translations of the majority of these works had been available since at least the mid-twelfth century. By taking seriously Grosseteste's Aristotelian formation and command of Euclid the structure and scope of *On the Sphere* become clearer. These intellectual engagements speak also to Grosseteste's own intellectual development. *On the Sphere* sits in continuity with elements of his first treatise *On the Liberal Arts*, which gave pride of place to the art of astronomy.[11] The continuities are important (Ch. 6, §2), and the later treatises should be read with knowledge of its predecessor in mind. Nevertheless, *On the Sphere* also marks a transition away from the liberal arts framework and towards natural philosophy, a transition mirrored across the thirteenth-century university curriculum.

Whether or not Grosseteste intended the treatise for a classroom is, strictly speaking, impossible to establish, though the evidence of its later transmission, and in a limited context illustration (Ch. 10), implies a work frequently consulted, which is consistent with a schools' text. The invitation Grosseteste repeats 'to imagine' the geometry set down in the treatise and the structures and movements it describes also indicates some form of student audience. This is honoured in the creation of the Virtual Celestial model which accompanies this volume in an online format and also in the sequence of diagrams to explain key aspects of the treatise (Ch. 4). What can be known or suggested of Grosseteste's

---

[10] James McEvoy, 'The Chronology of Robert Grosseteste's Writings on Nature and Natural Philosophy', *Speculum*, 58 (1983), 614–55; at 617 McEvoy presents Grosseteste in *On the Sphere* as 'still far from being the complete Aristotelian scholar'.

[11] Grosseteste, *De artibus liberalibus*, §§11–13.

whereabouts during the probable period of composition (Chs. 1, 2) certainly does not preclude a notion of the treatise as a teaching text, though for whom remains unknown. *On the Sphere* can, then, be placed in a broader context in terms of Grosseteste's career.

While the evidence is patchy, and this is not at all unusual, strong arguments can be made for a continued association with Hereford diocese, as advanced by Southern and Joseph W. Goering, and deepened in this volume (Ch. 1, §3). *On the Sphere* was written in years of political upheaval, military campaigns, and the first movement to royal recovery in England, and in the context of renewed and vigorous reform of the church led by a confident papacy. In this context it is also possible to place some of Grosseteste's other early writings, the treatises on pastoral care, and the obligation of the church to educate its clergy and laity as to their Christian responsibilities, the consequences of sin, mechanisms for restitution, and the economy of salvation. Taking these works into consideration adds a different dimension to thinking about Grosseteste's shorter scientific works and *On the Sphere* in particular (Ch. 2, §2). Although the treatise is counted as the third work in Grosseteste's scientific canon, he had also written, quite probably, two or three treatises on pastoral care over the same period. *On the Sphere* is the product of an experienced writer, in his forties, and, as far as the evidence shows, with a supportive network and patron in Hugh Foliot, archdeacon, and, by 1219, bishop of Hereford.

On the Sphere was part of a larger movement within medieval European astronomical learning and was composed in a period of considerable intellectual change. Longer inheritances from ancient Greek and medieval Islamicate thinking, made available through Latin translation, transformed the way in which astronomy was understood, taught, and practised (Ch. 5). Grosseteste's contribution to this movement was, however, independent and distinct in its structure, scope, and intentions, and to that extent the treatise is both unusual and misunderstood. Grosseteste would go on to explore the calculation of time, the elements, celestial and meteorological phenomena, the formation of the universe, and the action and activity of light and rays.[12] *On the*

---

[12] See McEvoy, 'The Chronology', 617.

8 THE SCIENTIFIC WORKS OF ROBERT GROSSETESTE

Sphere establishes the physical arena in which those other interests would play out.

As with the first volume in this series more complex mathematical discussion is presented throughout in the form of inset boxes as well as tables and figures. This is an instantiation of the interdisciplinary approach adopted for the elucidation of Grosseteste's scientific works, which flows from the collaborative reading symposia which constituted the primary forum for encountering and understanding the text. All authors of *Mapping the Universe* have brought their particular disciplinary expertise to the task of collaborative writing as well as the experience of translating and sharing their insights in the reading sessions. Writing together is a different exercise to reading, and a very different one over the compass of a monograph-length study. It is, however, a methodology that the Ordered Universe Research Project has championed as a means to draw out the richness in breadth and depth of Grosseteste's thought, the wider culture and networks in which this is to be understood, and the implications of medieval exploration of the world around them for modern questions of the same phenomena.

The current volume, unlike that which it succeeds and those that will follow, is dedicated to a single treatise. Since this is a jointly-written study, in its parts, and, more distinctively, as a whole, the individual chapters are listed as for a conventional monograph study. Most chapters are the product of collaborative writing and the contributions of individual authors are indicated in the front matter. Internal references to *On the Sphere* are made in-text in what follows and indicated by paragraph number for both Latin edition and modern English translation. Cross-references to pertinent discussion in other parts of the volume are used throughout, these are again in-text, and are indicated by chapter and section numbers. A cross-reference within a chapter is indicated by section number alone. Names of ancient and medieval authors have been retained in their conventional English-language forms. With respect to Arabic-language authors these are named by a simplified transliteration from Arabic, or where appropriate, most commonly in discussion of their reception by medieval Latin authors, in their conventional Latinate forms (Albumasar, Alfraganus).

INTRODUCTION 9

What follows is an elucidation of a remarkable treatise, from the edition and translation, to the explanation of astronomical terms, analysis of its striking features, for example trepidation, and its larger structures and arguments. To this are added consideration of the diagrammatic tradition, the longer conceptual frameworks which Grosseteste inherited, directly and indirectly, and the historical context in which he and his writing can be placed. The dialogue between now and then finds sharper focus in the dialogue between the authors of the volume, from their individual perspectives, and a wider collaboration between natural sciences and humanities. It is in this spirit that the volume was conceived and produced, and it is in the same spirit that it is offered to wider readership.

1

Dating *On the Sphere* and Locating Robert Grosseteste

Robert Grosseteste's treatise *On the Sphere* sets out, according to its opening lines, to describe both the shape of the *machina mundi*, the 'world machine', that is the universe, and the positions, shapes, and movements of the higher bodies (DS §1). While the treatise is not a comprehensive account of astronomy (see Intro. and Chs. 6 and 7), its compass is, nevertheless, wide-ranging, mapping both particular and general features of the celestial arena looking from the earth outwards and upwards. The intellectual arc of the treatise will be explored in later chapters. In what follows, the particular issues connected with dating will be considered, alongside an analysis, which extends into the subsequent chapter, of Grosseteste's activities in the historical record and his other intellectual interests over the same period. This includes, most notably, the literatures of pastoral care and mathematics, the latter with respect to the contents of Bodleian Library, MS Savile 21, copied, it can be closely argued, by Grosseteste himself. The material presented here takes both a wider and in some specific areas a more local perspective, matching the scope of Grosseteste's treatise: the contemplation of the cosmos and the emphasis laid on this activity from an individual human perspective.

1 Dating *On the Sphere*

While it is difficult to make a precise dating for the *On the Sphere*, sound arguments can be made for the decade between 1215 and 1225, and plausible suggestions for the narrower range of 1217–21. At the beginning of the fourteenth century Nicholas Trevet noted that Grosseteste had composed the treatise, alongside the *Compotus*, a commentary on

DATING *ON THE SPHERE* AND LOCATING ROBERT GROSSETESTE   11

Aristotle's *Posterior Analytics* and other philosophical works, while master of arts. This places *On the Sphere* within the first of the three periods by which Grosseteste defined his career: cleric, master of theology and priest, and then bishop.[1] Direct medieval record for *On the Sphere* offers no further help to its dating, though its survival in a high number of manuscripts speaks to a wide dissemination and reception (see Ch. 3, §1). There is also no explicit evidence for dating within the treatise itself. While a watertight chronology for Grosseteste's scientific canon is neither possible nor sensible to construct given the lack of corroborating evidence, comparison between the works can be used to suggest lines of intellectual development notwithstanding the difficulties that attend this type of analysis. If placed between 1215 and 1225 *On the Sphere* postdates *On the Liberal Arts* and *On the Generation of Sounds*, and, indeed, it clearly draws on, and expands positions set out in the first of these earlier treatises (see Ch. 6, §2).[2] It also seems to predate the *Compotus* on the basis of the source use noted below.[3] This leaves the principal suggestions for the date of *On the Sphere* rooted in the sources used by Grosseteste. A scholarly consensus also places the treatise after *c.*1215 and, as advocated by Cecilia Panti, before *c.*1220.[4]

Central to the suggested dating of *On the Sphere* is the absence of a particular source, namely *On the Motions of the Heavens* by al-Biṭrūjī

---

[1] Nicholas Trivet, *Annales ex regum Angliae, 1135–1307*, ed. Thomas Hog (London: English Historical Society, 1845), 243: 'Qui, cum esset magister in artibus super librum Posteriorum compendiose scripsit. Tractatus etiam de Sphaera et de Arte compoti, multaque alia in philosophia utilia edidit'. On the inadequacies of Hog's edition, see: Frank A. C. Mantello, 'The Editions of Nicholas Trevet's *Annales ex regum Angliae*', *Revue d'histoire des textes*, 10 (1982 for 1980), 257–75; on Trevet see: James G. Clark, 'Trevet, Nicholas (*b.* 1257x65, *d.* in or after 1334)', *Oxford Dictionary of National Biography* (Oxford: Oxford University Press, 2004): http://www.oxforddnb.com/view/article/27744 (accessed 23 February 2021). For Grosseteste's schema, see James Ginther, *Master of the Sacred Page: A Study of the Theology of Robert Grosseteste ca. 1229/30–1235* (Aldershot: Ashgate, 2004), 1, n. 2, the source is *Sermon* 31 (the sermons are unedited, here Ginther uses London, British Library MS Royal 7.E.ii, fol 344rb).

[2] See *Knowing and Speaking*, 11, 14–18, 35, 224–5.   [3] Grosseteste, *Compotus*, 14–19.

[4] The most detailed discussion is that by Panti, *Moti, virtù e motori celesti*, 68 and 69–87. See also Southern, *Grosseteste*, 145 n. 7; McEvoy, *Philosophy*, 506, and his 'The Chronology of Robert Grosseteste's Writings on Natural and Natural Philosophy', *Speculum*, 58 (1983), 614–55 at 618: 'the work need not have been composed more than a few years later than 1215'; Richard C. Dales, 'Robert Grosseteste's Scientific Works', *Isis*, 52 (1961), 381–402 specifically excludes *On the Sphere* from consideration. A. C. Crombie, *Robert Grosseteste and the Origins of Experimental Science 1100–1700* (Oxford: Oxford University Press, 1953), 48, suggests *c.*1220 following Baur who proposed 1215–30, *Die Philosophischen Werke*, 64.

12 THE SCIENTIFIC WORKS OF ROBERT GROSSETESTE

(Latinized as Alpetragius), composed in about 1185 and translated into Latin by Michael Scot. This translation was completed, according to its colophon, in August 1217.[5] The basics of al-Biṭrūjī's argument were familiar to Grosseteste by the time he came to write the *Compotus*. Al-Biṭrūjī's treatise is referred to in its first chapter in a discussion on the motion of the planetary spheres in relation to the length of the year. In the course of this Grosseteste noted that Aristotle and Alpetragius describe such motion without recourse to epicycles:

> And according to [Aristotle] a planet has no intrinsic movement other than the movement of its sphere; and according to him there is no movement on an eccentric or on an epicycle.
>
> And Alpetragius recently devised a mode [of reasoning] and explained how it is possible to save the forward movements, stops, and retrograde movements of the planets and inflections and reflections and all the appearances, by using Aristotle's mode [of reasoning], and without an eccentric and an epicycle.[6]

Here, Grosseteste made explicit the differences between Aristotelian cosmology and its astronomical interpretation by al-Biṭrūjī, and the cosmology espoused by Ptolemy, in whose system epicycles played a central explanatory function for planetary patterns of movement (see Ch. 9).[7] The distinction between the two systems is not highlighted in *On the Sphere*, where Ptolemy's *Almagest* is mentioned as an authority in conjunction with its critical development by Thābit ibn Qurra (Latinized as Thebit) (DS §49), and where it lies behind the discussion of solar and lunar movement. Given the subject matter explored in *On the Sphere*, it seems plausible that had Grosseteste known al-Bitrūjī's

---

[5] Alpetragius (al-Biṭrūjī), *De motibus caelorum*, ed. F. J. Carmody (Berkeley: University of California Press, 1952). The colophon in two versions is transcribed by Philipp Nothaft in Grosseteste, *Compotus*, 16 n. 60.

[6] Grosseteste, *Compotus*, c.1, 64–5: 'Et secundum ipsum planeta non habet motum alium proprium a motu spere sue. Et motus in excentrico et epicycle nihil est secundum ipsum. / Et Alpetragius nuper adinvenit modum et explanavit, quomodo possible est salvare processus et stationes et retrogradationes planetarum et inflexiones et reflexiones et omnia apparentia per modum Aristotelis et absque excentrico et epicyclo'. Translation lightly emended by Sigbjørn Sønnesyn.

[7] See Panti, *Moti, virtù e motori celesti*, 79–82.

DATING *ON THE SPHERE* AND LOCATING ROBERT GROSSETESTE   13

work he would have used it explicitly. That it makes no appearance lends credence to the notion that Grosseteste's *Compotus* is subsequent to *On the Sphere* and that the latter was composed before 1217–*c.*1221. The wider time-frame allows for the dissemination of, and Grosseteste's encounter with, Michael Scot's translation, in which connection a visit to Paris in the early 1220s is probable and links well with the composition of the *Compotus* (see below §2.2).[8]

A second source whose relation to Grosseteste's *On the Sphere* brings implications for dating is the treatise of the same name by John of Sacrobosco. So little is known about Sacrobosco's life that it is difficult to come to any unequivocal judgements on the dates of the works ascribed to him: the *Algorismus*, *Compotus*, a tract on the quadrant, and *On the Sphere*.[9] That said, a position on which of the two treatises *On the Sphere* came first is important to establish since it bears on the sources available to either author. While Ludwig Baur placed Grosseteste's version before that of Sacrobosco, good arguments can be, and have been, made for the reverse, from George Sarton, Lynn Thorndike, and Cecilia Panti, rehearsed and elaborated in this volume (see Ch. 5, §4.2; Ch. 7, §§4 and 6).[10] These arguments include the length of Grosseteste's treatise, which is about half that of Sacrobosco's, and that, while the two have broad similarities in structure and content, Grosseteste develops different lines of thought, uses different terminologies, and adopts a different format.[11] Sacrobosco shows no knowledge of the trepidation of the equinoxes, in contrast to Grosseteste, or the treatise attributed to Thebit *On the Motion of the Eighth Sphere* in which it was formulated, which, as will be shown below, Grosseteste may himself have copied. The style of Sacrobosco's text, as Thorndike pointed out, is also closer to twelfth-century pedagogy than the thirteenth.[12] Sacrobosco's text is littered with classical quotations; on the rare occasion that Grosseteste included such he truncated, oddly, verses cited in full by

---

[8] This will be explored in more detail Volume III of this series.
[9] Thorndike, *The* Sphere *of Sacrobosco*, 3–4.
[10] Baur, *Die Philosophischen Werke*, 64; George Sarton, *Introduction to the History of Science*, 2.2 (of 5 parts in 3 vols) (Baltimore: Carnegie Institute of Washington, 1931) 584; Thorndike, *The Sphere of Sacrobosco*, 10–14; Panti, *Moti, virtù e motori celesti*, 69–75.
[11] Thorndike, *The* Sphere *of Sacrobosco*, 10–11, 13; Panti, *Moti, virtù e motori celesti*, 69–70.
[12] Thorndike, *The* Sphere *of Sacrobosco*, 5.

14   THE SCIENTIFIC WORKS OF ROBERT GROSSETESTE

Sacrobosco from Virgil and Ovid (DS §16).[13] Absent too from Grosseteste's treatise are mnemonic verses for the reader, an absence which indicates different audiences for the two works. That Sacrobosco's treatise was more successful as a teaching text is borne out by its far greater number of surviving manuscripts and its extensive, later, commentary tradition.[14]

If Sacrobosco's treatise *On the Sphere* preceded that by Grosseteste, then its composition can be posited as before *c*.1215. On the face of it this is not incompatible with what little is known of Sacrobosco's life, but Thorndike and Olaf Pedersen present later dates ranging from *c*.1220 to *c*.1230. In the case of the first date, *c*.1220, this suggestion was made tentatively, and was founded on the notion that Grosseteste's *On the Sphere* might be placed at *c*.1224, a date linked, incorrectly as it now seems, to his rectorship of the Franciscans at Oxford.[15] However, Thorndike's suggestion was merely to show a plausible framework. His arguments for the earlier dating of Sacrobosco work equally well if the composition of both works is pushed slightly earlier then *c*.1220, as suggested by Grosseteste's non-use of al-Bitrūjiī.

The second dating suggestion for Sacrobosco's *On the Sphere*, that of *c*.1230, was made by Pedersen, and largely on two grounds.[16] First, the relation between Sacrobosco's *Compotus*, which can be dated to around 1232 or 1235 on the grounds of a calculation for the incarnation and a reference to the present year, and the suggestion that it postdated his *On the Sphere*.[17] Second, the author and character of the earliest commentary on Sacrobosco's treatise, the arguments around the dating of which also assume that Sacrobosco was active at the University of Paris. A connection to Paris is certainly plausible, although what weight should be given to the evidence is more problematic. Sacrobosco, was,

---

[13] Panti, *Moti, virtù e motori celesti*, 71–3.

[14] Thorndike, *The* Sphere *of Sacrobosco*, 18–41, 74. On the remarkable reception of Sacrobosco's *On the Sphere*, see the work of 'The Sphere: Knowledge System Evolution and the Shared Scientific Identity of Europe': https://sphaera.mpiwg-berlin.mpg.de, accessed 12 April 2021; project outputs include *De sphaera of Johannes de Sacrobosco in the Early Modern Period: The Authors of the Commentaries*, ed. M. Valleriani (London: Springer Nature 2020).

[15] Thorndike, *The* Sphere *of Sacrobosco*, 14.

[16] O. Pedersen, 'In Quest of Sacrobosco', *Journal for the History of Astronomy*, 16 (1985), 175–220.

[17] Pedersen, 'In Quest of Sacrobosco', 187–9.

DATING *ON THE SPHERE* AND LOCATING ROBERT GROSSETESTE   15

seemingly, buried in the church of St Mathurin, and Bartholomew of Parma noted in his commentary on *On the Sphere* written at the end of the thirteenth century that Sacrobosco wrote it while at the University of Paris.[18] He was not mentioned, however, by any contemporaries. The earliest commentary on Sacrobosco's treatise has been taken to indicate a Parisian milieu for the treatise as the city is named in the closing sections.[19] The commentary is attributed to Michael Scot, who died in 1235.[20] It is also replete with references to Aristotle's natural philosophy, which, if it were to have been produced in Paris, suggest a date of either before the 1210 ban on the public teaching of these works or after its repeal in 1231.[21] Nevertheless, the attribution to Michael Scot is not certain.[22] Amongst other things the commentary makes no mention of Thebit or trepidation, which while consistent with Sacrobosco's text, might have been expected in a commentary by Michael Scot; if he was not the author of the commentary then the suggestions for dating disappear. And, even if he was the author all that is established, strictly speaking, is that Sacrobosco's *On the Sphere* existed *by c.*1231. Pedersen adopted this date with more enthusiasm, drawing close to an argument made by Thorndike that the composition of Sacrobosco's *Compotus* and *On the Sphere* in close chronological proximity was more plausible than a longer gap between the two.

Pedersen's arguments for the dating of Sacrobosco's *On the Sphere* focus, not unreasonably, on Sacrobosco's works. The relationship to Grosseteste's *On the Sphere* is not, however, covered in any great detail, offering a similar judgement to Southern's that it is difficult to tell which author might have used the other.[23] Nevertheless, Pedersen also states

[18] Thorndike, *The Sphere of Sacrobosco*, 2, for Bartholomew's remarks, and 28–9 for further information on his commentary.

[19] Thorndike, *The Sphere of Sacrobosco*, 21.

[20] Lynn Thorndike, *Michael Scot* (London: Nelson, 1965), 38.

[21] Pedersen, 'In Quest of Sacrobosco', 192; for the 1210 prohibition, see *Chartularium Universitatis Parisiensis*, i, ed. Heinrich Denifle and Émile Chatelain (Paris: Delalain, 1889), Pars prima, 70, no. 11.

[22] Pedersen, 'In Quest of Sacrobosco', 192; Thorndike, *The Sphere of Sacrobosco*, 23. See more recently C. A. Musatti, 'Alcune considerazioni sulla paternità del commento alla Sphaera di Giovanni Sacrobosco attribuito a Michele Scoto', in Pina Totaro and Luisa Valente (eds.), *Sphaera. Forma immagine e metafora tra Medioevo ed età moderna*, Lessico intellettuale europeo, 117 (Florence: Olschki, 2012), 145–65.

[23] Southern, *Grosseteste*, 145 n.7.

16   THE SCIENTIFIC WORKS OF ROBERT GROSSETESTE

that, 'In general there seems to be no serious reason to quarrel with Thorndike's impression that Sacrobosco's work is the earlier.' The implications of the date assigned to Grosseteste's treatise in this connection are not considered. On the logic of Sacrobosco's text being dated to c.1230 and prior to Grosseteste's, this would put the date of composition for the latter into the 1230s. This is not coherent with the evidence and suggestions made for Grosseteste's text marshalled above. If the argument for proximity between Sacrobosco's *Compotus* and *On the Sphere* is relaxed, since it is not demonstrable anyway, and the first commentary on *On the Sphere* is not attached too closely to Michael Scot or chronological proximity to the main text, it is possible to offer more coherent suggestions. Were Sacrobosco to have composed his *On the Sphere* before 1215 this would leave a gap of twenty-years or longer until his *Compotus*; if between 1215 and 1220 then between fifteen and twenty years. Either way there is nothing to suggest that this was not case and such a gap of years is neither implausible nor unlikely. Grosseteste's own works were composed over a considerable span, the shorter scientific works from perhaps as early as 1195 to perhaps as late as 1229/30. There is much, then, to recommend the suggestion that Sacrobosco's *On the Sphere* lies in the first two decades of the thirteenth century. If the Paris connection is sustained it might be suggested further that Grosseteste's period in France during the period of the interdict over England provided ample opportunity for his familiarity with, or acquisition of, the earlier treatise.[24]

A further area for consideration in dating Grosseteste's *On the Sphere* is in a sense ancillary to the principal arguments around the use of al-Biṭrūjī and the relation to Sacrobosco, but with important implications and possibilities. It is the evidence of Bodleian Library, MS Savile 21, discussed in particular by S. Harrison Thomson and Richard Southern in terms of Grosseteste's autograph and his interests in astrology.[25] A full assessment of MS Savile 21 is to be found in the following chapter (Ch. 2, §1). Here it is worth noting that the relevant sections of MS Savile 21, those associated with what is probably Grosseteste's handwriting, can be dated to 1215/16. The

---

[24] *Knowing and Speaking*, 203–22.
[25] Harrison Thomson, *Writings*, 30–3; Southern, *Grosseteste*, 107.

DATING *ON THE SPHERE* AND LOCATING ROBERT GROSSETESTE   17

dating occurs in two horoscopes, one for the vernal equinox of 1216, one for a conjunction of Saturn-Mars in October that same year, but these are accompanied by a range of other texts on astronomy and mathematics, including *On the Motion of the Eighth Sphere*, attributed to Thebit, and translated into Latin by Gerard of Cremona.[26] As noted above, Grosseteste's *On the Sphere* discusses the trepidation of the equinoxes where Sacrobosco's does not, and it is highly likely that Grosseteste's knowledge was drawn from *On the Motion of the Eighth Sphere*. MS Savile 21 gives the distinct probability that Grosseteste knew and copied this text in 1215/16. This offers further circumstantial suggestion for an earlier date for Sacrobosco, and the possibility that Grosseteste's interest in the text formed part of his preparation for *On the Sphere*, which might then be placed somewhat closer to 1215/16 or thereabouts.[27]

2 Historical Context of the Period *c*.1210–*c*.1220

Before moving to the more specific evidence for Grosseteste's activities during the years suggested for the composition of *On the Sphere*, it is important to cast some sense of the wider contexts in which his life and work are to be placed. Within the ambit of Grosseteste's own experience this decade saw the continued incursion of western Europeans into Greece and the Aegean, and the institution of the territories ruled directly from Venice or Genoa, as well as the establishment of the principal Frankish lordships, following the fall of Constantinople to a Latin crusader army in 1204. These included the Latin empire in Constantinople, the kingdom of Thessaloniki, the megaskyrate of Athens and Thebes, and

---

[26] Editions in *The Astronomical Works of Thabit b. Qurra*, ed. Francis J. Carmody (Berkeley: University of California Press, 1960) and José María Millás Vallicrosa, 'El "Liber de motu octave sphere" de Tābit ibn Qurra', *Al-Andalus*, 10 (1945), 89–108. Additional commentary and English translation in Otto Neugebauer, 'Thâbit ben Qurra "On the Solar Year" and "On the Motion of the Eighth Sphere"', *Proceedings of the American Philosophical Society*, 106 (1960), 264–99. See in addition C. Philipp E. Nothaft, 'Criticism of Trepidation Models and Advocacy of Uniform Precession in Medieval Latin Astronomy', *Archive for History of Exact Sciences*, 71 (2017), 211–44.

[27] The possibility that Bodleian Library, MS Savile 21 postdates *On the Sphere* should also be borne in mind, which would place the composition of the treatise to 1215 or even slightly before, see Panti, *Moti, virtù e motori celesti*, 78.

18   THE SCIENTIFIC WORKS OF ROBERT GROSSETESTE

the principality of Achaia.[28] As political units they were volatile, fissiparous, and unstable; the claims of the Latin empire extended over the whole of what was known as 'Romania', the Latin Aegean, but in practice were almost never exercised outside Constantinople.[29] Latin conquest of Greece would have direct implications for Grosseteste's later intellectual interests. From the 1230s onwards Grosseteste studied Greek, and very probably in the company of Master John of Basingstoke a translator of Greek, who had been active in the lordship of Athens, and others, including Nicholas the Greek.[30] As bishop of Lincoln Grosseteste translated works by John of Damascus, Pseudo-Dionysius, Aristotle, the *Testaments of the Twelve Patriarchs*, and parts of a Byzantine lexicon, the *Suda*. Grosseteste's skill in this arena impressed even Roger Bacon.[31]

The conquest of Constantinople by Latin forces was the product of what is known now as the Fourth Crusade. Constantinople was certainly not the original objective which was rather Jerusalem, captured after nearly a century of Latin Christian rule, by Ṣalāḥ al-Dīn in 1187. The Third Crusade reinforced what remained of the kingdom of Jerusalem, based at Acre, but failed to re-take the Holy City.[32] Failure and misdirection of crusades to Jerusalem led, in part, to the papacy claiming the primary role in its organization as articulated in Canon 71 of the Decrees of the Fourth Lateran Council (the gathering setting out a platform for church reform and pastoral care which would dominate Grosseteste's clerical life (see Ch. 2, §1)).[33] Enacted energetically by Innocent III

---

[28] Peter Lock, *The Franks in the Aegean: 1204–1500* (Longman: London, 1995), 6–8.

[29] Lock, *The Franks in the Aegean*, 6.

[30] See Southern, *Grosseteste*, 9, 185–6; John of Basingstoke's accomplishments in Greek are recorded in Matthew Paris, *Chronica majora*, s.a. 1252, ed. Henry Richards Luard, 7 vols. (London: Longman, Green, Longman and Roberts, 1872–83), v. 285–7.

[31] Meridel Holland, 'Robert Grosseteste's Greek Translations and College of Arms MS Arundel 9', in James McEvoy (ed.), *Robert Grosseteste: New Perspectives on His Thought and Scholarship* (Turnhout: Brepols, 1995), 121–47; Roger Bacon, *Opus tertium*, in *Opera quaedam hactenus inedita*, ed. J. S. Brewer (London: Longman, Green, Longman and Roberts, 1859), 91; Roger Bacon, *Compendium studii philosophiae*, in *Opera quaedam hactenus inedita*, 474.

[32] Amongst a vast literature, see Jonathan Phillips, *The Crusades, 1095–1204*, 2nd ed. (London: Routledge, 2014), chs. 12–13; and his *The Life and Legend of the Sultan Saladin* (New Haven, CT: Yale University Press, 2019).

[33] *Concilium Lateranense IV a. 1215*, Canon 71, in *Conciliorum Oecumenicorum Decreta*, ed. J. Alberigo, J. A. Dossett, P. P. Joannou, C, Leonardi, P. Prodi, 3rd ed. (Bologna: Istituto per le Scienze Religiose, 1973), 230–71, at 267–71. For the implementation of the conciliar decrees, see Jeffrey M. Wayno, 'Rethinking the Fourth Lateran Council of 1215', *Speculum*, 93 (2018), 611–37, esp. 628.

DATING *ON THE SPHERE* AND LOCATING ROBERT GROSSETESTE   19

(r. 1198–1216) it fell to his successor Honorius III (r. 1216–27) to call and administer the Fifth Crusade. With a wide-ranging recruitment across Christendom, the military effort was directed for the main part at Ayyubid Cairo, the principal power in the region.[34] The early stages of the crusade from 1217 involved an abortive siege of Jerusalem by Christian forces including King Andrew of Hungary, before a shift, with the arrival of fresh German, Italian, English, and French forces, along with the papal legate Pelagius, to Egypt in 1218. The successful siege and capture of the Egyptian Mediterranean city of Damietta, during which St Francis of Assisi was given permission to meet and preach to the Sultan al-Malik al Kāmil, proved illusory.[35] A later advance on Cairo by the crusader army collapsed, leading to the retreat and capitulation of the crusaders by 1221. The Sultan's terms for their lives were the surrender of Damietta and withdrawal from Egypt.

English nobles joined the Fifth Crusade armies, although without royal direction, and only after the settling of the baronial rebellion in 1217.[36] Supporters of both sides in the rebellion took part in the crusade, including Ranulf, earl of Chester, a key supporter of the royalist cause, as also William, earl of Derby, alongside the leader of the rebels Robert Fitzwalter and other prominent figures such as Henry de Bohun, earl of Hereford, and Saer, earl of Winchester.[37] Nor was the English contribution unimportant. As Christopher Tyerman notes they were identified by contemporaries as a distinct group, and the earl of Chester, amongst others, was influential in the higher command of the crusade army. Enthusiasm for crusade continued even after the ignominious results. Henry III's tutor, Philip of Aubigné, despite an attempted prohibition by Pope Honorius, managed to join the crusade in April 1221, and although

---

[34] The most detailed single account remains James M. Powell, *Anatomy of a Crusade, 1213–1221* (Philadelphia: University of Pennsylvania Press, 1986). See also E. J. Mylod, Guy Perry, Thomas W. Smith, and Jan Vandeburie (eds.), *The Fifth Crusade in Context* (London: Routledge, 2017).

[35] John Tolan, *Saint Francis and the Sultan: The Curious History of a Christian-Muslim Encounter* (Oxford: Oxford University Press, 2009).

[36] Christopher Tyerman, *England and the Crusades* (Chicago: University of Chicago Press, 1988), 95–9.

[37] Tyerman, *England and the Crusades*, 97; Guy Perry, 'A King of Jerusalem in England: The Visit of John of Brienne in 1223', *History* 100 (2015), 627–39, at 629–30.

20   THE SCIENTIFIC WORKS OF ROBERT GROSSETESTE

he arrived only to see the end of the affair this did nothing to diminish his enthusiasm for the cause.[38] His letter to Ranulf of Chester describing the failures of the expedition was probably instrumental in provoking fund-raising for John of Brienne, the titular king of Jerusalem.[39] Philip took up the cross again in 1228, in company with Peter des Roches, bishop of Winchester, on the Holy Roman Emperor Frederick II's campaign which, temporarily, negotiated Christian rulership of Jerusalem. Philip died, and was buried, in the Holy City in 1235.[40] Crusade was, then, an important feature of the broader political and spiritual landscape for Grosseteste's contemporaries.

Crusade was present in other theatres and with other valences as well. Innocent III had summoned a crusade against Albigensian heretics in and around the County of Toulouse in southern France in 1209. The campaign lasted until 1229. By 1216 Simon of Montfort had emerged as the dominant crusade leader and as an independent lord of some considerable standing having taken titles and lands by conquest, from Béziers and Carcassone to the County of Toulouse itself.[41] Astonishing as they were, these gains evaporated almost as quickly as they had been acquired. Simon himself was killed before the walls of Toulouse in 1218 and his eldest son Amaury was unable to sustain his inheritance, ceding all claims to Louis VIII of France in 1224. The crusade itself quickly evolved into a far more complex re-structuring of temporal and spiritual control in the region.[42] Capetian kings extended their authority into the region; the Count-Kings of Aragon, saw their trans-Pyrennean authority eroded following the death of Peter II in 1213 at Muret, fighting against Simon of Montfort; and the kings of England contended with perturbations in their lordship in Gascony.[43] While it is unlikely that Grosseteste had any direct connection to the region or the activity of

[38] David Carpenter, *Henry III 1207–1258* (Newhaven: Yale University Press, 2020), 23.
[39] Perry, 'A King of Jerusalem', 630–1.   [40] Tyerman, *England and the Crusades*, 98–9.
[41] G. E. M. Lippiat, *Simon V of Montfort & Baronial Government 1195–1218* (Oxford: Oxford University Press, 2017), 2–4.
[42] Dan Power, 'Who Went on the Albigensian Crusade?', *English Historical Review*, 128 (2013), 1047–85, at 1047.
[43] T. N. Bisson, *The Medieval Crown of Aragon* (Oxford: Oxford University Press, 1986), 39–40, 68–9; N. Vincent, 'England and the Albigensian Crusade', in Björn Weiler and I.W. Rowlands (eds.), *England and Europe in the Reign of Henry III (1216–1272)* (Ashgate: Aldershot, 2002), 67–97.

DATING *ON THE SPHERE* AND LOCATING ROBERT GROSSETESTE 21

crusade the on-going fighting against the Cathars certainly would have been known to him; heresy was something on which he would, as bishop, pronounce very fiercely, and of which he would provide a famous description in his deathbed speech as recorded by Matthew Paris: 'Heresy is a judgement chosen according to human understanding, contrary to Sacred Scripture, publicly taught and obstinately defended'.[44]

Closer to home, the conflict between the English crown and some amongst its leading baronial subjects from the end of John's reign, inter-meshed with existing struggles with the French king, Philip II (r. 1180–1223) and his son, Prince Louis, the future Louis VIII (r. 1223–6), provides an essential background for any assessment of Grosseteste's activities during these years. The events of this period were of considerable moment, many with social, political, economic, and intellectual ramifications for the rest of the century. They encompass the reconciliation of John and Innocent III in 1213, including the homage of the former to the latter, the lifting of the interdict over England and Wales in 1214, the collapse of John's campaigns against Philip II in the same year, and English baronial rebel-lion leading to Magna Carta in 1215.[45] Although rapidly repudiated by John and swiftly annulled by the pope, Magna Carta would become for Grosseteste's generation and beyond a powerful statement of a changed political world, in which kingship could be constrained and the liberty of the king's subjects emphasized.[46] In the immediate circumstances John's disavowal of Magna Carta provoked further rebellion seeking to replace Plantagenet rule with Capetian in the person of Prince Louis.

John's death in October 1216 left his nine-year-old son Henry as his successor.[47] He also left an inheritance under considerable strain with a

---

[44] Matthew Paris, *Chronica majora*, s.a. 1253, v. 400.

[45] C. R. Cheney, *Pope Innocent III and England* (Stuttgart: Anton Hiersemann, 1976), 332–7.

[46] The literature on Magna Carta is vast. For recent discussion, see David Carpenter, *Magna Carta* (London: Penguin, 2015); James Holt, *Magna Carta*, 2nd ed. (Cambridge: Cambridge University Press, 1992, repr. 2015); Nicholas Vincent, *Magna Carta: A Very Short Introduction* (Oxford: Oxford University Press, 2012); A. Musson, *Medieval Law in Context: The Growth of Legal Consciousness from Magna Carta to the Peasants' Revolt* (Manchester: Manchester University Press, 2001); P. A. Brand, *The Making of the Common Law* (London: Hambledon Press, 1992); Stephen Church, *King John: England, Magna Carta and the Making of a Tyrant* (London: Macmillan, 2015), 214–34.

[47] Detailed discussion of the minority is to be found in the magisterial studies of David Carpenter, *The Minority of Henry III* (London: Methuen, 1990), summarized in Carpenter, *Henry III*, 1–57.

22 THE SCIENTIFIC WORKS OF ROBERT GROSSETESTE

substantial part of the south-east of England controlled by Louis and the English rebels, led by Robert Fitzwalter, including London though not Dover, the castle of which held out under Hubert de Burgh.[48] The royalist party acted swiftly to crown the young Henry III, supervised by the papal legate Guala, and to appoint the seventy-year-old Earl William Marshal as regent and guardian for Henry.[49] Magna Carta was revised and re-issued in 1216, demonstrating a commitment to the liberties it entailed, as much as these had been rejected by John.[50] The year 1217 saw the military defeat of the rebels, at the battle of Lincoln and in a naval confrontation off the coast of Sandwich, Kent.[51]

A settlement followed negotiated at the peace of Kingston and Lambeth. This was lenient on the rebels, some of whom, as noted above, joined their erstwhile opponents on crusade. Louis was allowed to return home, his English followers to receive those lands they held at the beginning of the struggle. Louis's supporters amongst the clergy were made to seek absolution from the pope.[52] There followed a complex series of processes by which the king's minority council sought to restore the resources of the crown, its revenues and rights. Re-asserting the power of the crown against, for example, over-mighty sheriffs, ostensibly the king's local agents but in practice often running their localities for their own purposes, was a central element of this programme. So too the re-establishment of the mechanisms of royal justice. The gradual expansion of the young king's household is one measure for the pace of recovery: in 1217 he provided robes at Christmas for seven household knights, in 1220 this figure was twenty-five.[53] William Marshal's death in 1219 forced a change in governance with the emergence of a triumvirate comprising the Justiciar Hubert de Burgh, Peter des Roches, bishop of Winchester, and Pandulf the Papal Legate.[54] This lasted until July 1221 when Pandulf's legation came to end and the Peter was removed as Henry III's tutor and guardian by October 1221, though the evidence suggests that his contact with the royal household had been diminished

---

[48] Carpenter, *Minority*, 21.
[49] Carpenter, *Minority*, 13–16.
[50] Carpenter, *Minority*, 21–4.
[51] Carpenter, *Minority*, 35–44.
[52] Carpenter, *Minority*, 44–5.
[53] Carpenter, *Minority*, 227.
[54] Carpenter, *Minority*, 128–262.

DATING *ON THE SPHERE* AND LOCATING ROBERT GROSSETESTE   23

from January.[55] Henry was fourteen in October 1221, not quite in his majority, but no longer a child. The minority was never formally brought to an end and Henry would take a more active political role from 1223 onwards. For the time being Hubert dominated with Stephen Langton, archbishop of Canterbury, playing an increasingly important role, and the recovery of the royal estate continued.

A similar process of reconstruction following John's reign and the rebellion was undertaken with respect to church governance and the revivification of papal legal processes and diocesan administration. These issues would touch Grosseteste directly (see §3.3). Institutions of learning in England were affected in similar ways as well. The University of Oxford had re-opened in 1214 after its voluntary closure in 1209 in response to the *Suspendium clericorum*, that is, the hanging of two or three scholars by secular justice from which they were immune by dint of their clerical status, and royal aggression towards the body of scholars.[56] The re-opening was carried out as part of the settlement of the interdict, in this case involving a separate award, the Legatine Ordinance, administered by Nicholas of Tusculum. This award gives some evidence for a growing corporate identity of the masters of the university, as a guild, with recompense from the town for their part in the *Suspendium*, and the clerical status of the university enshrined in the ordinance.[57] It is within the award too that the first reference to a chancellor for the university can be identified, an appointment made by the diocesan bishop, in this case of Lincoln. Later thirteenth-century Oxford masters would claim to elect their chancellor and that the bishop's role was to confirm their choice. That bishops did not see their role in quite the same way is shown in the dispute between the masters and Bishop Oliver Sutton in 1295 in which Grosseteste's putative role as chancellor or *magister scholarum*

---

[55] Carpenter, *Minority*, 256–60; Nicholas Vincent, *Peter des Roches: An Alien in English Politics, 1205–1238* (Cambridge: Cambridge University Press, 1996), 195–208.

[56] *Knowing and Speaking*, 218–22.

[57] See G. Pollard, 'The Legatine Award to Oxford in 1214 and Robert Grosteste', *Oxoniensia*, 39 (1974), 62–72. For further references, see §3.2 below. For more general discussion, Alan B. Cobban, *The Medieval English Universities: Oxford and Cambridge to c.1500* (Berkeley: The University of California Press, 1988), 44–50; R. W. Southern, 'From Schools to University', in Jeremy I. Catto (ed.), *The History of the University of Oxford*: Vol. 1, *The Early Oxford Schools* (Oxford: Oxford University Press, 1984), 1–36; and M. B. Hackett, 'The University as Corporate Body', in Catto, *History of the University of Oxford*, 37–98.

24 THE SCIENTIFIC WORKS OF ROBERT GROSSETESTE

was mentioned. Whether Grosseteste should be associated with the chancellorship, and, just as importantly, when, as discussed below (§3.2 and in Appendix 1). While there is no doubt that collective identity was important to the university from 1214 too much should be not read into its early manifestations in light of later developments.

Indeed, the decade after 1214 has little evidence for scholastic activity at Oxford, though what there is points to a slow re-building of organizational structures emerging from the disruption of the later years of King John's reign to more stability and permanence.[58] From 1225 onwards the more distinctive elements to Oxford's structures and academic interests took shape, a process in which Grosseteste was involved, although by strict account of the evidence only from his appointment as lector to the Franciscans of the city in c.1230.[59] Oxford and Cambridge emerged from the early thirteenth century as the only two universities in the kingdom of England; the vibrant Cathedral schools of Grosseteste's youth, at York, Exeter, possibly Hereford, and above all at Lincoln, did not develop further.[60] Many reasons for this can be adduced: the size of the population, the comprehensive nature of studies offered at Oxford and Cambridge, and the growing power and presence of the two universities which, positively and negatively, discouraged other similar foundations.

In the period in which *On the Sphere* was composed, however, this shift in the evolution of English institutions of higher learning was only just beginning. Monastic houses continued to add to their libraries and sustain connections with the secular schools; secular Cathedrals remained centres of learning and training for clergy especially in light of the emphasis placed on this role in the Fourth Lateran Council. Lincoln, for example, seems to have retained some reputation as a centre for learning in the years following the death of William de Montibus in 1213, whose theological acumen had brought this fame to the city.[61]

---

[58] Southern, 'From Schools to University', 34.

[59] Michael Robson, 'Robert Grosseteste and the Franciscan School at Oxford (c.1229–1253)', *Antonianum*, XCV (2020), 345–82; Giles E. M. Gasper, 'How to Teach the Franciscans: Robert Grosseteste and the Oxford Community of Franciscans c.1229–35', in Lydia Schumacher (ed.), *The Early English Franciscans* (Berlin: De Gruyter, 2021), 57–75.

[60] Cobban, *Medieval English Universities*, 26–34.

[61] Joseph W. Goering, *William de Montibus (c.1140–1213): The Schools and the Literature of Pastoral Care* (Toronto: Pontifical Institute of Mediaeval Studies, 1992), 26.

DATING *ON THE SPHERE* AND LOCATING ROBERT GROSSETESTE 25

Grosseteste's scholarly supporter for his first post at Hereford, Gerald of Wales, spent the last decade of his life, until about 1223, predominantly in Lincoln, although he may have died in Hereford.[62] English intellectual pre-eminence was rapidly focusing on Oxford and Cambridge, but not exclusively or at an even pace. Many other plausible locations for scholarly endeavour existed in England, including, in the case of Grosseteste, Hereford. Outside of England the appeal of study at the University of Paris remained strong and a path that Grosseteste would tread, it is almost certain, in the years to come soon after the composition of his treatise on the uses of astronomy.

3 Where Was Grosseteste?

Into this context Grosseteste's activities, insofar as they can be followed, can be placed. That he was in France during the later years of the interdict has been suggested previously with respect to his deathbed scene as recorded by Matthew Paris and Grosseteste's apparent recollection of sermons preached against Christian money-lenders, usurers, from Cahors.[63] When he returned to England is not possible to ascertain but at or around the return of the bishops to the kingdom and the lifting of the interdict is plausible. Grosseteste would have come home to the complex circumstances within the kingdom provoked by the rapprochement between John and Innocent. These alone make it likely that his return to England was for a number of years. While there are good contextual reasons to posit further visits to France, and to Paris in particular, travel would have been constrained during the course of the conflicts, between first John and then the Minority Council, against the rebels and Prince Louis. At least until the end of 1217, then, Grosseteste is likely to have been in England. A Parisian residence of some sort might be suggested for the early 1220s on the basis of Grosseteste's mention of the meridian of that city in his *Compotus*, and on the Parisian reception

---

[62] Robert Bartlett, 'Gerald of Wales (*c.*1146–1220x23)', *Oxford Dictionary of National Biography* (Oxford: Oxford University Press, 2004); online ed., Oct. 2006, http://www.oxforddnb.com/view/article/10769 (accessed 26 April 2021).

[63] *Knowing and Speaking*, 203–14.

26 THE SCIENTIFIC WORKS OF ROBERT GROSSETESTE

of his treatise *On Comets*.[64] A more extended visit around 1225, coinciding with Grosseteste's appointment to his first benefice, is also likely and would fit his later familiarity with Parisian approaches to theology and a seeming friendship with William of Auvergne.[65]

The period in which *On the Sphere* was composed, from about 1213/14 to about 1218 at the earliest, and probably c.1221 or so at the latest, would seem to place Grosseteste in England. This provokes further speculation as to the circumstances in which it was written and for whom it was intended. More evidence can be marshalled for Grosseteste's other activities and location within this period which confirm his presence in England on a number of different occasions, and in particular connection to the diocese of Hereford. This, as Southern and Goering point out, makes sense especially if the suggestion that Hugh Foliot had acted as Grosseteste's patron from 1198 is taken up. The circumstances of his Hereford activities are worth probing in a little more detail, but before this, consideration needs to be given to older historiographical positions on the offices claimed for Grosseteste, his family, and a tradition which places Grosseteste squarely in Oxford during this period, and as the university's first chancellor in 1215.

3.1 Archdeacon at Large and a Family Affair?

Older traditions place Grosseteste in a number of different clerical roles during this period.[66] Wharton identified him as archdeacon of Chester in 1210, for which there is no corroborating evidence.[67] The *Fasti Ecclesiae Anglicanae* of 1854 lists him as archdeacon of Leicester, to which post he

---

[64] Giles of Lessines, *De essentia, motu et significtione cometarum*, c. 3, ed. Lynn Thorndike, *Latin Treatises on Comets Between 1238 and 1368 A.D.* (Chicago: The University of Chicago Press, 1950), 103–84 at 116–17; Panti, *Moti, virtù e motori celesti*, 135–6.

[65] The evidence for Grosseteste in Paris in the early 1220s will be addressed in Volume III of this series, that of the mid-1220s in Volume IV; Grosseteste and Oxford will be covered in Volume V.

[66] These were reviewed in J. C. Russell, 'The Preferments and "Adiutores" of Robert Grosseteste', *The Harvard Theological Review*, 26 (1933), 161–72. As representative of the older historiographical traditions, see Samuel Pegge, *The Life of Grosseteste* (London: John Nichols, 1743), 19–23, summarized in Francis Stevenson, *Robert Grosseteste, Bishop of Lincoln* (London: MacMillan and Co. 1899), 25–9.

[67] Henry Wharton, *Anglia Sacra*, 2 vols (London: Richard Chiswel, 1691), i. 457.

DATING *ON THE SPHERE* AND LOCATING ROBERT GROSSETESTE   27

was appointed, certainly, in 1229, but also as archdeacon of Northampton in 1221.[68] He was not.[69] More intriguing is the identification of Robert Grosseteste as the archdeacon of Wiltshire, in Salisbury diocese, from 1214, and as the rector of Calne in the same diocese.[70] The actual archdeacon and incumbent at Calne, was not Robert, but Richard Grosseteste.[71] Richard, a chaplain to Herbert Poor, bishop of Salisbury, 1194–1217, had been archdeacon from at least 1199 until his replacement in 1223.[72] While surnames are uncommon in English society until 1200, the matching style gives pause for thought.[73]

Whether Richard and Robert were related is an open question. Details on Robert Grosseteste's family are patchy, with the contemporary information dating from 1232 onwards. He had a sister, Ivette (Juetta), who predeceased him, and there is evidence for other family members.[74] Adam Marsh asked Grosseteste in 1249 to support two of his relatives in their university studies, and later writing.[75] Two other Grossetestes appear in his episcopal register, John installed in a benefice in 1237 and Robert in 1245.[76]

---

[68] *Fasti Ecclesiae Anglicanae*, compiled by John le Neve, corrected and continued by T. Duffus Hardy, 3 vols. (Oxford: Oxford University Press, 1854), ii. 55.

[69] 'Archdeacons: Northampton', in *Fasti Ecclesiae Anglicanae 1066–1300*: Vol. 3, *Lincoln*, ed. Diana E Greenway (London, 1977), 30–2. British History Online: http://www.british-history.ac.uk/fasti-ecclesiae/1066-1300/vol3/pp30-32 (accessed 28 April 2020).

[70] Russell, 'Preferments and "Adiutores"', 166–7.

[71] 'Archdeacons: Wiltshire', in *Fasti Ecclesiae Anglicanae 1066–1300*: Vol. 4, *Salisbury*, ed. Diana E Greenway (London, 1991), 33–7. British History Online: http://www.british-history.ac.uk/fasti-ecclesiae/1066-1300/vol4/pp33-37 (accessed 27 April 2020); 'Prebendaries: Calne', in *Fasti Ecclesiae Anglicanae 1066–1300*: Vol. 4, 57–9.

[72] C. L. Kingsford and B. R. Kemp, 'Poor [Pauper], Herbert (d. 1217), Bishop of Salisbury', *Oxford Dictionary of National Biography*, 23 September 2004: https://www.oxforddnb.com/view/10.1093/ref:odnb/9780198614128.001.0001/odnb-9780198614128-e-22524 (accessed 3 May 2020).

[73] See the English Surnames Series, for example, Richard McKinlet, *Norfolk and Suffolk Surnames in the Middle Ages* (London, Chichester: Phillimore, 1975).

[74] Robert Grosseteste, *Episcopi quondam Lincolniensis Epistolae*, ed. Henry Richards Luard, Rolls Series (London: Longman, Green, Longman and Roberts, 1861), Letter 8, addressed to Ivette, dated to shortly after 1 November 1232; Robert Grosseteste, *The Letters of Robert Grosseteste, Bishop of Lincoln*, trans. F. A. C. Mantello and Joseph W. Goering (Toronto: University of Toronto Press, 2010), 75–7. Adam Marsh, *The Letters of Adam Marsh*, ed. and trans. C. H. Lawrence, 2 vols. (Oxford: Oxford University Press, 2006), Letters 12 and 53, 130–3, 148–9. See also N. M. Schulman, 'Husband, Father, Bishop? Grosseteste in Paris', *Speculum*, 72 (1997), 330–46. The implications of Schulman's suggestion for Grosseteste as married and with children will be discussed in Volume III in this series.

[75] Marsh, *Letters*, Letter 35,1.100–5, at 102–5.

[76] *Robert Grosseteste as Bishop of Lincoln, The Episcopal Rolls, 1235–1253*, ed. Philippa M. Hoskin (Woodbridge: Boydell for The Lincoln Record Society, 2015): John nos. 629, 665, 933; Robert nos. 1661, 1720–1.

28   THE SCIENTIFIC WORKS OF ROBERT GROSSETESTE

Later sources include an anecdote that Grosseteste's sister married one of his chamberlains, by whom she had become pregnant, and older traditions insist that she was a nun.[77] The lowliness of Grosseteste's background is stressed in sources from the later thirteenth and early fourteenth centuries but with no further specificity; Richard of Bardney's *Life of Grosseteste* from the early sixteenth century invokes Grosseteste's mother, but this work has no, or limited, evidential status in this respect.[78]

While the possibility of a familial connection between Richard and Robert may not be demonstrable, but it should not be ruled out. There are points of comparison, particularly connected to clerical networks and patronage, which can be explored. For example, Herbert, the son, in all probability, of Richard Ilchester, bishop of Winchester, had connections to Lincoln, where he was cathedral canon by 1167, and probably archdeacon of Northampton in the early 1170s. Later archdeacon of Canterbury, Herbert also gained canonries in Salisbury and Wells, in addition to Lincoln. He was elected bishop of Lincoln in 1186 by the chapter but was refused the office by Henry II. In 1194 he was elected, and appointed, as bishop of Salisbury. There are interesting connections to Robert Grosseteste's life and career here, not least the links to Lincoln, and to Wells (it was Hugh of Wells, bishop of Lincoln, previously archdeacon of Wells, who gave Robert his first recorded benefice, at Abbotsley). Connections are even stronger between Robert Grosseteste and Herbert Poor's successor at Salisbury, his younger brother Richard.[79] As dean of Salisbury, Richard Poor spent the years of the interdict in Paris, making a meeting with Grosseteste possible. Many years later, as both bishop of Salisbury and then of Durham, he settled a dispute between Grosseteste and the Abbey of Reading over payments from the benefice at Abbotsley.[80]

---

[77] S. Gieben, 'Anecdota Lincolniensia. La preghiera mattutina del vescovo; La debolezza umana della sorrella Ivetta; L'eretica che non voleva bruciare', in P. Maranesi, ed., *Negotium Fidei. Miscellanea di studi offerti a Mariano D'Alatri in occasionne del suo 80° compleanno* (Rome: Bravetta, 2002), 127–44.

[78] *Lanercost Chronicle, 1201–1346*, ed. Joseph Stevenson (Edinburgh: Bannatyne Club, 1839), s.a. 1235, 44–5; Richard of Bardney, *Vita Roberti Grosthed*, ed. Henry Wharton, *Anglia Sacra*, ii. cc. III–VI, 326–7.

[79] Philippa Hoskin, 'Poor [Poore], Richard (d. 1237), Bishop of Salisbury', *Oxford Dictionary of National Biography*, 23 September 2004: https://www.oxforddnb.com/view/10.1093/ref:odnb/9780198614128.001.0001/odnb-9780198614128-e-22525 (accessed 17 May 2020).

[80] Giles E. M. Gasper, 'Robert Grosseteste at Durham', *Mediaeval Studies*, 76 (2014), 297–303.

DATING *ON THE SPHERE* AND LOCATING ROBERT GROSSETESTE   29

Grosseteste's links to Hereford are demonstrable (§3.3 below) though their exact form over an extended period of time is less easy to identify. That he was also linked, in some way, to the close-knit ecclesiastical hierarchy associated with the secular (not monastic) cathedrals of Salisbury, Lincoln, and the mixed secular/monastic arrangements of Bath and Wells is also worth noting in terms of the networks to which he might have belonged. This association comes to bear particularly on his appointment to Abbotsley and will be outlined in Volume IV of this series. For the present purpose, while the antiquarian tradition was mistaken in its attribution of multiple clerical offices to Grosseteste, the emergence of Richard Grosseteste allows for a little more speculation on a family connection. A fraternal relationship is a possibility, an avuncular one perhaps more likely, although the difference in date of death between the two, *c*.1223 and 1253, and appointment to office does not necessarily indicate a difference in date of birth. Both possible relationships would conform to prevailing models in clerical families.[81] The implications of such a relationship are less easy to draw out. As the case of Gerald of Wales and his nephew shows (Ch. 2, §2.3.1), such ties could be complicated and destructive as much as generous, and nepotism did not attract universal approval.[82] There is no evidence for any contact between Richard and Robert, but there is not anyway a great deal of consistent evidence for the latter's career from *c*.1214 to *c*.1225. Grosseteste's pre-episcopal career invites particular questions, not least how someone of his evident intellectual interests and talents was supported, which require careful and open-minded consideration; the range of his possible activities should not be interpreted too rigidly.

3.2 Chancellor of Oxford?

One of the longer-running debates about Grosseteste's career is his disputed role as the first chancellor of the University of Oxford. The circumstances reveal much about the development of the university,

---

[81] Julia Barrow, *The Clergy in the Medieval World: Secular Clerics, their Families and Careers in North-Western Europe, c.800–c.1200* (Cambridge: Cambridge University Press, 2015), 113–57.
[82] Hugh Thomas, *The Secular Clergy in England, 1066–1216* (Oxford: Oxford University Press, 2014), 190–208.

30   THE SCIENTIFIC WORKS OF ROBERT GROSSETESTE

its historiographical, and differing assessments of Grosseteste's intellectual interests and their institutional instantiation. Southern and Goering concur in the correction of alternative analyses associated both with scholarship on Grosseteste's life and works and on the establishment of, and evidence for, corporate identity at the university, which place Grosseteste as chancellor some time not earlier than 1214 or later than 1221.[83] The issue of the chancellorship turns on the interpretation of a statement made by Oliver Sutton, bishop of Lincoln (1280–99) in 1295 that:

> ... blessed Robert, formerly bishop of Lincoln, who had held such an office while he was teaching at the aforementioned university, said at the start of his episcopacy that his immediate predecessor as bishop of Lincoln had not allowed the same Robert to be called chancellor, but Master of the Scholars [Magister Scholarum][84]

The statement (see Appendix 1), which supplies no date for the episode, indicates that Grosseteste, in 1235, recalled that he had been allowed to use the title *magister scholarum* rather than *cancellarius* by bishop Hugh of Wells. How Oliver Sutton came to know the anecdote is open to interpretation; institutional memory perhaps, connected to his uncle, Henry of Lexington, bishop of Lincoln (1253–8) in succession to Grosseteste. Henry was prebendary of Calne in Salisbury (a prebend held previously by Richard Grosseteste) by 1237 and became dean of Lincoln in 1246 in the immediate aftermath of Grosseteste's claim to rights of visitation over the cathedral chapter.[85] Another possible route is through Adam

---

[83] Daniel A. Callus, 'Robert Grosseteste as Scholar', in D. A. Callus (ed.), *Robert Grosseteste, Scholar and Bishop* (Oxford: Oxford University Press, 1955) 1–69, 7–10, esp. at 9. See *Snappe's Formulary: Extracts from a Formulary attributed to John Snappe and other Records relating to Oxford University*, ed. H. E. Salter, Oxford Historical Society, 1st ser. 80 (Oxford: Oxford University Press, 1923), 319. What follows draws heavily on Southern, Grosseteste, pp. xxix–xxxii; and Joseph W. Goering, 'Where and When Did Grosseteste Study Theology?', in James McEvoy (ed.), *Robert Grosseteste: New Perspectives on his Thought and Scholarship* (Turnhout: Brepols, 1995), 17–51 at 47–50. See also Southern, 'From Schools to University', 32–6.

[84] *The Rolls and Register of Bishop Oliver Sutton 1280–1299*: Vol. V, *Memoranda May 19, 1294– May 18, 1296*, ed. Rosalind M. T. Hill, Lincoln Record Society, 60 (Printed for the Lincoln Record Society in Hereford: Hereford Times Ltd, 1965), 60: '... beatus Robertus quondam episcopus Lincolniensis qui huiusmodi officium gessit dum in Universitate predicta regebat, in principio creationis sue in episcopum, dixit, proximum predecessorum suum episcopum Lincolniensem non permisisse quod idem Robertus vocaretur cancellarius, sed magister scholarum.'

[85] 'Prebendaries: Calne', 57–9; 'Deans', in *Fasti Ecclesiae Anglicanae 1066–1300*: Vol. 3, 5–12; Grosseteste, *Epp.* 121 and 122. Southern, *Grosseteste*, 264–5.

DATING *ON THE SPHERE* AND LOCATING ROBERT GROSSETESTE 31

Marsh, who wrote to Grosseteste as bishop (the letter is undated), concerned for Oliver Lexington (Sutton) and asking for promotion to a benefice for the young scholar. Adam, who describes Oliver as 'specially dear to me in Christ [*michi in Christo specialiter dicto*]', could easily have been the source of the anecdote.[86]

The question of Grosseteste's title came up in a dispute between Sutton and the masters of Oxford, as to whether the latter elected their chancellor, or whether he was nominated by the former.[87] The example of Grosseteste as a chancellor elected by the masters in defiance of the rights and privileges of the bishop was put forward by Sutton. The early history of the chancellorship of the university is complicated and dates to the period of the Legatine Ordinance in 1214, a treaty, in effect, between the town, the scholars, and the bishop of Lincoln (Hugh of Wells) in whose diocese Oxford lay.[88] The appointment of a chancellor for the university is one possibility explored in the ordinance, and one which would match Parisian practice. That the office emerged reasonably quickly is clear from the record of a Master Geoffrey de Lucy as university chancellor in 1215.[89] The early chancellors usually did not hold office for as long or as continuously as their successors, so de Lucy's tenure in 1215 does not automatically preclude Grosseteste having held the position. That said, it is also possible, as Southern points out, that de Lucy had held the position since 1214 and possibly until 1220.[90]

One line of interpretation for Grosseteste's tenure of the office from an institutional perspective, is that the lesser title of *magister scholarum* rather than *cancellarius* indicates a date before William de Lacey, that is from 1214–15.[91] This on the basis that the bishop had some caution about the development of the role of chancellor with respect to his own rights of appointment, explaining the difference in description of the role. However, the fact that Sutton's original comment gives no date for

---

[86] Marsh, *Letters*, Letter 14, i.34–7.
[87] Southern, *Grosseteste*, p. xxx; see also R. M. T. Hill, 'Oliver Sutton, Bishop of Lincoln, and the University of Oxford', *Transactions of the Royal Historical Society*, 4th ser., 31 (1949), 1–16.
[88] C. H. Lawrence, 'The Origins of the Chancellorship at Oxford', *Oxoniensia*, 41 (1976), 316–23.
[89] Mary Cheney, 'Master Geoffrey de Lucy, an Early Chancellor of the University of Oxford', *The English Historical Review*, 82 (1967), 750–63.
[90] Southern, *Grosseteste*, xxxi.   [91] Pollard, 'The Legatine Award', 62–72.

32   THE SCIENTIFIC WORKS OF ROBERT GROSSETESTE

Grosseteste's tenure makes this suggestive rather than demonstrated, and an argument made to fit Grosseteste's candidacy into the list of early chancellors. The proposition makes the additional assumption that Grosseteste was both suitable for, and able to fulfil, the role at that point. It is here that the interpretation of Grosseteste's career as a whole comes into play, and in particular the version offered by Fr Daniel Callus, followed and elaborated by James McEvoy.[92] This involves a second historiographical debate over Grosseteste's formal training in theology, of what that was likely to have consisted, and when it took place. Callus, writing before Pollard, Cheney, and Lawrence's more detailed investigations of the office, took Sutton's statement to show that Grosseteste had held the office of chancellor, and that the alternative title offered was down to a period of transition as the status of the role was settled, a process complete by 1221.[93] Callus noted too that for Grosseteste to have been chancellor he must have been regent in theology.[94] Although not defined in statute, medieval university chancellors were generally from the higher faculties of theology or law, as opposed to the faculty of liberal arts.[95]

On this model then, Grosseteste had undertaken formal theological study by 1214. Callus's construction of the career (which does recognize the fragmentary nature of the evidence), moves Grosseteste from Lincoln to Oxford, followed by the period in Hereford, after which it is suggested he returned to Oxford 'to resume his studies'.[96] He would then have been affected by the closure of the university between 1209–14, left for France, and returned to take up the role of chancellor. The sojourn in France is linked then to the commencement of theological studies, as the most logical step in Grosseteste's career, and this allows him to return with sufficient qualification to act as chancellor or equivalent.[97] Grosseteste's theological training and, presumably, teaching extends then from 1214 onwards.

---

[92] D. A. Callus, 'The Oxford Career of Robert Grosseteste', *Oxoniensia*, 10 (1945), 42–72; Callus, 'Robert Grosseteste as Scholar', 6–10; McEvoy, *Philosophy*, 8–10; James McEvoy, *Robert Grosseteste* (Oxford: Oxford University Press, 2000), 22–9.
[93] Callus, 'Grosseteste as Scholar', 8–9.   [94] Callus, 'Grosseteste as Scholar', 8.
[95] Goering, 'Where and When', 49.   [96] Callus, 'The Oxford Career', 45.
[97] Callus, 'The Oxford Career', 50.

DATING *ON THE SPHERE* AND LOCATING ROBERT GROSSETESTE   33

Flaws in this model were pointed out by Southern, on several occasions, and by Goering. A significant problem is one that Callus acknowledged, namely the lack of any definite evidence to tie Grosseteste to Oxford at any point, strictly speaking, before his appointment as lector to the Franciscans in *c.*1230. Nor, as Goering has shown, is there any evidence for Grosseteste's theological training, or for works being the product of an early Parisian education before 1215 or in the decade following; instead his theological study seems likely to have begun, in Paris, in about 1225.[98] Other evidence points in the same direction: Nicholas Trevet in his early fourteenth-century account of Grosseteste's life states that he was a master of arts when he completed the commentary on *Posterior Analytics*. This is dated now to *c.*1225, and there is no reason why this should be regarded as inconsistent with Trevet's statement.[99]

While there is now a general scholarly consensus that the scientific *opuscula*, *Compotus*, commentaries on *Physics* as well as on *Posterior Analytics* formed a more consistent basis of Grosseteste's intellectual interests in the period *c.*1195 until at least 1225. Lack of evidence for formal training in theology does not preclude theological interests.[100] As Goering and Mantello have shown, Grosseteste developed a particular interest in penitential literature, part of pastoral care.[101] The coterminous development of these writings, which include *On the Temple of God* for the period under scrutiny with the scientific *opuscula*, is explored in the next chapter.

Taken together the lack of evidence for connection to Oxford, or for formal training in theology would seem to militate strongly against the notion that Grosseteste was chancellor, or *magister scholarum* in Oxford, in or around 1214/15. Nevertheless, unless Sutton's remark was not to be taken at face value, or was even intended as humorous, Grosseteste does appear to have held the latter office at some point. For Southern, this took place in the

---

[98] Goering, 'Where and When', 26, 41–3.

[99] Goering, 'Where and When', 23–4; Nicholas Trivet, *Annales ex regum Angliae*, 242.

[100] For the current scheme of Grosseteste's writings, see Cecilia Panti, 'Robert Grosseteste and Adam of Exeter's Physics of Light: Remarks on the Transmission, Authenticity and Chronology of Grosseteste's Scientific Opuscula', in John Flood, James R. Ginther, and Joseph W. Goering (eds.), *Grosseteste: Intellectual Milieu* (Toronto: Pontifical Institute of Mediaeval Studies, 2013), 165–90, tabulated at 185.

[101] Goering, 'Where and When', 27–35.

34 THE SCIENTIFIC WORKS OF ROBERT GROSSETESTE

later 1220s, and as an act of defiance by the university, intent on asserting what they considered their right to elect a chancellor. This presupposes a sense of corporate identity which is better understood as a more extended evolution from the Legatine Ordinance.[102] Hugh of Wells, on this account, then refused to give the proper title in response, preserving his own rights as he perceived them. This works better with the chronology and the evidence for Grosseteste's career. However, there are additional problems, not least that Hugh of Wells seems to have been well-disposed in other matters to Grosseteste. An alternative interpretation is offered by Goering, who focuses on the particularity of the title *magister scholarum*.[103] If Grosseteste were still a master of arts, and not a regent master of theology or law, then this might explain Hugh's decision to award a different title. As Goering points out, there is a clear distinction to be held in mind between the reasons why Bishop Sutton invoked the case of Grosseteste, and the decision that Hugh of Wells made in insisting on a different title to *cancellarius*.[104]

The positions taken on the chancellorship and the beginnings of Grosseteste's theological training are important for an assessment of where Grosseteste was to be found in the period from *c*.1214 to *c*.1221, and the sort of scholarship in which he might be expected to have engaged. Such evidence as exists points towards Hereford, and it is in this rather different context that the setting for *On the Sphere* can be imagined plausibly. The importance of Hereford to the first stages of Grosseteste's career, as discussed in Volume I of this series, is undeniable.[105] The extent to which his connections with the city, diocese, and region continued are important to explore in any reconstruction of his possible career. A strong suggestion can be made, following Southern and Goering, for Grosseteste's association with the diocese of Hereford, at least until some point in the early 1220s, and probably supported by Hugh Foliot, archdeacon of Shropshire and later himself bishop (1219–34).[106]

---

[102] Southern, *Grosseteste*, xxxi.   [103] Goering, 'Where and When', 48–50.

[104] An issue for a later 'chancellorship' for Grosseteste is that he is difficult to fit into the sequence of chancellors of the later period, see Hackett, 'The University as Corporate Body', 45–7. The change in nomenclature might make this an easier task.

[105] *Knowing and Speaking*, 18–35.

[106] Southern, *Grosseteste*, 66–9; Goering, 'Where and When', 20, 26; *Knowing and Speaking*, 210–12.

3.3 Grosseteste at Court: Legal Activities

While there is no evidence for Grosseteste's association with either of the two next successors to William de Vere (d. 1198) as bishop of Hereford, Giles de Braose (1200–15) and Hugh Mapenor (1216–19), a continuing relationship with Hugh Foliot and Hereford diocese can be demonstrated. Grosseteste was appointed as papal judge-delegate by Pope Innocent III in company with Archdeacon Hugh between 1213 and 121. He also featured a summons to appear before the royal justices in 1220 and again in 1221 for having heard a lay case in an ecclesiastical court which involved Hugh, as bishop, in its wider circumstances. Finally, and more briefly, Grosseteste was accorded a high status as a witness in a charter presenting a priest in charge to the parish of Culmington, six miles north of Ludlow, an appointment authorized by bishop Hugh.[107] Whether Grosseteste was in Herefordshire permanently or intermittently is not clear, and it should be noted that although no other evidence for his location has come to light no certain conclusions can be drawn as to Grosseteste's whereabouts. He could, quite easily, have been elsewhere, although any suggestion here would have to take into account the political upheaval in England and Wales during the First Baron's War, rebellion, and its aftermath.

3.3.1 Papal Judge-Delegate

The first evidence to consider is Grosseteste's appointment as a papal judge-delegate by Innocent III, not later than Innocent's death in 1216, and, given the complications of the interdict, probably after 1213/14. Developing over the twelfth century and notably from the pontificate of Alexander III (1159–81), the system of judges-delegate was fully established in that of Innocent III (1198–1216). One of the most striking features of the twelfth-century Latin church was the rapid growth of the papal judicial system, fuelled by an increasing volume of appeals in the first instance to the pope. Church law was separate from secular law during the Middle Ages, the latter also evolving rapidly, in the case of

---

[107] Since the current focus is on the period up to *c.*1221 the implications of the Culmington charter will be explored in more detail in Volume III of this series.

36 THE SCIENTIFIC WORKS OF ROBERT GROSSETESTE

twelfth-century England into common law.[108] Jurisdictional separation should be seen in the broader context of appeals for the freedom of the church, *libertas ecclesiae*, from secular influence. This was the watchword and programme of reform from the mid-eleventh century onwards. Henry II of England's dispute with Archbishop Thomas Becket, with the public murder of Becket in his cathedral in 1170 which shocked contemporaries at its finale, is perhaps only the most famous example of challenges to ecclesiastical privilege. The nature of that challenge, and the articulation of how the freedom of the church was to be understood, was debated and was contested throughout the twelfth and thirteenth centuries (and beyond), becoming a central component in the conceptual frameworks of political and spiritual life.

Judicial appeal to Rome emerged from broader movements for church reform and instigated a sophisticated and detailed series of processes and institutions. Prominent amongst these was the mandating of judges to act for the pope outside Rome. As early as the pontificate of Paschal II (1099–1118), delegation was used, very often to diocesan bishops, in cases where local knowledge was essential. Appeals expanded dramatically by Alexander III's pontificate with hundreds of petitions per year, a trend which continued until the mid-thirteenth century.[109] Partly as a result delegation of cases increased, alongside the evolution of the papal chancery and its official instruments. For example, the *Audientia litterarum contradictarum*, the office concerned with the issuing of mandates, was probably established under Innocent III. The new papal offices led, as Jane Sayers points out, to an 'increase in the number of men who were directly concerned in the judicial administration of the church both at the centre and in the provinces.'[110] Rising numbers of delegations drove a deepening of the pool of delegates; by the 1180s abbots and lesser church dignitaries, for example archdeacons, were acting in this capacity.

---

[108] See John Hudson, *The Formation of the English Common Law: Law and Society in England from the Norman Conquest to Magna Carta* (London: Longman, 1996); Patrick Wormald, *The Making of English Law: King Alfred to the Twelfth Century: Vol. 1, Legislation and its Limits* (Oxford: Blackwell, 1999).

[109] Jane E. Sayers, *Papal Judges Delegate in the Province of Canterbury 1198–1254* (Oxford: Oxford University Press, 1971), 13.

[110] Sayers, *Papal Judges Delegate*, 19.

DATING *ON THE SPHERE* AND LOCATING ROBERT GROSSETESTE   37

Alongside these changes came shifts in the study and practice of the law of the church, canon law. The collection compiled by Gratian in Paris in the 1140s, the *Decretum* or the *Concordance of Discordant Canons*, did much to provide a more uniform model for the church, and the collection was soon and extensively glossed. From the 1170s the evidence suggests a shift in emphasis to collections of decretals (papal judgments or decrees on points of canon law), with more detailed discussion of specific points of canon law from the pope. From compilations made at local centres, and privately, larger, Christendom-wide decretal collections were put together. There was an English element to this activity. Schools for canon law at Worcester, Canterbury, and Exeter in the 1160s and 1170s show evidence of decretal collection from examples sent to their vicinity. A number of English scholars were active at Bologna in the later twelfth and early thirteenth centuries, and English collections were used as models at this point for Bolognese compilations. The first official collection of church law (the better known collection by Gratian was not officially sanctioned) was commissioned in 1210 by Innocent III, the *Compilatio tertia* of Petrus Collivachinus of Benevento, which paved the way for extended treatment of the *Decretals* by Raymond of Peñafort in 1234.

The circumstances of Grosseteste's earlier life can be placed into the developments of canon law, in procedure and practice. While he did not, so far as is known, attend Bologna, or Paris, for legal training, he grew up in an England of the 1180s and 1190s where local centres for such training were established, and where compilations of papal judgements were created. Moreover, since Grosseteste can be connected to Lincoln from the later 1180s, it is possible that he encountered Master Vacarius, a notable teacher of law, active in the city at this point.[111] The activities of Vacarius have been the subject of some scholarly debate; that Grosseteste might, conceivably, have been taught by him would fit with Gerald of Wales's recommendations to William de Vere.[112]

---

[111] *Knowing and Speaking*, 17–18; Peter Landau, 'The Origins of Legal Science in England in the Twelfth Century: Lincoln, Oxford and the Career of Vacarius', in Martin Brett and Kathleen Cushing (eds.), *Readers, Texts and Compilers in the Earlier Middle Ages* (Aldershot: Ashgate, 2009), 165–82.

[112] *Knowing and Speaking*, 17–18; Landau, 'Origins of Legal Science'; R. W. Southern, *Scholastic Humanism and the Unification of Europe*: Vol. II, *The Heroic Age* (Oxford: Blackwell, 2001), 155–6.

38 THE SCIENTIFIC WORKS OF ROBERT GROSSETESTE

The papal judicial system in England was severely disrupted during the interdict, making its reconstruction a priority from 1213 onwards. John's submission to Pope Innocent III, including offering the kingdom as a papal fief to the papal nuncio Pandulf in May 1213, created the circumstances for a reconciliation between the king and the English church.[113] This was marked by the return of the archbishop of Canterbury, Stephen Langton, in July, accompanied by the papal legate Nicholas of Tusculum, at John's request. The process of bringing the interdict to an end began, turning to a considerable degree on the compensation to be paid by the king to the bishops. The final compensation was far less than that demanded originally, and the interdict was declared to be lifted at a council at St Paul's in late June 1214.[114] Nicholas of Tusculum's legation lasted from September 1213 to December 1214, during which period he delegated cases in his capacity as legate to deal with the backlog of appeals, part of a programme to restore papal administration within the kingdom.[115] Nicholas's successors as legate included Cardinals Guala (2016–18) and Pandulf (1218–21) both of whom continued to delegate cases in England. Often this involved lesser clerics, and, sometimes, quite trivial cases.[116]

It is in this context that Grosseteste's mandate, issued between 1213 and 1216, to act as papal judge-delegate by Innocent III should be placed.[117] The case in question had been brought to Rome by William, son of Ralph of Hallow, against the prior and abbey of Worcester over the legality of William's dismissal as the prior's butler. The verdict was recorded in the cartulary of Worcester Cathedral Priory.[118] Grosseteste's fellow judges-delegate were H., the archdeacon of Salop, and R. the rural dean of Sapey.[119] 'Sapey' must refer to either Upper or Lower Sapey, the

---

[113] S. T. Ambler, *Bishops in the Political Community of England, 1213–1272* (Oxford: Oxford University Press, 2017), 64–5; Vincent, *Peter des Roches*, 90–3; Cheney, *Pope Innocent III and England*.

[114] Vincent, *Peter des Roches*, 93.   [115] Sayers, *Papal Judges Delegate*, 28–9.

[116] Sayers, *Papal Judges Delegate*, 31–2.

[117] C. R. Cheney and Mary G. Cheney, *The Letters of Pope Innocent III (1198–1216) Concerning England and Wales* (Oxford: Oxford University Press, 1967), no. 1156B, p. 189; Southern, *Grosseteste*, 67.

[118] *The Cartulary of Worcester Cathedral Priory (Register I)*, ed. R. R. Darlington, Pipe Roll Society Publications, 76 = ns 38 (1968), 72–3, no. 129, see 73, n. 2 for dating.

[119] *Cartulary of Worcester Cathedral Priory*, 72.

DATING *ON THE SPHERE* AND LOCATING ROBERT GROSSETESTE   39

former in Worcestershire, the latter Herefordshire, both in the north-east of Hereford diocese and to the north-west of Worcester and Hallow.[120] The identity of the rural dean in question is not known. H., the archdeacon of Salop (Shropshire), was Hugh Foliot. Although the county of Shropshire was divided between the dioceses of Hereford and Lichfield-Coventry, and both dioceses included a Shropshire archdeaconry, Hugh, and Hereford must be meant here.[121]

More can be gleaned from the details of the case. All three judges-delegate were secular clergy, that is non-monastic, and Grosseteste is identified only as *magister*; he had no ecclesiastical benefice at the time. This, as Goering notes, 'suggests that his qualifications as judge... derived from sources other than diocesan administrative office'.[122] The case itself was resolved with William agreeing to resign his right to the butlership to the prior and monks, in addition to his conferring land on them, and receiving in return support from the monastery for his mother, father, and himself, but not his heirs. The judges conclude that: 'We, therefore, the resignation into our hands having been made by the aforementioned William, absolve the oft-mentioned prior and monks, by the apostolic authority invested in us in this part, from the charge of the William mentioned and his heirs concerning the aforementioned office of butler...'.[123] Grosseteste, in his forties, was

---

[120] It is worth noting also that the bishop of Hereford had tenants, the de Beauchamp family, in Upper Sapey, Acta of Hugh Foliot in *English Episcopal Acta VII: Hereford 1079–1234*, ed. J. Barrow (Oxford: Oxford University Press for the British Academy, 1993), 216, no. 280.

[121] The case and those hearing it have been the occasion for a slight misunderstanding in secondary discussion. Goering ('Where and When', 20) suggests that Grosseteste was involved in two different cases connected to Worcester, one alongside the archdeacon of Salop (Lichfield) before 1216, citing Cheney and Cheney, *Letters of Pope Innocent III*, 189, and one alongside Hugh Foliot 1213–16, citing Southern, *Grosseteste*, 67. In fact these are the same case, and involve only Hugh as archdeacon of Shropshire in Hereford diocese. Sayers (*Papal Judges Delegate*, 217) rightly reports the three judges as 'the archdeacon of Salop, Master R. Grosseteste, and the rural dean of Sapey', the notice in Cheney and Cheney explicitly invokes Hugh as the archdeacon of Salop and the description of the case matches that given in the Worcester *Cartulary*. Darlington noted with respect to the entry in the *Cartulary* that half of the original cirograph is extant as Worcester Cathedral MS B 406, though faint and in parts illegible (*Cartulary of Worcester Cathedral Priory*, 73).

[122] Goering, 'Where and When', 21.

[123] *Cartulary of Worcester Cathedral Priory*, 73: 'Nos igitur facta a prefato W[illelmo] in manus nostras dicta resignatione, sepedictos priorem et monachos auctoritate apostolica qua fungebamur in hac parte absoluimus ab inpetitione dicti W[illelmi] et heredum suorum super prefato offitio pincerne...'.

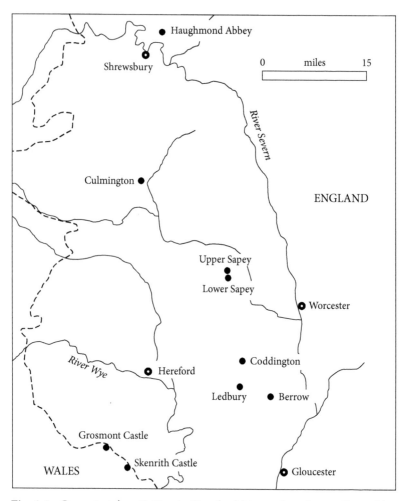

Fig. 1.1. Grosseteste's activities in Herefordshire and environs. Created by Rosie Taylor.

entrusted with a delegated papal case, even though he was unbeneficed and a secular cleric; and in connection to Hugh Foliot, here in his capacity as archdeacon.[124] Although limited, the evidence shows Grosseteste as an active participant in the re-establishment of church

---

[124] Grosseteste was not ordained as a priest until 1225 and his installation as rector of Abbotsley was described there as of 'a clerk in deacon's orders', Southern, *Grosseteste*, 69.

DATING *ON THE SPHERE* AND LOCATING ROBERT GROSSETESTE   41

order in England in the years following the interdict, and in a capacity which fits his expertise in law praised by Gerald of Wales in his letter of recommendation to Bishop William de Vere of Hereford, *c.*1195, and his continued relationship with Hugh Foliot.

3.3.2 A Brush with Royal Justice

In this connection it is possible to connect Grosseteste more firmly still to Hugh on the latter's election as bishop of Hereford in 1219 in records of two quite different incidents. The first shows Grosseteste acting again in a legal context, the second with appointments within the diocese. To take the first instance, Grosseteste appeared in a case that came before the royal justices in late August 1220. The case concerned actions taken at least a few months earlier by Grosseteste himself, Master Robert de Cotinton' (probably Coddington, some three miles to the north of Ledbury, although the similarly named villages in south Cheshire and Nottinghamshire should also be noted) and Gilbert the chaplain of Ledbury. The three men, acting as ecclesiastical judges, had overreached their authority when they heard a case which should have gone before the royal justices, concerning as it did an issue of lay service, and were summoned to answer for themselves. None of the three attended in August. Had they been laymen with landholding their property could have been held in distraint, that is seized by the sheriff to force sureties that they would attend a future court date. Since this was not the case, the sheriff was authorized to instruct the bishop of Hereford, Hugh Foliot, to distrain Grosseteste and his companions and compel their attendance at court on Michaelmas Day, 29 September, five weeks away.[125]

That they did not do so emerges in a notice of the subsequent hearing, in which more detail is recorded of the original complaint. This involved Roger le Fevre, and Roger Cook and his wife Juliana.[126]

> Roger le Fevre presented himself to the court on the fourth day against Roger Cook and his wife Juliana concerning the case of why they, in contravention of the prohibition etc., had pursued a case in the court of

---

[125] *Curia Regis Rolls*, 4 and 5 Henry III (London: HMSO, 1952), 171.
[126] *Curia Regis Rolls*, 4 and 5 Henry III, 328.

42 THE SCIENTIFIC WORKS OF ROBERT GROSSETESTE

Christianity [an ecclesiastical court] over the lay service of the same Roger in Bergh' [Berrow, Worcestershire] etc.; and against Master Robert Grosseteste and Master Robert of Coddington and Gilbert chaplain of Ledbury concerning the case of why they had heard the case in an ecclesiastical court etc.[127]

Juliana and Roger Cook did not attend the court, and neither did the ecclesiastical judges. This time the sheriff was instructed to compel them to attend three weeks after St Hilary's feast-day, 14th January, that is, early February, 1221. The bishop of Hereford was again ordered to distrain the three clerics, or now, to be summoned by the sheriff to explain why he had not done so. No further mention of the case is to be found.

The implications of this episode for Grosseteste's activities require a little more scrutiny. It seems reasonably clear that Grosseteste and his companions had been reported, presumably on appeal from Roger le Fevre, to the royal justices, for hearing the original case in an ecclesiastical court. So appraised, the justices had instructed the sheriff accordingly. At face value the details of the case offer no particular insights as to why Grosseteste was involved. Nevertheless, there is a little more to be gleaned about the local situation, and more still about the broader context of legal and political reform between late 1220 and early 1221, and the roles of the sheriff and bishop of Hereford. First, the local aspects. It may be a coincidence of names but it is worth noting perhaps a grant of land at Colcombe near Hereford from Hugh Foliot between 1219 and 1229x30, quite possibly on the earlier side, to one Matthew Cook.[128] While no firm connection can be made between Matthew and Roger Cook, the possibility that they were related, and the family were in

---

[127] *Curia Regis Rolls,* 4 and 5 Henry III: 'Rogerus le Fevr' optulit se quarto die versus Rogerum Cocum et Julianam uxorem ejus de placito quare ipsi contra prohibitionem etc. secuti sunt placitum in curia Christianitatis de laico feodo ipsius Rogeri in Bergh' etc., et versus magistrum Robertum Grosseteste et magistrum Robertum de Cotinton' et Gilibertum Capellanum de Ledebir' de placito quare ipsi tenuerunt placitum in curia Christianitatis etc.' For Berrow, see 'Parishes: Berrow', in *A History of the County of Worcester,* Vol. 3 (London, 1913), 257–61. British History Online: http://www.british-history.ac.uk/vch/worcs/vol3/pp257-261 (accessed 1 January 2020).

[128] Acta of Hugh Foliot in *English Episcopal Acta VII: Hereford 1079–1234,* 285–6, no. 357. The original grant had been made to Richard of Kent by Giles de Braose 1200x1215, Acta of

DATING *ON THE SPHERE* AND LOCATING ROBERT GROSSETESTE 43

some way favoured by Hugh Foliot, can be put forward. This might further explain why the case was heard by one of Hugh's trusted familiars.

The wider circumstances are also instructive; the period in which the case was heard was one of considerable political fluidity and concerns the triumvirate who held power after the resignation and then death of William Marshal in 1219. William had restored the exchequer and tried, as far as possible, to ensure that the appeal for royal justice articulated in Magna Carta was answered, restoring the judicial bench at Westminster and initiating a general eyre, a legal circuit of royal judges.[129] Serious issues remained to be resolved: the power of Llewelyn ap Iorweth in Wales and the March, the ongoing impoverishment of royal revenue and resources, the question of the king's majority and the problems this might pose for those holding royal offices (whether they would keep them or not), and the existence of powerful sheriffs and castellans not easy for the royal government to direct.[130]

As previously mentioned, the triumvirate, established between April and June 1219, consisted of an uneasy alliance between the frequently geographically dispersed Hubert de Burgh, Peter des Roches, and Pandulf, who had replaced Guala as Legate in December 1218. Hubert rapidly strengthened himself against Peter des Roches, the former's role as justiciar ensuring that he controlled the royal seal. The triumvirate met in Hereford between 28th June and 3rd July 1219, probably, as Carpenter states, 'to bring influence to bear at the election of the new bishop of Hereford', that is Hugh Foliot.[131] Meeting with the king's council at Hereford, Hubert offered important financial incentives to the sheriff of Hereford, Walter de Lacy, in the form of a suggestion of a short respite for presenting his account as well as arranging a gift from the king for Walter's wife of 100 oaks. Hubert's support for Walter may well have been related to a long-running dispute over the Three Castles (Grosmont, Skenfirth, and Whitecastle) in Gwent. Granted to Hubert by

---

Giles de Braose in *English Episcopal Acta VII: Hereford 1079–1234*, 216, no. 280; Richard sold the grant to Hugh Foliot, who granted it to Walter of Hope, who resigned it. Hugh then made the grant to Matthew.

[129] David Crook, *Records of the General Eyre* (London: HMSO, 1982).

[130] Carpenter, *Minority*, 108–27.   [131] Carpenter, *Minority*, 148.

44 THE SCIENTIFIC WORKS OF ROBERT GROSSETESTE

John in 1201, the king had given them over to William de Braose in 1204; in 1218 Hubert asserted his rights against the senior heir of the de Braose family, Reginald. Walter, as lord of Ewyas Lacy in the south-west of Herefordshire, was able to support Hubert's claim.[132] A prominent family on the Welsh March, the de Lacys also held Irish lordships in Meath. Walter's attempts to succeed to and retain his estates exemplify the vicissitudes of royal patronage during John's reign with open conflict between them in 1210. The rebellion offered Walter a route back to favour, with his English and Irish lands restored by 1216. He was also made sheriff, which he held until 1223, and acted as custodian for Hereford diocese during the period between Giles de Braose's death and Hugh Mapenor's appointment as bishop.[133]

A key moment for the triumvirate in its re-assertion of royal authority was the second coronation of the king on 17th May 1220 by Archbishop Stephen Langton in Westminster Abbey. Important steps were taken in the following months to bring defiant sheriffs and castellans to order. A great council which followed in August 1220 instituted a tax, a carucage (2 shillings on every yoked plough), which although patchy in its financial return was at least a mechanism for asserting some measure of control over the sheriffs. While problems in Poitou and with Llewelyn ap Iorweth in Wales continued, a number of royal castles had been secured by early 1221. Within the triumvirate, however, the balance of power had shifted, much to the detriment of Peter des Roches. The fact that in October 1221 Henry III would come into his majority, seems to have occasioned a wider debate in first months of 1221 about any future role Peter might play as guardian. Peter's departure on pilgrimage for Santiago de Compostela at Easter 1221 was probably connected to these discussions.[134] He was accompanied on this pilgrimage, by Hugh Foliot, bishop of Hereford.[135]

---

[132] Carpenter, *Minority*, 138–9 and 149. Brock Holden, *Lords of the Central Marches: English Aristocracy and Frontier Society, 1087–1265* (Oxford: Oxford University Press, 2008), 204–5.

[133] M. T. Flanagan, 'Lacy, Walter de (d. 1241), Magnate', *Oxford Dictionary of National Biography*, 23 September 2020: https://www.oxforddnb.com/view/10.1093/ref:odnb/9780198614128.001.0001/odnb-9780198614128-e-15864 (accessed 8 March 2020).

[134] Carpenter, *Minority*, 239–43; Vincent, *Peter des Roches*, 197–8.

[135] *Annales de Dunstaplia*, in *Annales monastici*, ed. Henry Richards Luard, 5 vols (London: Longman, Green, Longman, Roberts and Green, 1864), iii. 68.

DATING *ON THE SPHERE* AND LOCATING ROBERT GROSSETESTE   45

Grosseteste's summonses to answer to the royal justices between September 1220 and February 1221, can be placed, then, into this broader context. He heard the case at a point where royal authority, through the triumvirate, was consistently being re-asserted if inconsistently enforced. A significant element in this process was the provision of royal justice. The council meeting in 1219 at Hereford reveals the support for Walter de Lacy by Hubert de Burgh. That Hugh Foliot accompanied Peter des Roches on pilgrimage might also suggest some form of favoured or privileged relationship. In this case, some of the circumstances of the legal case become more explicable, notably the extent to which the sheriff was instructed to pursue the case, and the extent to which the bishop seems to have protected Grosseteste and his companions, by not distraining the clerics.[136] In fact, in de Lacy's absence in Ireland from 1220, his shrieval responsibilities would have been assumed, in all probability, by his deputy. There is no evidence that any further action was taken against the three clerics, or against Juliana and Roger Cook, after January 1221. Absence of evidence should not be taken to indicate, necessarily, that nothing happened, but it might be possible to speculate further, that the case was dropped as a result of shifting power dynamics affecting Hugh Foliot. Peter des Roches, and presumably Hugh, returned to England in July 1221, the former in company with the first mendicants, the Dominicans, to enter England. Given a decision to leave the kingdom by his patron, the question whether Grosseteste did so or not as well, becomes intriguing. A visit to Paris in 1221, for example, would fit reasonably well with the evidence for the chronology of his works, especially that connecting his treatises *On Comets* and the *Compotus*, to that city (see above §3).[137]

3.3.3 Haughmond Abbey
Whatever the case, as noted above, the final evidence for Grosseteste's activities in the period is found in the cartulary of Haughmond Abbey, Shropshire (in the medieval diocese of Coventry-Lichfield), compiled in

---

[136] Holden, *Lords of the Central Marches*, 204–5.
[137] This will be explored in more detail in Volume III of this series.

46 THE SCIENTIFIC WORKS OF ROBERT GROSSETESTE

the later fifteenth century.[138] Grosseteste's name appears as witness to the installation of Master John of Wroxeter to the church of Culmington, in Hereford diocese, approved by Hugh Foliot, now bishop, and presented by Abbot Osbert and the canons of Haughmond.[139] The date of the charter can be narrowed down: it postdates Hugh's appointment as bishop in 1219 and predates Abbot Osbert's replacement by Abbot William in 1226–7.[140] The other witnesses named are Master Thomas, precentor of Hereford and Master Richard of Hereford, the bishop's official. Thomas had been appointed precentor by November 1223, and in all likelihood not much before, putting the probable date range to 1223–6x7.[141] This is, then, several years after Grosseteste's appointment as a papal judge-delegate, and, in all probability, after the case involving the royal justices. The Haughmond charter is, therefore, evidence for his activities in the early 1220s. Haughmond was, by the 1220s, a significant and wealthy religious house of strict Augustinian Canons, the 'Canonici albi—White Canons', as noted by the late-twelfth- and early-thirteenth-century chronicler Gervase of Canterbury.[142] The community had consolidated and then expanded its property from the mid-twelfth century and enjoyed the particular patronage of the FitzAlan family. Culmington lay close to the north-eastern border of Hereford diocese with that of Coventry-Lichfield.

As archdeacon of Shropshire (Hereford diocese) since the 1180s, and then bishop, Hugh would have been familiar with the secular and ecclesiastical politics of the region. It is therefore all the more striking that Grosseteste should have been asked to witness the installation. That his name was recorded second, and before the bishop's official, a relatively recent post within the church at large and introduced to Hereford

---

[138] *The Cartulary of Haughmond Abbery*, ed. Una Rees (Cardiff: University of Wales Press, 1985); Jeffrey J. West and Nicholas Palmer, eds, *Haughmond Abbey: Excavation of a 12th-Century Cloister in its Historical and Landscape Context* (Swindon: English Heritage, 2014), esp. 6–27.

[139] *Cartulary of Haughmond Abbey*, 69, no. 271.

[140] *Cartulary of Haughmond Abbey*, 69, no. 271.

[141] *English Episcopal Acta VII: Hereford 1079–1234*, 262, no. 334.

[142] Gervase of Canterbury, *Mappa mundi*, in *The Historical Works of Gervase of Canterbury*, ed. William Stubbs Vol. 2 of 2 (London: Longman and Trübner, 1880), ii.436.

DATING *ON THE SPHERE* AND LOCATING ROBERT GROSSETESTE 47

only under Hugh, indicates, as Southern noted, a high status within the circles involved.[143] Hereford diocese, as noted above, was, in common with the rest of the English church, experiencing a period of re-construction after the years of the interdict and rebellion. In this sense Grosseteste at Haughmond was part of that process of re-building and stabilizing the diocese. The legal cases show Grosseteste to have been involved, in however small a part, in the reform of papal justice, of royal justice, and of diocesan jurisdiction. His relations with Hugh Foliot seem to have been strong, and Hereford seems to have provided a focus, so far as the evidence suggests, for his activities between *c.*1214 and *c.*1221, and perhaps a little later. Further dimensions can be suggested to those activities and for the context in which *On the Sphere* was composed. It is entirely possible, as the next chapter will explore, that between *c.*1215 and *c.*1225 Grosseteste was involved in the production of treatises and guides for clergy, and possibly directly to laity, on pastoral care, raising further questions and possibilities for how his career might be interpreted. In this there was more than a chronological overlap with the shorter scientific works, including *On the Sphere*. Grosseteste can be seen to have applied similar intellectual methods to the two genres with significant implications for his understanding of both.

---

[143] Southern, *Grosseteste*, 67.

2

Mathematics and Pastoral Care

Contextualizing Grosseteste's *On the Sphere*

The evidence for Grosseteste's career before the late 1220s is partial. While the lack of a detailed medieval life is perhaps a little surprising for a famous bishop of Lincoln who was put forward for canonization three times in the later thirteenth and fourteenth centuries with the customary dossier, a patchwork of evidence for the life of a medieval scholar is not at all unusual. Taken individually the records of his activity as papal judge-delegate, his summons to royal justice and witness to the charter of presentment to Culmington are inconclusive as to the shape and purpose of his activities. Taken together, and alongside the earlier evidence for his work in and around Hereford, a common thread emerges of commitment to the life of the church at a time of considerable and rapid reform of its procedures and structures locally and universally (Ch. 1, §3.3.1). The locus for that commitment seems to have been the diocese of Hereford, although that does not preclude Grosseteste's presence in other places, and in some sort of connection with archdeacon and then Bishop Hugh Foliot (Ch. 1, §3.3.3). The evidence of charter and summons indicate some form of active role in the processes of ecclesiastical re-ordering but there is additional evidence to suggest a more substantial contribution to the exercise and principles of church reform. Grosseteste's work in the practice, and apparent mis-practice, of canon law can be complemented by his work in the literature of pastoral care, which is essential, as noted by Goering, to consider in this respect.[1] It is quite possible that during the period *c.*1200–*c.*1221, that is, in the same year as

---

[1] Joseph W. Goering, 'Where and When did Grosseteste Study Theology?', in James McEvoy (ed.), *Robert Grosseteste: New Perspectives on his Thought and Scholarship* (Turnhout: Brepols, 1995), 26–35.

On the Generation of Sounds and *On the Sphere*, Grosseteste produced three works of *pastoralia*, *On the Way of Making of Confessions*, *Meditations*, and *On the Temple of God*. These are important for a fuller sense of his intellectual priorities and to contextualize further his scientific writing and possible audiences.

In addition to the textual evidence for Grosseteste's engagement with pastoral care a further body of material exists which gives insight into other aspects of his intellectual resources. This is the corpus of works that he appears to have copied himself now gathered in a Bodleian Library manuscript, MS Savile 21. The identification of Grosseteste's handwriting has been the occasion for some scholarly debate, which is assessed in what follows. The implications for Grosseteste's scientific vision are considerable, not least in demonstrating the range of primarily mathematical treatises that he copied, all of which would have been relevant to astronomical calculation. The materials in MS Savile 21 can be tied even more closely to the period of composition for *On the Sphere*. Two horoscopes which can also be identified as in Grosseteste's hand give specific dates, of 22 March and 4 October 1216. Quite what the horoscopes might mean for Grosseteste's interests is addressed later (Ch. 5, §2; Ch. 8, §2). The influence of the planets in their celestial spheres on what lies below is a central feature of ancient and medieval cosmology from which Grosseteste does not depart. In his first treatise, *On the Liberal Arts*, he discussed the role of astronomy in the calculation of the planetary conjunctions for the good practice of medicine, agriculture, and alchemy.[2] Nevertheless, it is perhaps worth noting that his negative views on judicial astrology, that is the forecasting of events, and the contradiction of human free will that this implies, were articulated clearly by the 1230s. Grosseteste's active role in pastoral care from *c.*1200 onwards might indicate that this was not a recent evolution of thought.[3] The horoscopes do, however, suggest that the rest of the mathematical material was collected at a similar time, and to this end allow a closer correlation with the conception and production of *On the Sphere*.

---

[2] Grosseteste, *De artibus liberalibus*, §§11–13; and *Knowing and Speaking*, 166–95.
[3] Robert Grosseteste, *Hexaemeron*, V.8–11, ed. Richard C. Dales and Servus Gieben (Oxford: Oxford University Press, 1982), 165–73.

50   THE SCIENTIFIC WORKS OF ROBERT GROSSETESTE

1 MS Savile 21: Contents and Script

MS Savile 21 is a composite volume which preserves a number of twelfth-and thirteenth-century treatises on astronomy/astrology and mathematics. The provenance of the booklets of which the current manuscript is comprised, their scribes, previous owners, and prior binding formats are largely unknown, although the final third of the present codex may have circulated as a unit even before the curatorial efforts of the seventeenth century that gave it its present form.[4] Two segments of the manuscript, namely, ff. 143r–160v and 161r–201r, have attracted particular attention as candidates to have been written or owned by Grosseteste.[5] The first and shorter of the two sections contains copies of the following texts. Items marked with an asterisk are those identified by S. Harrison Thomson as being in the hand of Grosseteste:[6]

I.i   ff. 143r–146v *Algorismus Jordani tam in integris quam fraccionibus demonstratus**

I.ii   ff. 146v–150r *Jordanus de fraccionibus**

I.iii   ff. 150v–151r *Thebith de proporcionibus**

I.iv   ff. 151v–153r *Thebith de figura Catha**

I.v   ff. 153v–155v *Tractatus patris Ascii Tebit filii Chore in motu accessionis et processionis* [commonly known as Pseudo-Thābit, *De motu octave sphere*]*

I.vi   ff. 155v–156r Three tables*

I.vii   ff. 156v–158r *Ars inveniendi eclipsim solis et lune**

I.viii   ff. 158r–160r Tables*

I.ix   f. 160v Horoscopes*

---

[4] F. Madan, H. H. E. Craster, and N. Denholm-Young, *A Summary Catalogue of Western manuscripts in the Bodleian Library at Oxford which have not hitherto been catalogued in the quarto series: with references to the Oriental and other manuscripts*, Vol. II, part ii, (Oxford: Oxford University Press, 1937), no. 6567, 1106–7.

[5] This identification was first made in Harrison Thomson, *Writings*, 'Grosseteste's Handwriting', *The Writings of Robert Grosseteste, Bishop of Lincoln 1235–1253* (Cambridge: Cambridge University Press, 1940), 30–3. The, identification has been supported, although cautiously by R. W. Hunt 'Appendix A', in D. Callus (ed.), *Robert Grosseteste, Scholar and Bishop* (Oxford: Clarendon Press, 1955), 132–41, at 133–4; and Southern, *Grosseteste*, 107, n. 35; and Panti, *Moti, virtù e motori celesti*, 75–6, 86.

[6] Harrison Thomson, 'Grosseteste's Handwriting', 30–3.

The longer section ff. 161r–199v contains:

II.i ff. 161r–169v Alexander de Villa Dei *Massa Compoti*
II.ii ff. 170r–176v Metrical computus
II.iii ff. 177r–181v Liturgical calendar
II.iv ff. 182r–200r *Experimentarius* (*Regula III*)
II.v ff. 200v–201r Astrological fragments*

Three issues are particularly pertinent in this encounter: the identification of Grosseteste's hand, the dating of those parts of the manuscripts so identified, and the implications of both the content and the collection of the materials for Grosseteste's interests and the resources to which he had access.

1.1 Grosseteste's Handwriting

Harrison Thomson was not alone in finding samples of Grosseteste's handwriting in surviving medieval manuscripts. M. R. James identified his hand among the marginalia of Cambridge, Corpus Christi College, MS 480 (CCCC MS 480), and the online catalogue description of Cambridge, University Library, MS Ff.1.24 (CUL MS Ff.1.24), attributes a number of marginal annotations to him. However, even a brief comparison of these samples shows that they are the work of several different scribes.[7]

A strong argument in favour of Harrison Thomson's identification of Grosseteste's handwriting in MS Savile 21 can be made following a detailed examination of the manuscripts in Plate 1 and other manuscripts that have been connected with Grosseteste. The first of these is a manuscript of biblical commentaries that he exchanged with the monks of Bury St Edmunds for a copy of the *Hexaemeron* of Basil. The *Hexaemeron* manuscript has not been identified, but the book of biblical commentaries is now Cambridge, Pembroke College, MS 7

---

[7] M. R. James, *A Descriptive Catalogue of The Manuscripts in the Library of Corpus Christi College Cambridge*, Vol. II of 2 vols. (Cambridge, 1912); Cambridge University Digital Library, 'Testaments of the Twelve Patriarchs (MS Ff.1.24)', https://cudl.lib.cam.ac.uk/view/MS-FF-00001-00024/1 (accessed 10 March 2021).

52 THE SCIENTIFIC WORKS OF ROBERT GROSSETESTE

(hereafter Pembroke MS 7).[8] One of the opening flyleaves preserves an inscription that records the transaction: 'Pledge by *Magister* Robert Grosseteste for the *Hexaemeron* of Basil'.[9]

It might be expected that the recipient of the gift, that is the monk in charge of the library at Bury St Edmunds, would have written such an inscription, particularly with the form *memoriale* rather than the donor. There are, however, other uses of *memoriale* in the first half of the thirteenth century where the word clearly means 'pledge', which makes Grosseteste's authorship of the donation inscription grammatically plausible. Beryl Smalley noted that in *The Statutes of the Priors of Durham* pertaining to Thomas de Melsonby (prior 1234–44) issued in the year 1235, there is a stipulation that the books belonging to the prior and convent should not be loaned, unless the borrower offered a pledge (*memoriale*) of equal value.[10] Furthermore, as will be set out below, the hand that wrote the inscription occurs in other manuscripts that have been associated with Grosseteste and his circle, and which have no connection with the community of Bury St Edmunds. The date of the exchange of manuscripts cannot be determined with any great precision, but presumably it took place before 1235, and not between the spring of 1229 and early November 1232 since Grosseteste refers to himself as a *magister* rather than bishop, or archdeacon.[11]

---

[8] The details of the exchange are discussed in Harrison Thomson, *Writings*, 25–6, but although A. G. Little ('Review of S. Harrison Thomson The Writings of Robert Grosseteste, 1940', *The English Historical Review*, 56 (1941), 306–9 at 308) agreed with Harrison Thomson that the inscription is in Grosseteste's hand, he noted that Harrison Thomson's reconstruction of the transaction 'is open to question'.

[9] 'Memoriale magistri Roberti grossetesti pro exameron basilii', Cambridge, Pembroke College, MS 7, f. ii v.

[10] B. Smalley, 'A Collection of Paris Lectures of the Later Twelfth Century in the MS. Pembroke College, Cambridge 7', *The Cambridge Historical Journal*, 6 (1938), 103–13 at 104: 'Quod nullus Liber de Armariolo accomodetur sine memoriali:- Item statutum est per eosdem ut nullus Liber accomodetur alicui per Librarium, vel per alium, nisi receperit memoriale aequipollens; nisi fuerit ad instanciam domini Episcopi'. See J. Raine (ed.), *Historiæ Dunelmensis scriptores tres, Gaufridus de Coldingham, Robertus de Graystanes, et Willielmus de Chambre*, Surtees Society, 9 (London and Edinburgh: J. B. Nichols & Son, 1839), xliii, from Durham, Cathedral Library, MS B.IV.26, f. 5r, available online via the Durham Priory Library Recreated project.

[11] In the earliest of his surviving letters, written, according to its most recent editors, between 1225 and 1228, Grosseteste refers to himself as a *magister*. Upon being granted the positions of archdeacon and canon of Lincoln Cathedral between spring 1229 to November 1232, Grosseteste describes himself in his letters for the period as an archdeacon, but he reverted to

CONTEXTUALIZING GROSSETESTE'S *ON THE SPHERE*   53

This hand appears again in in CUL MS Ff.1.24, a tenth-century manuscript of early Christian texts in Greek which was in Grosseteste's possession by *c.*1240, and which includes the text of the *Testaments of the Twelve Patriarchs*, a work Grosseteste would later translate from this manuscript.[12] A selection of the pertinent annotations in CUL MS Ff.1.24 includes: f. 25r (see Plate 1); f. 42v (see Plate 1); f. 107v (see Plate 3), and f. 229v (see Plate 3) and the headings added to ff. 203r–261r that name the patriarch whose testament can be found on the page below. The notes on ff. 25r and 42v were added to the *Book of Chronicles*, books I and II, the one on f. 107v relates to the *Hypomnestikon biblion Ioseppou* and the one on f. 229v comments on the *Testaments of the Twelve Patriarchs*. The first two marginal notes are attributed to the 'hand of Robert Grosseteste' in the online catalogue, but they are strikingly different and it is not easy to see how they might all be attributed to the same person even when taking into account that medieval people frequently varied the formality of their handwriting according to circumstance.

The Latin annotations in CUL MS Ff.1.24 merit disambiguation given the longstanding recognition of Grosseteste as an owner and user of the manuscript and the current scribal designations supplied in the online catalogue entry and description. The note in the lower margin of f. 25r, written in plummet, that is with a lead point rather than ink, is described as being 'in the hand of Robert Grosseteste', but it is in fact the sole contribution of an unknown scribe, and describes a missing portion of the main text in Latin. The note in the outer and lower margin of f. 42v, also described as being 'in the hand of Robert Grosseteste', discusses the etymology of 'palm trees' φοίνικας, but it is not at all like the hand of the annotation on f. 25r, nor the donation note in Pembroke MS 7,

---

the title of *magister* in a letter written in early November 1232 having renounced these clerical positions. In his letters written after being elected bishop of Lincoln in March 1235 he uses his episcopal epithet to describe himself. See Robert Grosseteste, Letters 1–11, in *Epistulae*, ed. Henry Richards Luard (London: Longman, Green, Longman, and Roberts, 1861), 1–54, and *The Letters of Robert Grosseteste, Bishop of Lincoln*, trans. F. A. C. Mantello and Joseph Goering (Toronto: University of Toronto Press, 2010), 22, 35–86, but see footnote 94 below.

[12] 'Catalogue Entry: Testaments of the Twelve Patriarchs (MS Ff.1.24)', Cambridge University Digital Library, https://cudl.lib.cam.ac.uk/view/MS-FF-00001-00024/1 - note that the record includes unpublished notes by M. R. James. On the translation, see M. de Jonge, 'Robert Grosseteste and the Testaments of the Twelve Patriarchs', *The Journal of Theological Studies*, 42 (1991), 115–25 at 118.

54 THE SCIENTIFIC WORKS OF ROBERT GROSSETESTE

and is clearly not the work of the same scribe. The annotator of f. 42v also added notes to ff. 45r, 53v, and is probably responsible for messy erasures on ff. 87r and 92r; this scribe, writing in a distinctively large and unsteady hand, also added an annotation to another Greek manuscript associated with Grosseteste CCCC MS 480, f. 9v (Plate 1), and so may have been a member of Grosseteste's circle as Grosseteste's interest in Greek developed.

The annotations on CUL MS Ff.1.24 ff. 107v, 229v (Plate 3) and the headings added to *The Testaments of the Twelve Patriarchs* closely resemble the aspect, ductus, and letterforms of the donation inscription in Pembroke MS 7, f. ii v (Plate 2).[13]

The distinctive o+r letter combination that is present in Pembroke MS 7 in *memoriale* is also present in CUL MS Ff.1.24, f. 229v in *laboribus,* as is the e with the long and high tongue, and the p with the backward flick to its descender. The only way for these handwriting specimens to match is if they are samples of Grosseteste's handwriting.

The identification of Grosseteste's handwriting in Pembroke MS 7 and CUL MS Ff.1.24 provides a sufficient foundation for the assessment of the writing in the Savile manuscript. Close comparison of the relevant portions of Pembroke MS 7, CUL MS Ff.1.24, and Savile MS 21 shows that Grosseteste copied Savile MS 21 ff. 143r–160v. The idiosyncratic, and therefore diagnostic features of his handwriting can be described as follows (and see Plate 4):

a is teardrop-shaped and has a single compartment; ascenders (with the exception of d) are roughly perpendicular to the baseline;

d is wavy, extravagant, and tilted at a 45° angle relative to the baseline; the final minim of h, m, and n, is often pulled downward and backward below the baseline for decorative effect, and p has a backward flick to its descender as noted above; two forms of r are used, but where r follows o there is a preference for a 2-shaped r that sits close to the bowl of the o;

s differs depending on the location of the graph within the word and the graphs around it, but there is a preference for round-s in the initial position with the bottom part of that letter sometimes dropping below the baseline, and tall s with a deep hook and a foot that drops below the

---

[13] There are also notes in his hand on ff. 108v and 215v.

CONTEXTUALIZING GROSSETESTE'S *ON THE SPHERE*   55

baseline in the middle of words; *et* is 7 rather than &, and the downstroke of the 7 is usually crossed with a straight diagonal stroke ascending up and to the right; the standard abbreviation stroke ‾ is straight and relatively long, and all of the standard abbreviation sigla and conventions are in frequent use.

Although the letterforms mentioned above are sufficient to ascribe a portion of MS Savile 21 to Grosseteste's pen, further diagnosis is possible by comparing the Hindu-Arabic numerals present in both Savile 21 and CUL MS Ff.1.24. Grosseteste used Hindu-Arabic numerals in the tables found on MS Savile 21 ff. 155v–156r, 158r–160v, but he preferred Roman numerals in the prose extracts that are interspersed with the tables. In CUL MS Ff.1.24 the numerals were added to keep a count of each patriarch in *The Testaments of the Twelve Patriarchs* alongside their names, perhaps to allow for more efficient navigation of the manuscript as he was translating the text.

The use of Hindu-Arabic numerals was still somewhat rare in England in the first half of the thirteenth century, and there was still a relatively large degree of variation in how people wrote the numbers 2, 4, 5, and 7 in particular, so the similarity of these glyphs, or features, is further evidence that Grosseteste wrote MS Savile 21.[14]

1.2 Dating the Grosseteste Sections of MS Savile 21

Internal evidence allows for a relatively narrow date range to be assigned to MS Savile 21 ff. 143r–160v (Part I) by virtue of the two horoscopes mentioned below, both of which are found on f. 160v. The diagrams on f. 160v both have captions explaining the dates to which they correspond according to the Arabic *hijrī* calendar system. The first, upper, chart is for the Sun's entry into Aries and is annotated:

---

[14] For an introduction to the use of Hindu-Arabic in western Europe, see G. F. Hill, *The Development of Arabic Numerals in Europe: Exhibited in Sixty-Four Tables* (Oxford: Oxford University Press, 1915); C. S. F. Burnett, *Numerals and Arithmetic in the Middle Ages* (Farnham: Ashgate, 2010); and C. S. F. Burnett, 'The Palaeography of Numerals', in F. T. Coulson and R. G. Babcock (eds.), *The Oxford Handbook of Latin Palaeography* (Oxford: Oxford University Press, 2020), 24–36.

56 THE SCIENTIFIC WORKS OF ROBERT GROSSETESTE

Hec figura anni 611 annis arabum, 11 mensibus 2 diebus, 48 minutis, 22 die martis post meridiem hora jovis [This is the horoscope of the year [starting after] 611 Arabic years, 11 months and 2 days, 48 minutes; 22nd day of March, hour of Jupiter in the afternoon].

This translates to 3 Dhū al-Ḥijjah in the year 612 of the Hijra, which was 23 March 1216, not 22 March, as stated. The reason the Islamic date above indicates 23 March is that the day in the Toledan Tables begins from the previous noon.

The second, lower, chart pertains to a conjunction of Saturn and Mars with the inscription:

Hec est figura coniunccionis saturni et iovis annis arabum 612 perfectis, mensibus 5, diebus 20, horam[reading *horis*] equalibus 21, minuta 58 [This diagram is for the conjunction of Saturn and Jupiter; 612 Arabic years completed, 5 months and 20 days, 21 equal hours, 58 minutes].[15]

The chart in fact shows a conjunction of Saturn and Mars, rather than Jupiter, in Libra and accords with the inscription added into the upper chart:

Coniunccio saturni et martis annis arabum perfectis 612, mensibus 5, die vicesimo primo sexti mensis in libra 3 gradus 50 minuta [Conjunction of Saturn and Mars; 612 Arabic years completed and 5 months, the 21st day of the 6th month, Libra 3 degrees, 50 minutes].

The corresponding dates for the lower chart are 13 Jumādá II of the following Arabic year to the previous diagram, which corresponds to 4 October 1216. Taken together these dates allow 1216 to be posited as an important point for Grosseteste's active engagement with the texts that he copied down.

---

[15] This was wrongly transcribed by Harrison Thomson, *Writings*, 32, reading 'aquarii[?]' for 'equalibus'.

CONTEXTUALIZING GROSSETESTE'S *ON THE SPHERE*   57

1.3 Resources for Study

The material gathered by Grosseteste in MS Savile 21 ff. 143–60 would fit very well for preparation of *On the Sphere* and also for the slightly later *Compotus*. The contents are of clear relevance for research into astronomy and its applications, and any calculations involved. They include recent mathematical works by Jordanus; three short treatises attributed to Thābit, one of which is *On the Motion of the Eighth Sphere*; calendar tables relating to the Arabic calendar (and to calendar conversions); materials on eclipses; a note on horoscopes and tables for the 'zodiacal houses' (§1) as well as the two horoscopes.[16] The last folio contains the two horoscopes as noted above. If these texts were indeed copied out by Grosseteste himself then it is clear that he was working on up-to-date aspects of the subjects represented. Moreover, the mathematical works of Jordanus were themselves very advanced in relation to the standard treatments of the subject used in schools in the early thirteenth century.[17] It may not be coincidental that Jordanus not only set out explanations and practical expositions of many mathematical fields relevant to astronomy but also considered the 'science of the stars [*Scientia astrorum*]' to be an integral part of mathematical study. The fact that he distinguished between the understanding of the stars in relation to future events on the one hand, and the 'mathematical science of the stars' on the other, and considered the former to be an area for judgement rather than mathematical precision, would, perhaps, have appealed also to Grosseteste.[18]

---

[16] For Thābit's works, see *The Astronomical Works of Thābit b. Qurra*, ed. F.J. Carmody (Berkeley: University of California Press, 1960). Helpful discussion is provided by F. J. Ragep, 'Al-Battānī, Cosmology and the Early History of Trepidation in Islam', in Josep Casulleras and Julio Samsó (eds.), *From Baghdad to Barcelona: Studies in the Islamic Exact Sciences in Honour of Prof. Juan Vernet*, Anuari de Filologia, XIX, 2 vols (Barcelona: Instituto 'Millás Vallicrosa' de Historia de la Ciencia Arabe, 1996), i 267–98.

[17] Alexander Neckam's *Sacerdos ad altare*, ed. C. J. McDonough, CCCM 227 (Turnhout: Brepols, 2010), c. XII, *De arismetica et musica* specifies Boethius and Euclid as the fundamental works, to be followed by Boethius on music. The comments make it clear that arithmetic is here a subject in the service of music. Alexander Neckham, *Sacerdos ad altare*, 196–7. An Oxford statute of the early fourteenth century still specified study of Boethius and Euclid but added Sacrobosco's *De sphera* and an unspecified *algorismus*. See *Statuta antiqua universitatis oxoniensis*, ed. S. Gibson (Oxford: Oxford University Press, 1931), 32–3.

[18] On Jordanus' work, and this division of the science of the stars, see J. Høyrup, 'Jordanus de Nemore, Thirteenth-Century Innovator; an Essay on Intellectual Context, Achievement and Failure', *Archive for the History of the Exact Sciences*, 38 (1988), 307–63.

58   THE SCIENTIFIC WORKS OF ROBERT GROSSETESTE

The collection of mathematical and astronomical material is intriguing in its own right. It sheds some light onto Grosseteste's personal resources for study, and the books and booklets that he may have acquired in the earlier part of his life. Grosseteste's library, as Richard Hunt pointed out, is not straightforward to identify.[19] There is no surviving catalogue, nor does his will survive, although it seems that the Franciscans had some responsibility for his own writings and books by others that he owned. Adam Marsh wrote to the dean of Lincoln, Richard Gravesend in May 1254, a few months after Grosseteste's death, on the subject stating that he, Adam, would consult with the provincial minister, William of Nottingham, as to the course of action to take.[20] Nicholas Trevet recorded in the early fourteenth century that Grosseteste had bequeathed all of his books to the Oxford Franciscans.[21] There they remained, in the convent, rather than the student, library, and a good number can be identified from marginal notes and extracts made by others, in particular the redoubtable Thomas Gascoigne.[22] The friars' library seems to have been in a process of dispersal from the later fifteenth century though at what pace is difficult to tell. John Leland's famous description of the cobwebbed, decrepit, state of the library might, as Jeremy Catto suggests, indicate change over the short term just as much as the long.[23] In a letter to Thomas Cromwell, Leland singled out the books of Bishop Robert as having been stolen by Franciscan visitors to the Oxford house.[24]

Most of the books that can be identified as having belonged to Grosseteste are of speculative or pastoral theology, or biblical exegesis. Material from the period before his lectorship to the Oxford Franciscans is difficult to single out. Amongst the possibilities are a copy of Aristotle's *Physics* with notes by Grosseteste which was included in the Greyfriars

---

[19] R. W. Hunt, 'The Library of Robert Grosseteste', in Callus, *Robert Grosseteste*, 121–45.
[20] Adam Marsh, Letter 75, *The Letters of Adam Marsh*, ed. and trans. C. H. Lawrence, 2 vols (Oxford: Oxford University Press, 2006), i. 184.
[21] Nicholas Trivet, *Annales ex regum Angliae, 1135–1307*, ed. Thomas Hog (London: English Historical Society, 1845), 243.
[22] On Gascoigne, see Southern, *Robert Grosseteste*, 313–15.
[23] Jeremy Catto, 'Franciscan Learning in England, 1450–1540', in James Clark (ed.), *The Religious Orders in Pre-Reformation England* (Woodbridge: Boydell, 2002), 97–104 at 97.
[24] Hunt, 'The Library', 132.

CONTEXTUALIZING GROSSETESTE'S *ON THE SPHERE* 59

library, according to a later comment by William of Alnwick, lector in 1316–17.[25] Another intriguing note, which may be connected to the *Dicta* although the geometrical formulations of *On the Sphere* might also be recalled, is found in Corpus Christi College, Oxford, MS 251, f. 83v: 'The square of a circle: Let there be a semicircle around the straight line AB... to which is subtended a squared side. I found this demonstration in Oxford in a certain note [*cedula* for *schedula*] of the Lord of Lincoln.'[26] Another contender for a book from Grosseteste's earlier years is a copy of Abū Maʿshar (work not specified) which he had glossed himself as noted by John Lathbury in his mid-fourteenth-century treatise *On Lamentations.*[27] Abū Maʿshar's *Introduction to Astronomy* was a source for Grosseteste's first treatise *On the Liberal Arts.*[28] To these suggestions can be added the collection of works that he copied, now preserved in MS Savile 21. All of these indications help in an incremental manner, an assessment of Grosseteste's interests and resources at the time he wrote *On the Sphere*.

The dating of the sections of MS Savile 21 ff. 143r–160v while not providing a fixed point for Grosseteste's whereabouts does give some more specific chronology to his intellectual activities. In terms of the date for *On the Sphere* it is plausible to regard the copying of treatises by Jordanus and those attributed to Thābit as preparatory for Grosseteste's own work, placing composition after 1216. This fits the broader historical context for England with 1216–17 dominated by political turmoil and military action (Ch. 1, §2). What Grosseteste's copying of the particular contents of MS Savile 21 also reveal is a scholar willing, and able, to engage with new material and approaches. As will be seen later (Chs 6 and 7) this instinct was married to a sustained critical encounter with Aristotle's natural philosophy which had been available to Latin scholars, at least in the translations from Greek by James of Venice, since the mid-twelfth century. A very similar situation can be suggested for

---

[25] *The Friars' Libraries*, ed. K. W. Humphreys (London: The British Library in association with The British Academy, 1990), 224.

[26] *Friars' Libraries*: 'Quadratura circuli. Esto circa lineam rectam AB semicirculus... cui subtenditur latus quadrata. Hanc demonstracionem inueni Oxon' in quadam cedula domini Lincol'.'

[27] *Friars' Libraries*, 225.　　　[28] *Knowing and Speaking*, 193–5.

60 THE SCIENTIFIC WORKS OF ROBERT GROSSETESTE

Grosseteste's activities and learning in a different, although not entirely unrelated, sphere, with which he was occupied at the same time. This is the literature of pastoral care to which he made a notable contribution, characterized by the same balancing of contemporary and more established sources. Grosseteste's involvement in *pastoralia*, as part of the generation of clerics who would go on to enact and champion the vision of church reform set down in the great oecumenical council of 1215, Lateran IV, provides a larger context still in which to place assessment of his career and of the composition of *On the Sphere*.

2 Pastoral Care

Pastoral care and its literature play an important role in identifying the shape of Grosseteste's career and offer, too, a different set of insights into his intellectual formation. Goering and Frank Mantello point out in a series of articles, summarized by the former in a separate essay the enduring interest of *pastoralia* to Grosseteste.[29] This material is particularly useful for the period from 1199 to *c.*1225 when his whereabouts are less easy to pin down. Nevertheless, while Grosseteste's interests in this arena demonstrably reach into his early career it is not easy to be precise on exactly how far back they extended. There are few indications, for example, that he came under the direct tutelage of William de Montibus at Lincoln, the foremost master for pastoral care literature in England who had returned to the city from Paris at some point in the 1180s. Even if at Lincoln, Grosseteste at this point would have been studying in arts rather than theology and William taught to a rather narrower,

---

[29] J. Goering and F. A. C. Mantello, 'The Early Penitential Writings of Robert Grosseteste', *Recherches de théologie ancienne et médiévale*, 54 (1987), 52–111; J. Goering and F. A. C. Mantello, 'The *Meditaciones* of Robert Grosseteste', *Journal of Theological Studies*, N.S. 36 (1985), 118–28; J. Goering and F. A. C. Mantello, 'The *Perambulauit Iudas...* (*Speculum confessionis*) Attributed to Robert Grosseteste', *Revue Bénédictine*, 96 (1986), 125–68; Goering, 'Where and When'. What follows draws heavily particularly on Goering's position in these matters, which formed a major part of his scholarly oeuvre. See, by contrast, Leonard Boyle, 'Robert Grosseteste and the Pastoral Care', *Medieval and Renaissance Studies*, 8 (1979), 3–51, repr. in his *Pastoral Care, Clerical Education, and Canon Law, 1200–1400* (London: Variorum, 1981), 3–9, where Grosseteste's first interest in the cure of souls is dated firmly to the rectorship of Abbotsley in 1225.

CONTEXTUALIZING GROSSETESTE'S *ON THE SPHERE*   61

more cloistered, and more theologically able audience, than the wider, parochial, audience to whom Grosseteste addressed his works in this genre.[30] Nevertheless, he is not likely to have been unacquainted with William with whom he probably witnessed a charter for the bishop of Lincoln, Hugh of Avalon, 1189x92, and in view of William's role as public preacher on penance.[31] It is worth returning also to the letter of recommendation written to Bishop William de Vere of Hereford by Gerald of Wales, on Grosseteste's behalf. Gerald emphasizes his subject's skill at law, in medical care, and in the liberal arts. He goes on, however, to lay particular stress on Grosseteste's integrity of faith:

> But while most masters of the abovementioned faculties for the most part are prone to be of dubious faith, you will know him as character-ized by faith beyond all other virtues and gifts of the soul with which he excels, and conspicuous by his fidelity.... you will find him a man in whom your spirit may find rest.[32]

Pastoral care, as Goering points out, was a subject that William de Vere seems to have taken seriously during his reign at Hereford.[33] Guy, prior of the Augustinian house at Southwick, praised William for his interest in penitential literature, and, in the 1190s, sent to him a copy of his latest work in this area, which laid heavy emphasis on confession.[34] Gerald of Wales himself produced in his *Gemma ecclesiastica—The Jewel of the Church* a two-fold manual on clerical morality and the sacraments, designed for clergy in Wales articulated 'in plain ordinary language without ornament as in commonplace judgments, in order that it may

[30] Joseph W. Goering, *William de Montibus (c 1140–1213), The Schools and the Literature of Pastoral Care* (Toronto: Pontifical Institute of Mediaeval Studies, 1992), 12–13, 42–57, 98–9.

[31] *Knowing and Speaking*, 17; Goering, 'Where and When', 28.

[32] Gerald of Wales, Letter XVIII in *Symbolum electorum, pars prima*, in *Giraldi Cambrensis opera*, ed. J. S. Brewer, 8 vols, Rolls Series, 21 (London: Longman, 1861–91), i. 249: 'Cum enim praedictarum facultatum artifices fide plerumque vacillare soleant, praeter caeteras virtutes et animi dotes quibus excellit, fide noveritis ipsum et fidelitate conspicuum.... et in quo plurimum quiescet spiritus vester invenietis.'

[33] Goering, 'Where and When', 28.

[34] D. A. Wilmart, 'Un opuscule sur la confession composé par Guy de Southwick vers la fin du XIIe siècle', *Recherches de théologie ancienne at médiévale*, 7 (1935), 337–52 at 338.

62   THE SCIENTIFIC WORKS OF ROBERT GROSSETESTE

be understood'.[35] This work he composed in Lincoln, 1194–9, during the extended academic retreat in which he seems to have met Grosseteste, before recommending him to Hereford.[36] All of these examples, taken as a piece, show that the circles to which Grosseteste can be connected in his earlier life placed some premium on the composition of *pastoralia*.

2.1 The Shaping of Pastoral Care

The concerns of Guy of Southwick, of William de Vere, Gerald of Wales, and slightly later, Grosseteste, respond to the much greater movement for pastoral care, common to the whole of medieval Europe, and which dominated discussion of the role, purpose, and function of the church.[37] This was especially the case from the middle decades of the twelfth century onwards, with new developments in theology and canon law, drawing on a range of different thinkers active for the most part in northern France, in the environs of Paris, or northern Italy around Bologna.[38] The tools were being forged for a church committed to the efficacy of its mission of salvation, from the creation of the glosses to the bible, and attempts to systematize questions of church doctrine by Hugh of St Victor and in Sentence Collections, notably the *Four Books of Sentences* of Peter Lombard. This process would also include, with the caveat that modern sub-disciplinary distinctions do not apply, the 'moral' or 'practical', approach to theology to be found in, for example, Peter Comestor, Peter the Chanter, and William de Montibus, and growing formalization of canon law under Master Gratian and his glossators, and local decretalists.[39] In pursuit of the latter the emphasis

---

[35] Gerald of Wales, *Gemma ecclesiastica*, Proemium, in *Giraldi Cambrensis opera*, ii.6: 'publicis admodum et planis absque ornatu tam verbis quam sententiis rudibus, solum ad intelligendum exposita'. English translation from Gerald of Wales, The Jewel of the Church, trans. John J. Hagen (Brill: Leiden, 1979), pp. 3–4.

[36] See *Knowing and Speaking*, 18–21.

[37] See Ronald J. Stansbury (ed.), *A Companion to Pastoral Care in the Late Middle Ages (1200–1500)* (Leiden: Brill, 2010).

[38] R. W. Southern, *Scholastic Humanism and the Unification of Europe*: Vol. I, *Foundations* (Oxford: Blackwell, 1995), 102–33, 243–318.

[39] Alexander Andrée, 'Editing the Glossa "Ordinaria" on the Gospel of John: A Structural Approach,' in Elisabet Göransson et al. (eds.), *The Arts of Editing Medieval Greek and Latin: A Casebook* (Toronto: Pontifical Institute of Mediaeval Studies, 2016), 1–20; Lesley Smith, *The Glossa Ordinaria: The Making of a Medieval Bible Commentary* (Leiden: E. J. Brill, 2009); Beryl

CONTEXTUALIZING GROSSETESTE'S *ON THE SPHERE*   63

placed on penance and confession in this period is distinctive, not least in a sacramental role, and this, alongside a similar stress on the importance of the Eucharist, shaped in particular and decisive ways the practice and belief of the medieval church in the later Middle Ages.[40] The whole complex of ideas surrounding pastoral care in this period were driven by a simpler injunction; the requirement of the church through its ordained members to ensure the spiritual health of their charges, in parishes, dioceses, provinces, and the whole of Christendom. The consequences for not doing so involved danger to the souls of those left uncared for, and equally for those who were delinquent in their office. None of this was new for the church; the cure of souls, according to Gregory Nazianzen (329/320–389/390) was 'the art of arts, and the science of sciences', the 'ars artium et disciplina disciplinarum' as rendered into Latin by Rufinus of Aquileia a few decades later.[41] The phrase was adopted enthusiastically by Pope Gregory the Great, and there is a long, and varied, history to the subject. Within the period of Grosseteste's lifetime, however, an urgency can be identified in the teaching of the arts of pastoral care and the role to be given to the parish priest.[42] This is represented not only by masters within the schools, but in the decrees of

Smalley, *The Study of the Bible in the Middle Ages*, 3rd ed. (Oxford: Blackwell, 1983); Paul Rorem, *Hugh of St Victor* (Oxford: Oxford University Press, 2009); and Hugh of St Victor, *De sacramentis*, PL 176, English translation, Roy J. Deferrari, *Hugh of St Victor: On the Sacraments of the Christian Faith* (Cambridge, MA: Harvard University Press, 1951); Philipp Rosemann, *The Story of a Great Medieval Book: Peter Lombard's 'Sentences'* (Toronto: Toronto University Press, 2007); on the distinctions which have been applied, with varying degrees of helpfulness, see Goering, *William de Montibus*, 36–42; John Baldwin, *Masters, Princes and Merchants: The Social Views of Peter the Chanter and His Circle*, 2 vols (Princeton: Princeton University Press, 1970); Atria A. Larson, *Master of Penance: Gratian and the Development of Penitential Thought and Law in the Twelfth Century* (Washington, DC: Catholic University of America Press, 2014); James A. Brundage, 'The Teaching and Study of Canon Law in the Law Schools', in Wilfried Hartmann and Kenneth Pennington (eds.), *The History of Medieval Canon Law in the Classical Period, 1140–1234: From Gratian to the Decretals of Pope Gregory IX* (Washington, DC: The Catholic University of America Press, 2008), 98–120.

[40] Joseph Goering, 'The Scholastic Turn (1100–1500): Penitential Theology and Law in the Schools', in Abigail Firey (ed.), *A New History of Penance* (Leiden: Brill, 2014), 219–38, at 229–33.

[41] Gregory Nazianzen, *Oration 2 'In Defence of his Flight to Pontus'*, 27, in Rufinus, *Orationum Gregorii Nazianzeni Novem Interpretatio*, ed. A. Engelbrecht, CSEL 46 (Vienna: Österreichische Akademie der Wissenschaften, 1910).

[42] Joseph W. Goering, 'The Internal Forum and the Literature of Penance and Confession', *Traditio*, 59 (2004), 175–227, reprinted and updated in W. Hartmann and K. Pennington (eds.), *The History of Courts and Procedure in Medieval Canon Law* (Washington, DC: The Catholic University of America Press, 2016), 379–428, at 381–7.

64 THE SCIENTIFIC WORKS OF ROBERT GROSSETESTE

the Third and Fourth Lateran Councils, in 1179 and 1215 respectively. Where the former speaks of the need for cathedrals to sustain teachers for the children of the poor and the clerics of that church, by the latter wider concerns are voiced for the impact of pastoral care.[43] Canon 21 includes the well-known injunction that all Christians should make an annual confession, coupled with an insistence on taking the Eucharist at least at Easter. This, however, is followed soon after by provision for competent practitioners.

To guide souls is a supreme art. We therefore strictly order bishops carefully to prepare those who are to be promoted to the priesthood and to instruct them, either by themselves or through other suitable persons, in the divine services and the sacraments of the church, so that they may be able to celebrate them correctly. But if they presume henceforth to ordain the ignorant and unformed, which can indeed easily be detected, we decree that both the ordainers and those ordained are to be subject to severe punishment. For it is preferable, especially in the ordination of priests, to have a few good ministers than many bad ones, for if a blind man leads another blind man, both will fall into the pit.[44]

This is exactly the constituency to which Grosseteste's early works on pastoral care were pitched. The proper training of priests is a well-known concern of his episcopacy, but one whose roots lie far deeper than 1235.[45]

---

[43] *Concilium Lateranense III a. 1179*, Canon 18, in *Conciliorum Oecumenicorum Decreta*, ed. J. Alberigo, J. A. Dossetti, P. P. Joannou, C. Leonardi, and P. Prodi, 3rd ed. (Bologna: Istituto per le Scienze Religiose, 1973), 211–25, at 220: 'magistro qui clericos eiusdem ecclesiae et scholares pauperes gratis doceat'.

[44] *Concilium Lateranense IV a. 1215*, Canon 27, in *Conciliorum Oecumenicorum Decreta*, 230–7, at 247: 'Cum sit ars artium regimen animarum districte praecipimus ut episcopi promovendos in sacerdotes diligenter instruant et informent vel per se ipsos vel per alios viros idoneos super divinis officiis et ecclesiasticis sacramentis qualiter ea rite valeant celebrare. Quoniam si ignaros et rudes de caetero ordinare praesumpserint quod quidem facile poterit deprehendi et ordinatores et ordinatos gravi decrevimus subiacere ultioni. Satius est enim maxime in ordinatione sacerdotum paucos bonos quam multos malos habere ministros quia si caecus caecum duxerit ambo in foveam dilabuntur. Quidam licentiam cedendi cum instantia postulantes ea obtenta cedere praetermittunt'. English translation from *Decrees of the Ecumenical Councils*, 2 vols., i: *Nicaea I to Lateran V*, ed. Norman P. Tanner (London: Sheed and Ward, 1990), available online at: https://www.papalencyclicals.net/councils/ecum12-2.htm (accessed 30 June 2020).

[45] Philippa M. Hoskin, *Robert Grosseteste and the 13-Century Diocese of Lincoln* (Brill: Leiden, 2019), 115–25, 128–52.

CONTEXTUALIZING GROSSETESTE'S *ON THE SPHERE* 65

Confession and penance lie, therefore, at the heart of the rapid and significant shift in emphasis in pastoral care, and its literature, in the thirteenth century, to which Grosseteste contributed.[46] They form part of a complex and organic relationship between canon law and theology. A metaphor which rapidly became commonplace depicted two spheres, or forums, where the external was the location of the law of the church, and the internal, that of conscience, served in particular by confession and penance.[47] The effects of this shift were universal. As Goering puts it: 'canon law was not just a system for lawyers, judges, and administrators, but was a body of jurisprudence that affected everyone, in the most intimate ways, in the confessional'.[48] As will be seen below, Grosseteste served in both the internal and external forum, appointed at least once as a papal judge-delegate in the diocese of Hereford, as well as composing works to guide priest-confessors.

2.2 Grosseteste's *Pastoralia* to *c*.1221

The body of work Grosseteste produced in this genre consists of six works principal works: *On the Way of Making Confession* (*De modo confitendi*), the *Meditations*, and the *Templum Dei*, the *Mirror of Confession* (*Speculum confessionis*), *God is* (*Deus est*), and the *God Known in Judea* (*Notus in Iudea Deus*).[49] Precise dating is impossible, but some progress can be made if, again, pre-conceived notions of Grosseteste's career are put to one side, for example, the notion that he

---

[46] L. E. Boyle, 'The Inter-Conciliar Period 1179–1215 and the Beginnings of Pastoral Manuals,' in F. Liotta (ed.), *Miscellanea Rolando Bandinelli Papa Alessandro III* (Siena, 1986), 45–5, and his collection of articles in *Pastoral Care*; Peter Biller and Alastair Minnis (eds.), *Handling Sin: Confession in the Middle Ages* (Woodbridge; Boydell & Bewer, 1999); Alexander Murray, 'Confession before 1215', *Transactions of the Royal Historical Society*, 6th ser., 3 (1993), 51–81, repr. in his *Conscience and Authority in the Medieval Church* (Oxford: Oxford University Press, 2015), 19–48.

[47] Goering, 'Internal Forum', in Hartmann and Pennington, *History of Medieval Canon Law*, 379–428.

[48] Goering, 'Internal Forum', 380.

[49] See footnote 5 above; and Siegfried Wenzel, 'Robert Grosseteste's Treatise on Confession, *Deus est*', *Franciscan Studies*, 30 (1970), 218–93; J. Goering and F. A. C. Mantello, '*Notus in Iudea Deus*: Robert Grosseteste's Confessional Formula in Lambeth Palace MS 499', *Viator*, 18 (1987), 253–73.

66   THE SCIENTIFIC WORKS OF ROBERT GROSSETESTE

began his magistracy in theology in the earlier 1200s, and by analysis of the sources used. The last two treatises appear to date from the episcopal period, the *Mirror of Confession* from around the second half of the 1220s, with the first three having good, or reasonable, claim to date to the period *c.*1200–*c.*1221.[50]

It is important to note here that Grosseteste would have required no formal, scholastic, training in theology to develop and sustain interest and expertise in penitential literature.[51] Grosseteste required no period of study at Paris, including any putative mastership in theology, for these endeavours, and as established in the previous chapter (Ch. 1, §3.2), hard evidence for Grosseteste's studies at Paris in the first decade of the thirteenth century is non-existent. None of the sources used in the pastoral care works was exclusive to scholastic theology. As Goering puts it: 'Any educated cleric, especially one like Grosseteste who was involved in ecclesiastical administration as well as scholastic teaching, could make full use of them without undertaking formal training in the schools of theology'.[52] Nor would it have been necessary for Grosseteste to have been a priest to write about the technicalities of confession. Grosseteste was still in deacon's orders in 1225 on receiving his benefice at Abbotsley, but he would have been in good company; an example of a well-known contemporary also not priested at the time that they wrote, was the later Innocent III.[53]

2.2.1 *On the Way of Making Confession (De modo confitendi)*

On the Way of Making Confession, 'a severely practical work' according to its editors, predates the reception of the canons of Lateran IV since it refers to the degrees of personal relationship within which marriage is prohibited as seven, not four, as the 1215 council ordered.[54] Moreover, Grosseteste makes use of the standard penitential sources to hand at the turn of the thirteenth century, with particular reliance on the collection

---

[50] Summary in Goering, 'Where and When', 29–33. The *Mirror of Confession* will be discussed in more detail in subsequent volumes in this series.

[51] Goering, 'Where and When', 27–9.   [52] Goering, 'Where and When', 29.

[53] Southern, *Grosseteste*, 69; Goering, 'Where and When', 27.

[54] Goering and Mantello, 'The Early Penitential Writings', 52; *Concilium Lateranense IV a. 1215*, Canon 50, in *Conciliorum Oecumenicorum Decreta*, 257–8.

CONTEXTUALIZING GROSSETESTE'S *ON THE SPHERE*   67

by Burchard of Worms (*c.*1020), and limited citation of patristic texts.[55] A quotation from the *Book on Penitence* (*Liber poenetentialis*) written at some point between 1208 and 1213 by the canon-penitentiary of St Victor in Paris, Robert of Flamborough, might indicate a date after which Grosseteste's treatise was composed.[56] However, it is quite likely, as Goering argues that this was a later addition to a text which enjoyed a complex reception, and whose contents were treated creatively.[57] A date before 1208, and certainly before 1213, can therefore be proposed. In other words, *On the Way of Making Confession* might be placed during the period of the interdict, and Grosseteste's visit to France, or, conceivably, before that, possibly as early as his association with William de Vere, 1195–8, or in the years following. In the context of his other writings this is at about the same point, then, as the composition of *On the Liberal Arts* and *On the Generation of Sounds*.

The structure of *On the Way of Making Confession* appears to have been at least two books, the first book being advice concerning confession, the second a treatise on penance, in addition to a collection of traditional penitential canons.[58] The second part is dependent on, and assumes knowledge of, the traditional canons, which, Grosseteste stresses, should be followed. The priest should not change the nature of the penance enjoined, though he could temper the length or intensity of the exercise, and the treatise offers examples of how this might be achieved.[59] The first part, though showing its debt to Burchard, adds something new. Grosseteste includes the first systematic list of practical questions for the confessor to pursue, a confessional interrogatory.[60] A series of questions to ask of the person confessing concern sins related to the seven vices and the sacraments. The longest versions incorporate a fuller analysis by Grosseteste of the ways in which sins can be identified and how they

---

[55] Goering and Mantello, 'The Early Penitential Writings', 65; Greta Austin, 'Jurisprudence in the Service of Pastoral Care: The "Decretum" of Burchard of Worms', *Speculum*, 79 (2004), 929–59.

[56] Robert of Flamborough, *Liber Poenitentialis*, ed. J. J. F. Firth (Toronto: Pontifical Institute of Mediaeval Studies, 1971).

[57] Goering, 'Where and When', 30.

[58] Goering and Mantello, 'The Early Penitential Writings', 63.

[59] Goering and Mantello, 'The Early Penitential Writings', 67–71.

[60] Goering and Mantello, 'The Early Penitential Writings', 64–5.

68 THE SCIENTIFIC WORKS OF ROBERT GROSSETESTE

manifest themselves.[61] A particularly good example of this is Pride, as noted by Goering and Mantello since, in this case, Grosseteste explains both its origins and the circumstances in which it flourishes, and therefore the best way in which the confession might be guided. This is seen clearly in the extracts below:

12. Concerning Pride

If he [the one confessing] has offended knowingly, which is the height of pride, and in this way pride is the first for those who move away from God and the last for those who return to him.

If he has ever attributed to himself the goods that he will have had.

If he believes that [the goods] have been given to him, but on account of his own merits, by God.

If he has boasted of having what he did not have.

If, while others remained unseen, he alone would want to be seen.

If he, being presumptuous concerning himself, has judged others.

If he was desiring of vainglory.

If he has sinned by hypocrisy.

If he has been luxuriating in fine clothing and such like.

If he has been disobedient to his parents.

If he has been disobedient to the church and its prelates.

If he has been disobedient to lords and authorities.

If a woman has been disobedient to a man concerning things that are not against God.[62]

---

[61] Goering and Mantello, 'The Early Penitential Writings', 65.

[62] Robert Grosseteste, *De modo confitendi*, Goering and Mantello, 'The Early Penitential Writings', 80–111, at 82–3, §12: '12. De *superbia* / Si scienter Deum offenderit, quod est magnae superbiae, et sic est / superbia prima recedentibus a Deo et ultima redeuntibus. / Si bona quae habuerit umquam sibi attribuerit. / Si crediderit sibi a Deo datum, sed tamen pro meritis suis. / Si se iactaverit habere quod non habuerit. / Si, ceteris despectis, singulariter voluerit videri. / Si de se praesumens alios iudicaverit. / Si inanis gloriae cupidus exstiterit. / Si per hypocrisim peccaverit. / Si in luxu vestium et huiusmodi. / Si inoboediens parentibus fuerit. / Si inoboediens ecclesiae et eius praelatis. / Si inoboediens dominis et potestatibus. / Si inoboediens mulier viro in his quae non sunt contra Deum.'

CONTEXTUALIZING GROSSETESTE'S *ON THE SPHERE* 69

The fuller description of pride provides the analysis on which the interrogatory proper is based. Pride is identified as emerging from natural causes (for, example from beauty, colour, or strength), by something the person confessing possesses by chance, and from what they have been given by grace (an excess of piety for instance). Examples of things possessed by chance include:

> knowledge, skill, riches, power, honour, familiarity; to have the summons of the wise, the rich, the powerful, of noble lords, and the society, dignity, favour and fame of such a sort, the love of great men, promotion, nobility, or refinement.[63]

Grosseteste ends with how pride emerges in the heart, from the mouth (for example in derision and mendacity), and in works (for example in disobedience), and by omission (for example in drinking).[64] The instinct within this part of the treatise to explain not only the 'quid' or 'what' of pride but also the 'why', the 'propter quid', is not dissimilar in general method to his scientific works.

2.2.2 *Meditations*

Grosseteste's second work of pastoral literature may well be the *Meditations*, a series of ten reflections on sin, penitence, and death. In part these can be seen in continuity to a longer monastic tradition, represented particularly by Anselm of Canterbury, a tradition transformed in the thirteenth century and later with vernacular versions for lay audiences, including sets of interrogations for deathbed confession.[65] A work of this sort from the early thirteenth century is the *Speculum*

---

[63] Goering and Mantello, 'The Early Penitential Writings',, 87, §22: 'Ex fortuitis, ut ex scientia, arte, divitiis, potestatibus, honoribus, familiaritatibus, vocatione sapientum, divitum, potentium, nobelium dominorum et huiusmodi societate, dignitate, favore, fama, magnatum amore, promotione, nobilitate, vel cultu.'

[64] Goering and Mantello, 'The Early Penitential Writings', 87–8, §§24–7.

[65] See Margaret Healy-Varley, 'Anselm's Afterlife and *De custodia interioris hominis*', in Giles E. M. Gasper and Ian Logan (eds.), *Saint Anselm of Canterbury and His Legacy* (Toronto: Pontifical Institute of Mediaeval Studies, 2012), 239–57; and Margaret Healy-Varley, 'The *Admonitio morienti* and *Meditatio ad concitandum timorem* in Vernacular Compilations', in Margaret Healy-Varley, Giles E. M. Gasper, and George Younge (eds.), *Anselm of Canterbury: Communities, Contemporaries, and Criticism* (Leiden: Brill, 2021), 240–61.

70 THE SCIENTIFIC WORKS OF ROBERT GROSSETESTE

religiosorum of Edmund of Abingdon written *c.*1214 for a monastic audience, and with which Grosseteste was familiar by the time he wrote the *On the Temple of God*.[66] By contrast, Grosseteste's *Meditations* are far shorter than any comparanda, and seem to be directed towards a non-monastic audience, probably a clerical one, though the possibility of a lay audience should not be dismissed too swiftly.[67]

In terms of dating, the treatment of the gifts that are given to those in heaven is useful. This was a popular topic in the early thirteenth century, and one to which Grosseteste returned in his disputation *On the Gifts* (*De dotibus*) of *c.*1230. The *Meditations* conform to teaching on the subject from *c.*1200, for example that of Stephen Langton, rather than the later discussions used in *On the Gifts*.[68] Moreover, the content and specific vocabulary of the *Meditations* is closely paralleled in *On the Temple of God*. The latter text repeats most of the concerns of the *Meditations*, that the sinner should contemplate the human condition, consequences of sin, the state of the blessed, and the damnation of the wicked, with the exception of an injunction to read Scripture.[69] Moreover, at the same point *On the Temple of God* uses the distinctive phrase, 'vile sperm' in the injunction 'Remember, human, and consider, that you were conceived in sin, and maybe even in mortal [sin], and from the repulsive matter of sperm'.[70] This phrase occurs in the *Meditations* in almost exactly the same form, and is attested in only one other source:

> 6. *You should meditate upon* the blessings of God: that is, *of creation*, from vile sperm, and beautiful life given by God, a noble creature in senses and member, and rational life; *of redemption* by the death of the Son; of the *communion of* sacraments. And in this way, you should move through to the individual daily [blessings] of support, of health,

---

[66] Goering and Mantello, 'The *Meditaciones* of Robert Grosseteste', 119.

[67] Goering and Mantello, 'The *Meditaciones* of Robert Grosseteste', 120–1.

[68] Goering and Mantello, 'The *Meditaciones* of Robert Grosseteste', 123–4.

[69] Grosseteste, *Templum Dei*, XXI.3–4. See Goering and Mantello, 'The *Meditaciones* of Robert Grosseteste', 121–2.

[70] Grosseteste, *Templum Dei*, XXI.6: 'Recordare homo et attende—quod in peccato conciperis, et forte in mortali, et ex uili materia spermatis.'

CONTEXTUALIZING GROSSETESTE'S *ON THE SPHERE* 71

and so forth. You were born ignorant, naked, an exile, and were endowed with virtue, enriched with possessions and raised up in dignity and order.[71]

In light of the close relation of the two treatises a strong suggestion can be made for the *Meditations* being produced at about the same time as, or before *On the Temple of God*, perhaps around 1215/16, although an earlier date is also possible.

2.3 Audience and Intention

Grosseteste's commitment to pastoral care from an early point in his career raises important questions as to whom his works were directed. Neither *On the Making of Confession* nor the *Meditations*, nor *On the Temple of God*, give any indication of their geographical provenance. As argued in Volume I of this series, a good case can be made for a continued association between Grosseteste and Hugh Foliot, archdeacon of Shropshire, in the years immediately following the death of William de Vere.[72] That he was then away in France during the interdict is also accepted, though the precise duration of his visit in unclear.[73] Such evidence as there is for his whereabouts after 1214 is reviewed in detail above (Ch. 1, §3). The *pastoralia* fit into this chronology well in terms of the sources Grosseteste appears to have used, notably his engagement after 1208x13 with the more recent penitential writings of Robert of Flamborough in Paris. It is possible then to posit Grosseteste as supplying material for clerical education, perhaps under the sponsorship of Hugh Foliot, first in the period 1200–8, that is to say, *On the Making of Confession*, with the *Meditations* and *On the Temple of God* following after the interdict.

---

[71] Robert Grosseteste, *Meditaciones*, 6, ed., Goering and Mantello, 'The *Meditaciones* of Robert Grosseteste', 127: 'Meditari debes Dei beneficia: scilicet creacionis ex vili spermate, vitam pulcram a Deo donatam, nobilem creaturam in sensibus et membris et vita racionabili; redempcionis per mortem Filii; communionis sacramentorum. Et sic discurre ad singula cotidiane sustentationis, salutis, et huiusmodi. Natum te imbecillem, nudum, et exulem, dotauit possessionibus, dignitate sublimauit et ordine'. See also p. 123 for discussion of 'vile sperm' and the only other usage, by Bernard of Clairvaux.
[72] *Knowing and Speaking*, 209–14.   [73] *Knowing and Speaking*, 204–8 and 214–18.

72 THE SCIENTIFIC WORKS OF ROBERT GROSSETESTE

The first audience of these works is very difficult, if not impossible, to identify. Indeed, the popularity of the treatises as indicated in the number of manuscripts shows that wherever and for whoever they were first produced, they enjoyed a rapid rise in circulation; for example, over twenty-four copies of *On the Making of Confession* survive.[74] Whether Hugh's archdeaconry, comprising the rural deaneries of Burford, Stottesden, Ludlow, Pontesbury, Clun, and Wenlock, provided the base for Grosseteste's activities is not clear; the larger churches of the Augustinian Canons at St Michael's, Chirbury, or St Laurence's in Ludlow are of the sort whose personnel might have been interested recipients of Grosseteste's expertise in pastoral care.[75] It is also not possible to know whether the advice contained in the treatises would have been disseminated in ways other than through manuscript copies, such as sermons or addresses at clerical gatherings or visitations. Hugh's own activities as archdeacon are also difficult to ascertain, though he is a frequent witness to episcopal charters.[76] That Grosseteste remained active in Hereford itself is also plausible. Though he may not have been appointed to the episcopal household of Giles de Braose, Hugh Mapenor, or even Hugh Foliot, nor installed in a cathedral canonry, and remained, as far as can be seen, without a benefice, this would be no impediment to being engaged with ecclesiastical education in the city, implementing the requirement perhaps of the Third Lateran Council (1179) that cathedrals should make provision for education of the clergy.[77]

Concern for education and pastoral care within the cathedral community in this period can be shown in a number of ways. As Julia Barrow has demonstrated, between 40 and 50 per cent of the canons of Hereford between 1160 and 1240 used the title of *magister* (master), which

---

[74] Goering and Mantello, 'The Early Penitential Writings', 59–63.

[75] M. J. Angold, G. C. Baugh, Marjorie M Chibnall, D. C. Cox, D. T. W. Price, Margaret Tomlinson, and B. S. Trinder, 'Houses of Augustinian Canons: Priory of Chirbury', in *A History of the County of Shropshire*: Vol. 2, ed. A. T. Gaydon and R. B. Pugh (London: Victoria County History, 1973), 59–62. *British History Online*, http://www.british-history.ac.uk/vch/salop/vol2/pp59-62 (accessed 6 July 2020); Michael A. Faraday, *Ludlow, 1085–1660: A Social, Economic, and Political History* (Chichester: Phillimore, 1991).

[76] Julia Barrow, 'Foliot, Hugh (d. 1234), Bishop of Hereford', *Oxford Dictionary of National Biography*, 4 October 2007, https://www.oxforddnb.com/view/10.1093/ref:odnb/9780198614128.001.0001/odnb-9780198614128-e-95044 (accessed 6 July 2020).

[77] *Knowing and Speaking*, 201.

CONTEXTUALIZING GROSSETESTE'S *ON THE SPHERE* 73

matches other non-monastic cathedrals such as Lincoln, with an even split between local and non-local masters.[78] A number amongst the chapter served as papal judges-delegate in the later twelfth and early thirteenth century.[79] In a similar vein the cathedral promoted the celebration of saint's cults, notably that of its dedicatee St Ethelbert, whose *Life* was composed by Gerald of Wales possibly in about 1195.[80] Gerald also offers some more detailed insights into the cathedral community and its interests in first decade and a half of the thirteenth century.

2.3.1 A Vignette from Gerald of Wales

Gerald's correspondence with a number of members of the Hereford Chapter in these years help to contextualize further the community around which Grosseteste is likely to have been active. A series of letters appended to Gerald's *Mirror of Two Men* composed in *c*.1218 as his record of a bitter dispute between him and his nephew (also Gerald) which erupted in 1208 and lasted for the next five years, are particularly useful.[81] The cause of the dispute was the nephew's refusal to honour an arrangement through which the older Gerald was to have received an income as a condition to his resigning to his nephew the archdeaconry of Brecon in 1203. Such an arrangement was in itself highly irregular since clerical office was not to be treated as if hereditary, leading Gerald of Wales to lay out the situation before Pope Innocent III who, Gerald notes, did not actively disapprove.[82] His nephew's refusal to support him, and the efforts, as he claimed, to defame him, meant that Gerald's correspondence drew in a wide range of other contacts, from Hereford and the southern March to Lincoln Cathedral, including William de Montibus.[83]

---

[78] Julia Barrow, 'The Canons and Citizens of Hereford *c*.1160–*c*.1240', *Midland History*, 24 (1999), 1–23 at 6.

[79] Barrow, 'The Canons', 7.

[80] Barrow, 'The Canons', 8; Robert Bartlett, *Gerald of Wales: A Voice of the Middle Ages* (Oxford: Oxford University Press, 1982; repr. Stroud: Tempus 2006), 177.

[81] Gerald of Wales, *Speculum duorum or A Mirror of Two Men*, ed. Yves Lefèvre and R. B. C. Huygens, trans. Brian Dawson, general editor Michael Richter (Cardiff: University of Wales Press, 1974), xxi.

[82] Gerald of Wales, *Speculum duorum*, xxxii–xxxiii, and Letter 7, 256/7.

[83] Gerald of Wales, *Speculum duorum*, xxxix–xlix.

74 THE SCIENTIFIC WORKS OF ROBERT GROSSETESTE

Something of the character of Hereford as an ecclesiastical centre is shown in the course of persistent accusations made by Gerald concerning the hospitality given by the chapter to his nephew. Hugh Mapenor as dean, William the Precentor (either William Foliot who held the office from 1186x95–1206x15 or his successor William of Kilpeck 1206x16 and 1220x3), and Canon Ralph Foliot were challenged directly on the matter.[84] A letter of *c.*1213 to the prior of Llanthony Secunda in Gloucester gives more detail about the role of chapter from the beginning of the dispute.[85] It was at Hereford, where the nephew was judging cases with the dean and precentor of Hereford and other unspecified clergy, that the bishop of St David's failed, according to Gerald, to represent properly the nephew's infamy. Despite the dean's observation that the nephew was 'appealing against the man who has given him everything he has and has reared him and almost created him from naught; without doubt this is a wicked plot he is executing', the bishop allowed the appeal.[86]

It is, however, the first letter, to Master Albinus, which is the most instructive for the educational environment in the diocese. He is addressed as canon of Hereford and as *dilectus* [beloved], a greeting undercut immediately by the remark, 'if only he were worthy of that love [et utinam digne diligendo]'. Albinus may have been part of the household of Giles de Braose on the latter's appointment as bishop of Hereford (1200–15) and rose to become chancellor of the cathedral

---

[84] Gerald of Wales, *Speculum duorum*, Letter 2, 160–7. 'Deans', in *Fasti Ecclesiae Anglicanae 1066–1300*: Vol. 8, *Hereford*, ed. J. S. Barrow (London, 2002), 7–13. British History Online, http://www.british-history.ac.uk/fasti-ecclesiae/1066–1300/vol8/pp7-13 (accessed 28 April 2019); *Acta*, lix and 304; 'Precentors', in *Fasti Ecclesiae Anglicanae 1066–1300*: Vol. 8, *Hereford*, 13–16; and *English Episcopal Acta VII: Hereford 1079–1234*, ed. J. Barrow (Oxford: Oxford University Press for the British Academy, 1993), 305; Julia Barrow, 'Athelstan to Aigueblanche, 1056–1268', in Gerald Aylmer and John Tiller (eds.), *Hereford Cathedral: A History* (London: Hambledon Press, 2000), 29–47, at 39, n. 102. This Ralph Foliot is not the archdeacon of Hereford who died 119/x9: 'Archdeacons: Hereford', in *Fasti Ecclesiae Anglicanae 1066–1300*: Vol. 8, *Hereford*, 23–5.

[85] The community of Llanthony Prima in the Black Mountains, founded in the second decade of the twelfth century, had been dispersed in 1135 and retreated first to Hereford and then to Gloucester where a cell was established, known as Llanthony Secunda. In 1205 the two Llanthony's were separated, though they were to be re-united, with the cell taking the role of the mother house at the end of the medieval period, in 1481.

[86] Gerald of Wales, *Speculum durorum*, Letter 7, 244/5: '"Contra eum igitur appellat, qui totum, quod habet, ei contulit ipsumque nutrivit et tanquam ex nichilo creavit: parvo procul-dubio fungitur consilio."'

CONTEXTUALIZING GROSSETESTE'S *ON THE SPHERE*   75

(1217x18–1225x7).[87] Gerald's letter to Albinus reveals a long-standing relationship, although the basis of the friendship is not specified. Gerald also mentions, in his first remonstration against the hospitality shown to his enemies at Hereford, a school administered by the chapter:

Although it is not surprising that you have been as kind to them as to anyone else, with your lecture-room and with the teaching of the school, which is open to all, annoying and good pupils alike, yet it is all the more surprising and is excusable by no possible argument, that you have admitted them under your roof, so we understand, as personal friends, day and night.[88]

The passage can be read as evidence for there being educational provision at a higher level, for clergy presumably, and a school for more elementary pupils, the latter open to all comers. Nicholas Orme suggests that Albinus was teaching a school 'of theology for older students, chiefly clergy or clerical trainees'.[89] Whatever the precise arrangements, it indicates the existence of educational structures within the city at the turn of the thirteenth century. The presence in the chapter of Nicholas, chancellor from 1187 to the mid-1190s, with the title *divinus*, and two *theologi*, Simon of Melun (fl. *c*.1190–1202) and Peter of Abergavenny (1201–19), adds some support to the theological interests of the chapter, whether harnessed in a school or not.[90] This would fit the expectations for provision of education for clergy in the Third Lateran Council, and would represent the sort of audience that might have been receptive to

---

[87] On Albinus's dates as chancellor, see 'Canons Whose Prebends Cannot Be Identified', in *Fasti Ecclesiae Anglicanae 1066–1300*: Vol. 8, *Hereford*, tabulated in 'Office Holders at Hereford Cathedral since 1300', compiled by G. Aylmer, J. Barrow, R. Caird, D. Lepine, and H. Tomlinson, in Aylmer and Tiller, *Hereford Cathedral*, appendix 2, 637–43 at 640. See also *English Episcopal Acta VII: Hereford 1079–1234*, lix and 307.

[88] Gerald of Wales, *Speculum duorum*, Letter 1, 156/7: 'Nec mirum tamen quod auditorium vestrum et scole doctrinam, que communis omnibus, tam discolis scilicet quam domesticis, esse solet, eis ut ceteris indulsistis; sed hoc revera non mediocriter admirandum nulloque prorsus fuco coloris excusandum, quod eos in hospitio vestro, sicut accecpimus, tanquam individuos nocte dieque socios'.

[89] Nicholas Orme, 'The Cathedral School before the Reformation' in Aylmer and Tiller, *Hereford Cathedral*, 546–78 at 567.

[90] Orme, 'The Cathedral School', 567. For Nicholas as *divinus*, see *English Episcopal Acta VII: Hereford 1079–1234*, 96–7, no. 142.

76 THE SCIENTIFIC WORKS OF ROBERT GROSSETESTE

Grosseteste's first interests in pastoral care. Hereford diocese in the first decade of the thirteenth century is both a possible and plausible venue for those interests to have been developed.

2.4 On the Temple of God (Templum Dei)

The final work amongst Grosseteste's treatises for pastoral care to be considered here, On the Temple of God, raises similar questions as to Grosseteste's activities during the period c.1215–c.1220. Depending on the dating of the text it may represent work carried out in or around Hereford, perhaps for the clergy of the diocese. It is further possible that the work was connected with Hugh Foliot, either as archdeacon, or, after October 1219, as bishop of Hereford himself. The effects of the interdict (1208–14) on the English church meant that most dioceses were in need of reform and reconstruction. Hugh's predecessor Hugh Mapenor (r. 1216–19) may well have experienced difficulties in restoring the diocesan finances, since he operated with a far smaller household then either his predecessor or successor.[91] He placed restrictions on both the mercantile and Jewish communities of Hereford, and was ordered by the sheriff to cease harassing the latter in 1218.[92] Hugh Foliot's period as bishop marks the first survival of extensive documents relating to the episcopal estate, which, as Barrow notes, implies that they were well administered.[93] The first bishop's official for Hereford, an administrative post used in a number of English dioceses from the later twelfth century, was appointed in Hugh's reign, probably between 1223 and 1226.[94] Hugh also worked hard to ensure that the chapter of the cathedral was sufficiently endowed so as not to be anxious about episcopal oversight.[95] Early in his reign cathedral canons are prominent as witnesses to episcopal acta. At the same time Hugh also supported the Grandmontine

[91] English Episcopal Acta VII: Hereford 1079–1234, xlviii and lx–lxi. What follows draws heavily on Barrow's assessment of the reign of Hugh Mapenor and Hugh Foliot.
[92] English Episcopal Acta VII: Hereford 1079–1234, xlviii.
[93] English Episcopal Acta VII: Hereford 1079–1234, l.
[94] Dating depends on the range identified for the charter recording the presentation of Master John of Wroxeter to the church of Culmington, by Haughmond Abbey (Ch. 1, §3.3.3).
[95] English Episcopal Acta VII: Hereford 1079–1234, xlix.

CONTEXTUALIZING GROSSETESTE'S *ON THE SPHERE* 77

order, and their establishment of a priory at Alberbury. It is also entirely possible that he assisted in the establishment of the Franciscans in Hereford before 1228.[96]

The dating of the treatise relies on a number of different factors. The most extensive study remains that of Goering and Mantello, whose conclusions are summarized in what follows.[97] A *terminus ad quem* would seem to be 1246, if Grosseteste's work is that referred to by Walter of St Edmund on his death in that year. For a *terminus a quo*, on the basis that Grosseteste's work can be shown to bear the influence of Robert of Flamborough's *Liber poenitentialis* (*Book of the Penitent/ Penitential*), then a date after its composition in 1208x1213 can be proposed.[98] Robert of Flamborough was canon-penitentiary at the Abbey of St Victor, so it is possible, and perhaps even likely, that Grosseteste may have encountered both the treatise and its author in the course of his sojourn in France during the interdict. The presence within Hereford diocese of a Victorine priory, founded originally at Shobdon 1135x4 and eventually re-located to Wigmore after some local perambulation by 1179, should also be recalled in this context.[99]

Arguments for an earlier dating for *On the Temple* focus around its relation to canon law, what is termed its psychological teaching, and stylistic features, both written and diagrammatic. The first evidence connected to legal material concerns the statement in the treatise that: 'Moreover, some say that prelates who study Civil Law or Medicine... are

---

[96] Barrow, 'Foliot, Hugh (d. 1234), Bishop of Hereford'; Barrow, 'Athelstan to Aigueblanche 1056–1268', 31–47 at 41. Whether the pre-1228 connections of the Franciscans to Hereford had any impact on Grosseteste's appointment as the first lector to the Oxford Franciscans in *c.*1229/30 is difficult to establish but worth contemplating, see Giles E. M. Gasper, 'How to Teach the Franciscans: Robert Grosseteste and the Oxford Community of Franciscans *c.*1229 35', in Lydia Schumacher (ed.), *Early Thirteenth-Century English Franciscan Thought* (Berlin: De Gruyter, 2021), 57–75 at 62–3.

[97] Grosseteste, *Templum Dei*, 4–8, 9–14.

[98] Robert of Flamborough, *Liber Poenitentialis*.

[99] William Dugdale, *Monasticon Anglicanum*, ed. John Caley, Henry Ellis, and Bulkeley Bandinel, Vol. 6 of 6 in 8 parts (London: T. G. March, 1846), part 1, 343–56. The first abbot of the house in *c.*1147–54/5 was Andrew of St Victor, English by birth, canon of St Victor and best known for his studies of the Hebrew bible, focusing on the literal interpretation. Andrew returned for a second stint as abbot in 1161–3 until his death in 1175. See Margaret Gibson, 'St Victor, Andrew of (*c.* 1110–1175), biblical scholar and abbot of Wigmore', *Oxford Dictionary of National Biography*, 23 September 2004, https://www.oxforddnb.com/view/10.1093/ref:odnb/9780198614128.001.0001/odnb-9780198614128-e-37116 (accessed 22 March 2020).

78 THE SCIENTIFIC WORKS OF ROBERT GROSSETESTE

excommunicated.'[100] If this is, as Goering and Mantello suggest, a reference to the prohibition of such study by Honorius III to the University of Paris in November 1219, then *On the Temple* may well have been composed after this point.[101] A second piece of evidence is the restriction on marriage in the canons of the Fourth Lateran Council to the fourth degree of relation. By contrast *On the Temple* records the second and third degrees of affinity as preventative of marriage.[102] While this might indicate a date before 1215, it might also reflect the speed at which knowledge of the Lateran IV canons travelled within post-1215 England. The new teaching, Goering and Mantello note, is evident in 'English synodal statutes as early as 1219'.[103] While suggestions on dating may not be precise, they are sufficient to rule out *On the Temple* as a product of Grosseteste's tenure as bishop of Lincoln (1235–53), or as lector to the Oxford Franciscans (*c.*1230–5) when such material would have been distinctly out of date. Goering and Mantello's suggestion that the understanding of the soul with which Grosseteste operates in *On the Temple* seems less sophisticated than that in the later *On the Intelligences—De intelligentiis* (*c.*1230) would also indicate an earlier date.[104]

A date closer to 1219 than later is, therefore, not implausible. That a clerical audience was intended for *On the Temple* probably lies behind its unusual schema in which the majority of the treatise is articulated in the form of lists, charts, and diagrams.[105] As an aid to reference it was, despite the evident difficulties in copying, a popular format adopted in a number of English treatises.[106] The form then suggests a didactic

---

[100] Grosseteste, *Templum Dei*, VII.4: 'Item dicunt quidam quod prelate audientes Leges uel Phisicam . . . excommunicati sunt'.

[101] Grosseteste, *Templum Dei*, 4.

[102] Grosseteste, *Templum Dei*, XVI.7, and foregrounded in VII.23.

[103] Grosseteste, *Templum Dei*, 4.

[104] Grosseteste, *Templum Dei*, 5. Robert Grosseteste's *De intelligentiis* (ed. Ludwig Baur, *Die philosophischen Werke des Robert Grosseteste, Bischofs von Lincoln* (Münster i. W.: Aschendorff, 1912)) is the second part of the first letter in Grosseteste's letter collection, addressed to Adam divided in the manuscripts into two distinct parts, Neil Lewis, 'Robert Grosseteste', *The Stanford Encyclopedia of Philosophy* (Summer 2019 ed.), ed. Edward N. Zalta, https://plato.stanford.edu/archives/sum2019/entries/grosseteste/ (accessed 16 July 2021); Grosseteste, *De intelligentiis*, 112–19; Robert Grosseteste, Letter 1, in *Epistulae*, 1–17. The dating of Letter 1 to 1225–8, in *The Letters of Robert Grosseteste, Bishop of Lincoln*, 22, relies on the notion that Grosseteste was a master in the schools at this point, for which there is, strictly speaking, no evidence; this might place the letter and the *De intelligentiis* a little later.

[105] Grosseteste, *Templum Dei*, 7.   [106] Grosseteste, *Templum Dei*, 8.

CONTEXTUALIZING GROSSETESTE'S *ON THE SPHERE* 79

function, the content that this was designed for the erudition of the clergy. Chapter 10 lists the knowledge and books with which a priest should be familiar:

The Office Books of the Church, such as the lectionary, the antiphonary, the order for baptism, the gradual and the missal.

A book of compotus so that he may know the moveable and immovable feasts.

The sacramentary so that he may know to minister with discernment at least in the sacraments.

A book of penitential canons so that he can discern between leprosy and leprosy (Deut. 17:8), that is between sins and imposed penances.

A homiliary of Gregory or some other saint, so that he may know to expound the Gospel to the people.[107]

While formal evidence for a school at Hereford is lacking until the mid-thirteenth century, some form of theological formation may well have been undertaken at the cathedral, as noted above (§2.3.1) from at least 1208.[108] It is plausible then, that Grosseteste created *On the Temple* in this sort of context. As far as dating is concerned, around 1219/20 would fit the early part of Hugh's episcopal career, which would also mark a suitable moment to engage with the fitness of the clergy of the diocese, as well as its finances. If this were to have been the case then *On the Temple* was written very probably shortly after *On the Sphere*, and before the *Compotus*.

On the Temple reveals little about the circumstances of its author, but there are some features worth noting. The examination of usury is detailed, and not in and of itself unusual, but Grosseteste's attendance at sermons preached by Eustace abbot of Fly, Stephen Langton, Robert of Courson, and James of Vitry between 1211 and 1213 against the usury of

---

[107] Grosseteste, *Templum Dei*, X.3: 'Libri Officii Ecclesie, ut lectionarius, antiphonarius, baptisterium, gradale, missale./Liber compoti, ut cognoscat festa mobilia et immobilia./Liber sacramentorum, ut sciat saltem in sacramentis discrete ministrare./Liber canonum penitencialium, ut sciat discernere inter lepram et lepram, siue inter peccata, et iniungere penitencias./ Liber omeliarum Gregorii uel alterius sancti, ut sciat exponere Euangelium populo.'

[108] Orme, 'The Cathedral School', 567.

80 THE SCIENTIFIC WORKS OF ROBERT GROSSETESTE

the Cahorsins might be recalled.[109] Usury also features as a form of greed, and it was a subject to which he would refer often as bishop.[110] Another point of interest occurs in discussion of simony (payment for spiritual office). This occurs, Grosseteste points out, in service, when you serve someone dishonestly, or honestly serve someone dishonest.[111] A more specific example follows almost immediately. Simony also occurs, 'if you serve a bishop on account of your legal knowledge and expertise in exchange for a certain level of regular stipend'.[112] Grosseteste may have had personal knowledge of the temptation such an arrangement might have provoked.[113]

That Grosseteste was engaged in *pastoralia* at the same time as his exploration of natural philosophy is, therefore, more than likely. Furthermore, a good case can be made that he saw the two areas of interest as overlapping in productive ways. *On the Temple* shows evidence, for example, of positions elaborated in his two earliest works, the scientific treatises *On the Liberal Arts* and *On the Generation of Sounds*. Reference to the tripartite division of soul into vegetative, sensible, and rational draws on Aristotelian natural philosophy, especially *On the Soul*, as demonstrated in *On the Generation of Sounds*.[114] In the same chapter of *On the Temple* which draws on the *Meditations* Grosseteste posits two ways of identifying contemplation, by *affectus* and by *aspectus*, that is, the mind's desire, and the mind's inner sight. This is the same pairing as to be found in *On the Liberal Arts*.[115] In *On the Temple*, Grosseteste sets up for the interaction of *aspectus* and *affectus* exactly the framework laid out in his earlier treatise.

---

[109] Grosseteste, *Templum Dei*, XIII.1–2; *Knowing and Speaking*, 204–8 for details on the preachers and on the Cahorsins.

[110] Grosseteste, *Templum Dei*, IX.4; on later uses, see Robert Grosseteste, *Dictum* 84; and Matthew Paris, *Chronica majora*, s.a. 1253, ed. Henry Richards Luard, 7 vols (London: Longman, Green, Longman, and Roberts, 1872–83), v. 401–7 for Grosseteste's deathbed comments of usury and heresy.

[111] Grosseteste, *Templum Dei*, XII.1.

[112] Grosseteste, *Templum Dei*, XII.3: 'Si pro sciencia et pericia iuris ministras episcopo pro certa taxacione stipendiarum'.

[113] See Goering, 'Where and When', 21.

[114] Grosseteste, *De generatione sonomrum*, §3; Grosseteste, *Templum Dei*, I.2, II.5, V.4, and see p. 5.

[115] Grosseteste, *De artibus liberalibus*, §2; Grosseteste, *Templum Dei*, XXI.3.

CONTEXTUALIZING GROSSETESTE'S *ON THE SPHERE* 81

Sight [*aspectus*] first looks; then it verifies what has been looked at or cognized, and when the fitting or harmful things have been verified within the mind or within sight, desire [*affectus*] covets to embrace the fitting, or retreats into itself to shun the harmful.[116]

Aspectus in *On the Temple* is connected to meditation of heaven, and comparison of life there to the fragility of earthly existence, and to the reading of scripture and the lives of the holy, through which to praise the Lord for his excellence. *Affectus* is used to love God strongly, wisely, and sweetly, and then to hurry towards heavenly rewards, spurning, tearfully, the transitory things of this world. In this way, then, *aspectus* looks to the heavens, engages actively with the promise of the life to come, compares it to mortal life on earth, distinguishing, in this way, between the fitting or harmful, and *affectus* moves quickly, then, to embrace the heavenly and shun the worldly.

Closer comparison of the basic structures of *On the Temple* and *On the Sphere* is also revealing. In both he moves from a macrocosmic to a microcosmic perspective. *On the Temple* begins by establishing the metaphor of the temple for the reception of Christ by sinful humanity guided by the clergy; as a building has three parts, foundation, walls, and roof, so too should the metaphorical temple of the priest.[117] Priests should have a two-fold temple, to match the two natures of Christ, divine and human, one for the soul and one for the body. The temple of the body has it foundations in the kidneys, responsible for growth and standing for temperance; it walls are the sides, back, and chest, which stand for fortitude against worldly, unfitting things; the roof is the head, housing the senses, and standing for prudence. The spiritual temple has for foundation the cognitive power of the soul, whose floor is the articles of faith; the walls are the potential power, whose principal ornament is hope; the roof is the affective power of the soul whose ornament is love. From this Grosseteste divides the parts into smaller segments in which the various aspects of pastoral care are located. This is also how *On the*

---

[116] Grosseteste, *De artibus liberalibus*, §2: 'Aspectus vero primo aspicit, secundo aspecta sive cognita verificat et, cum verificata fuerint apud mentem seu aspectum convenientia seu nociva, inhiat affectus ad amplexandum convenientia vel in se ipsum retrahit ut fugiat nociva.'

[117] Grosseteste, *Templum Dei*, I.1–3, diagrammatized at 4.

82 THE SCIENTIFIC WORKS OF ROBERT GROSSETESTE

Sphere is structured, as the rest of this volume will explore, moving from a description of the shape of the universe, the world machine, to the perspective of the individual observer (see Ch. 4).

On the Sphere and *On the Temple* also reveal a similar approach to sources, and other, contemporary works in their relevant field. Grosseteste takes the directions on pastoral care that Robert of Flamborough *Book on Pentitence* lays down, but in *On the Temple* explains *why* pastoral care matters. John of Sacrobosco's *On the Sphere*, as will be seen below (see Chs. 6 and 7), covers much, although not all, of the same material as Grosseteste, but with a quite different structure. Here again, Grosseteste's purpose is to move beyond an explanation of the *what* to a presentation of the *why*. His *On the Sphere* shows why astronomy has utility and purpose. In these ways, the essential methodology of Grosseteste's *On the Sphere* and *On the Temple* are identical. The *Compotus* plays a similar and important role here, bringing together science, astronomy, and pastoral care, and offering an important and original explanation for why the science of time-reckoning is essential.[118] Grosseteste's use, within *On the Temple*, of a framework set up in his scientific works points to a unitary mode which is characteristic of his thought. More specifically, it stresses an approach in which knowledge and virtue were conceived as being learnt together, as two sides of the same coin. In this, Grosseteste echoes his twelfth- and early thirteenth-century forebears, including John of Salisbury, and William de Montibus.[119] Where, however, the older thinkers tended to place the training of morals and intellect in the cloister, for Grosseteste, the arena for the theory and practice of virtue was the parish.

---

[118] Grosseteste, *Compotus*, 8–11. The *Compotus* wil be treated in more detail in Volume III of this series.

[119] Goering, *William de Montibus*, 39; Sigbjørn Sønnesyn, 'Word, Example, and Practice: Learning and the Learner in Twelfth-Century Thought', *Journal of Medieval History*, 46 (2020), 513–35.

3

On the Sphere

Manuscripts, Translation, Historiography, and Synopsis

1 Manuscript Transmission of *On the Sphere*

The special position *On the Sphere* holds among Grosseteste's treatises on nature is reflected in its manuscript transmission. Previous scholarship has comprehensively mapped out the dissemination of this treatise in space and time, but this dissemination has several noteworthy characteristics that deserve particular attention here.[1] While no direct correlation can be presupposed between Grosseteste's purpose in writing and subsequent copying and use of the text, the codicological reception history of *On the Sphere* provides information that can add depth and direction to interpretations of the treatise itself. It is therefore helpful for contextualizing the text itself to highlight features of the manuscript tradition that have particular significance for interpreting the text. The medieval attribution of *On the Sphere* to Grosseteste is clear from the manuscript tradition. Nicholas Trevet noted, at the beginning of the fourteenth century, that while Master of Arts, Grosseteste composed a treatise *On the Sphere* and one on the *Art of Compotus*, as well as the commentary on Aristotle's *Posterior Analytics*.[2] Towards the end of that

---

[1] For a conspectus of the manuscript copies of *On the Sphere*, see Panti, *Moti, virtù e motori celesti*, 211–44. The treatise was also presented in five printed editions from the early sixteenth century.

[2] Nicholas Trevet, *Annales ex regum Angliae, 1135–1307*, ed. Thomas Hog (London: English Historical Society, 1845), 243: 'Qui, cum esset magister in artibus super librum Posteriorum compendiose scripsit. Tractatus etiam de Sphaera et de Arte compoti, multaque alia in philosophia utilia edidit.'

84 THE SCIENTIFIC WORKS OF ROBERT GROSSETESTE

century *On the Sphere* was included by Henry of Kirksted in his list of works attributed to Grosseteste.[3]

The first and perhaps most striking feature of the manuscript dissemination of *On the Sphere* is the sheer number of surviving copies. The treatise is by far the most popular among Grosseteste's treatises on nature. The critical edition by Cecilia Panti lists forty-five manuscript witnesses and a collection of excerpts with a chronological range from the thirteenth to fifteenth centuries. Panti has uncovered even more copies since the edition was published in 2001.[4] As noted in the Introduction to this volume, this is more than three times the number of extant witnesses of Grosseteste's second most copied natural work, *On Light*, which exists in fourteen manuscript copies.[5] While the dissemination of Grosseteste's *On the Sphere* falls well short of the even more popular treatise of the same name by John of Sacrobosco, the popularity of Grosseteste's treatise suggests that this work provided a significant contribution to the extant astronomical corpus.

This popularity, moreover, seems to have arisen comparatively soon after the composition of the treatise. Sixteen manuscripts, a full third of the copies known when the critical edition was produced, date from *c.*1300 and earlier. These manuscripts are listed below with the sigla from Panti's edition and the corresponding sigla from the comprehensive list compiled for the Ordered Universe Project by Neil Lewis.[6]

1: Cambridge, Gonville and Caius College, 137 (Ca Panti; Cg1 Lewis)
2: Cambridge, University Library, Ff. 6. 13 (Cb Panti; Cu1 Lewis)
3: Vatican, Biblioteca Apostolica Vaticana, Pal. lat. 1414 (Cf Panti; Vp Lewis)

---

[3] Formally known as Boston de Bury; see R. H. Rouse, 'Boston Buriensis and the Author of the *Catalogus scriptorium ecclesiae*', *Speculum*, 41 (1966), 471–99.

[4] The manuscripts known to exist in 2001 are listed in Panti, *Moti, virtù e motori celesti*, 211–13. The additional manuscripts discovered since have not so far been analysed in print, and the present authors extend gratitude to Cecilia Panti for sharing the information based on her recent research.

[5] Panti, *Moti, virtù e motori celesti*, 208–9.

[6] Lewis's list of manuscripts containing Grosseteste's works of natural and theological philosophy can be accessed online at https://ordered-universe.com/manuscripts/.

MANUSCRIPTS, TRANSLATION, HISTORIOGRAPHY 85

4: Vatican, Biblioteca Apostolica Vaticana, Urb. lat. 1428 (Cg Panti; Vu Lewis)

5: Erfurt, Wissenschaftliche Allgemeinbibliothek, Amplon. Q. 351 (Eb Panti; Ea2 Lewis)[7]

6: Erfurt, Wissenschaftliche Allgemeinbibliothek, Amplon. Q. 355 (Ec Panti; Ea3 Lewis)

7: Firenze, Biblioteca Riccardiana, 885 (Fd Panti; Fr Lewis)

8: London, British Library, Add. 27589 (Lb Panti; La1 Lewis)

9: London, British Library, Egerton, 843 (Ld Panti; Le1 Lewis)

10: London, British Library, Harley 3735 (Lg Panti; Lh2 Lewis)

11: London, British Library, Harley 4350 (Lh Panti; Lh3 Lewis)

12: Oxford, Bodleian Library, Digby 191 (Od Panti; Od7 Lewis)

13: Oxford, Bodleian Library, Laud misc. 644 (Of Panti; Ol Lewis)

14: Paris, Bibliothèque Nationale de France, lat. 7195 (Pa Panti; Pb1 Lewis)

15: Paris, Bibliothèque Nationale de France, lat. 7413 (Pd Panti; Pb6 Lewis)

16: Verdun, Bibliothèque Municipale 25 item 4 (Vb Panti; Vmu Lewis)

This early dissemination offers a sharp contrast to the majority of the scientific opuscula to be published in this present series, the manuscript copies of which predominantly date from the late fourteenth century onwards. The treatise was explicitly attributed to Grosseteste's authorship from the beginning.[8] Ten of the sixteen earliest manuscripts contain ascriptions contemporary to the copying of the main text, while another two have ascription added in a different hand; no known manuscript copy attributes the text to another author.[9] As mentioned Nicholas Trevet included *On the Sphere* among the works Grosseteste composed. While the shorter scientific opuscula only seem to have attracted interest

---

[7] While Panti follows Baur's dating of this manuscript to the early fourteenth century in her list, the Erfurt catalogue dates the copy of *On the Sphere* in this composite manuscript to the late thirteenth; cf. Panti, *Moti, virtù e motori celesti*, 221, and Wilhelm Schum, *Beschreibendes Verzeichnis der Amplonianischen Handschriften-Sammlung zu Erfurt* (Berlin: Weidmann, 1887), 588.

[8] Panti, *Moti, virtù e motori celesti*, 67–8.   [9] Panti, *Moti, virtù e motori celesti*, 211–13.

86 THE SCIENTIFIC WORKS OF ROBERT GROSSETESTE

long after Grosseteste's lifetime, *On the Sphere* seems to have been an important foundation for Grosseteste's fame among his contemporaries as a man of learning.

The likelihood that the popularity of this text was a foundation for Grosseteste's authority rather than a consequence of it is borne out by the codicological context of the extant copies. While the minor scientific opuscula largely survive in later collections of Grosseteste's works, where the authority of the author is the gravitational centre around which the works are gathered, *On the Sphere* predominantly survives in an explicitly astronomical context. This suggests that the treatise was copied for the contribution it made to the astronomical corpus at least as much as for the authority Grosseteste's name carried. *On the Sphere* is included in the great fourteenth-century English collection of Grosseteste's works now preserved in the badly damaged British Library Cotton Otho D. X and its direct copy British Library Royal 6.E.5, as well as the fifteenth-century English collection now kept as Princeton, University Library, Garret 95; but the treatise is absent from the important continental collections now held in Florence (Biblioteca Marucelliana C. 163), Venice (Biblioteca Nazionale Marciana VI. 163), and Prague (Knihovna národního muzea, XII.E.5 and Naródní knihovna X.H.12).[10] The great majority of copies of *On the Sphere*, including all the earliest ones, are found in a codicological context focusing on the quadrivial arts of arithmetic, geometry, music, and astronomy, and the applications of these arts in computus and astrology.

A conspectus of the texts with which *On the Sphere* was frequently copied and bound offers interesting suggestions as to what function later readers and scribal directors saw the treatise as filling. The original codicological context is not always easy or even possible to ascertain; among the earliest copies listed above, Erfurt Amplon. Q.351, Florence Ricardiana 885, and Bodleian Digby 191 are composite manuscripts composed of texts copied over long periods of time, while British

---

[10] The British Library manuscripts are discussed in Panti, *Moti, virtù e motori celesti*, 228–9. The Florence, Venice, and Prague manuscripts are described in *Knowing and Speaking*, 52–3; Princeton Garret 95 is described in *Knowing and Speaking*, 227, cf. Don C. Skemer, *Medieval and Renaissance Manuscripts in the Princeton University Library*, Vol. 1 of 2 (Princeton: Princeton University Press, 2013), i. 203–11.

MANUSCRIPTS, TRANSLATION, HISTORIOGRAPHY   87

Library Egerton 843 originally formed a unit with Cambridge Trinity College Library 0.2.45.[11] Where the original codicological context has been preserved, however, clear trends can be discerned. Interestingly, Grosseteste's *On the Sphere* does not seem to have been in direct competition with Sacrobosco's treatise of the same name or Alfraganus's digest of Ptolemaic astronomy for inclusion in astronomical compilations. Eight of the sixteen oldest copies of *On the Sphere* contain both Grosseteste's treatise and Sacrobosco's *De sphera*, and one of these, Paris BNF lat. 7195, also contains Alfraganus.[12] Two manuscripts, Bodleian Laud misc. 644 and Paris BNF lat. 7413, contain Alfraganus but not Sacrobosco, while an additional two, Erfurt Amplon. Q.351 and Firenze Riccardiana 885, were later bound with copies of Alfraganus copied separately from Grosseteste's text.[13] Clearly, then, thirteenth-century and later compilers saw Grosseteste's text as filling a particular function not covered by Sacrobosco's and Alfraganus's textbooks.

The thirteenth-century compilations still integrally preserved do not always contain the same texts, but they cover a curriculum with a relatively consistent scope. Computistical works, including tables, canons explaining the tables, and more wide-ranging works of computistical theory are important elements. Grosseteste's *Compotus* often occurs together with *On the Sphere*, sometimes also with Sacrobosco's *De sphera*,

---

[11] For the Erfurt manuscript, see Schum, *Beschreibendes Verzeichnis*, 588; for the Florence manuscript, see Lynn Thorndike, 'Notes upon Some Medieval Astronomical, Astrological and Mathematical Manuscripts at Florence, Milan, Bologna and Venice', *Isis*, 50 (1959), 33–50, at 38–40; For Bodleian Digby 191, see W. D. Macray, *Bodleian Library Quarto Catalogues IX: Digby Manuscripts*, repr. with addenda by R. W. Hunt and A. G. Watson (Oxford: Bodleian Library, 1999), cols. 205–7; for BL Egerton 843, see Panti, *Moti, virtù e motori celesti*, 125–32.

[12] These eight are Cambridge, Gonville, and Caius 137 (see M. R. James, *A Descriptive Catalogue of the Manuscripts in the Library of Gonville and Caius College*, Vol. 1 of 2 (Cambridge: Cambridge University Press, 1907–8), i. 150–1); Cambridge, University Library Ff.6.13 (See Panti, *Moti, virtù e motori celesti*, 126–30); Vatican Urb. Lat. 1428 (Panti, *Moti, virtù e motori celesti*, 225); Erfurt Amplon. Q.355 (Schum, *Beschreibendes Verzeichnis*, 594–7); BL Add. 27589 (see Panti, *Moti, virtù e motori celesti*, 235–6); BL Egerton 843 (see previous footnote); BL Harley 3735 (see Panti, *Moti, virtù e motori celesti*, 216–17); Paris BNF lat. 7195 (See David Juste, 'MS Paris, Bibliothèque nationale de France, lat. 7195' (update: 20 April 2019), *Ptolemaeus Arabus et Latinus. Manuscripts*, http://ptolemaeus.badw.de/ms/99 (accessed 9 July 2021)).

[13] For Bodleian Laud misc. 644, see H. O. Coxe, *Laudian Manuscripts, Quarto Catalogues II*, repr. from the ed. of 1858–85, with corrections and additions, and a historical introduction by R.W. Hunt (Oxford: Bodleian Library, 1973), cols. 465–7 and p. 576; for Paris BNF lat. 7413, see Juste, 'MS Paris, Bibliothèque nationale de France, lat. 7413').

88   THE SCIENTIFIC WORKS OF ROBERT GROSSETESTE

computus, and his *Algorismus*. The *Theorica Planetarum* (see Ch. 6, §2) is also often found alongside the more general texts. Two manuscripts produced in France towards the end of thirteenth century, Bodleian Laud misc. 644 and Paris BNF lat. 7195, cover the entire quadrivium including music. All compilations contain combinations of theoretical expositions and practical guides to computistical calculations.

It seems clear, then, that comprehensiveness was not expected of any single astronomical text within the corpus. The corpus itself offered comprehensive if not entirely harmonious coverage (Ch. 5), and the various compilers could choose what they needed from the pool of available texts. This means that Grosseteste's *On the Sphere* should not, perhaps, be approached with set expectations to its scope and purpose. As will become clear in following chapters the treatise seems designed to fulfil a particular function within the larger distinct intellectual project, and consistent with the way it was subsequently disseminated in manuscripts.

2 Principles of This Translation

In line with the overall aims and purposes of this series, the translation of *On the Sphere* is intended to make Grosseteste's treatise accessible to those who have no Latin, and to assist those who have some. In the light of this, the present translation tends towards a literal, rather than free, rendering, while remaining attentive to the need for clarity. Where possible, one and the same English word has been used to translate one and the same word in Latin, to facilitate ease of navigation between the two. Complex conceptual frameworks are explained and explored in the chapters that accompany the text. As part of the Ordered Universe Project's iterative methodology the translation has been worked over in numerous collaborative reading groups, symposia, and workshops. As such it represents a sustained and interdisciplinary effort to understand and render Grosseteste's sometimes elliptical and taut mode of expression. It is to be hoped that, as a result, other readers will be stimulated to question and explore the text and its implications.

MANUSCRIPTS, TRANSLATION, HISTORIOGRAPHY 89

2.1 Particular Problems of the Present Translation

Grosseteste's terminology in *On the Sphere* reflects the wide variety of sources at his disposal, containing technical vocabulary translated into Latin from Arabic and Greek. This has resulted in unfamiliar and sometimes obscure terms being employed to denote familiar astronomical concepts and phenomena. This, in turn, leaves a challenge of translation, arising from a tension between comprehensibility and fidelity to Grosseteste's Latin. A few passages and terms have been particularly challenging to translate. These are listed here, with a brief exposition of the basis for the decisions made.

2.1.1 §1 and passim: *figura, situs, machina*

In keeping with his overall approach to astronomy, Grosseteste employed geometric language throughout *On the Sphere*. The present translation seeks to preserve, where possible, the technical distinctions and consistencies inherent in Grosseteste's geometric vocabulary, without thereby introducing potentially misleading connotations on the back of the English terms used. The term 'figura' has been rendered, as consistently as possible, as 'shape'. Throughout his works, Grosseteste used 'figura' both in a highly abstract and schematized sense and as a description of concrete material objects; 'shape' bridges this gap whereas the more intuitive 'figure' does not. The English 'form' has been reserved for the metaphysical uses of 'forma', which will play an important role in later volumes of this series.

The Latin term 'situs' resists translation into English using a single term. The rendering chosen here '[relative] position', sacrifices neatness to bring out the connotations of relationality inherent in the Latin. While 'locus' can mean 'place' or 'position' in a more absolute sense, 'situs' means place or location in relation to other things, the internal layout of a complex system or the position of something relative to such a system. As *On the Sphere* lays out the internal layout of the world machine, and the positions of higher bodies relative to one another and to the whole, the term 'situs' captures an essential aspect of Grosseteste's project. The full range of meaning the term carries should be kept in mind as the text is read.

90   THE SCIENTIFIC WORKS OF ROBERT GROSSETESTE

The word 'machina', borrowed into Latin from the Greek μηχανή, had a range of interrelated meanings, from 'scaffolding' and 'frame' through 'contrivance' or 'engine' to 'the structure or fabric of creation' (Ch. 1, §1). Grosseteste's usage here appears to denote the universe as a whole and not just the outer shell or frame, although the spherical shape of the outer shell is also the shape of the whole. A literal translation has been chosen here, but the mechanical connotations of the term in modern English do not seem to have been intended by Grosseteste.

2.1.2 §5 rationibus naturalibus et experimentis astronomicis

Grosseteste's reference to the two main argumentative routes by which the sphericity of the heavens and the earth can be established is made more difficult to translate by the connotations that the terms used have accumulated in the centuries following Grosseteste's lifetime. A prominent example here is the term 'experimentum', which, if translated simply as 'experiment', might appear to suggest that Grosseteste based his arguments on experimental science in something approaching the modern sense. The text itself does not support such an interpretation. What Grosseteste may have intended here is discussed below (Ch. 7, §2); for the present, it should be understood that the translation at this point 'natural reasoning and astronomical observations' represents a desire not to lead the reader to assumptions concerning the role of experiments and empiricism in Grosseteste's thought.

2.1.3 §13 cingulus signorum

This term has been rendered 'the girdle of signs', although 'belt' would have been equally appropriate from a purely linguistic perspective. The somewhat archaic 'girdle' has been chosen to emphasize that the term is not purely descriptive but in some ways functions like a proper name.

2.1.4 §20 cenith capitis

Grosseteste's normal expression for 'zenith' is 'cenith capitis', which literally translates as 'zenith of the head'. This term entered medieval Latin as a transliteration and translation of the Arabic term 'samt al-ra's', which means 'direction of the head'. The eventual Latin term 'cenith' represents a corruption of the Arabic 'samt'. While the

MANUSCRIPTS, TRANSLATION, HISTORIOGRAPHY   91

translation here uses 'zenith' on its own to avoid repeating the more cumbersome 'zenith of the head', it should be noted that Grosseteste never fully disconnected 'cenith' from its context in order to use it as a technical term on its own.

2.1.5 §31 sphera recta et obliqua

Grosseteste did not explain the meaning of the technical terms 'sphera recta' and 'sphera obliqua', and translating these terms introduces an additional layer of ambiguity. A clarification of the perceptual phenomena to which these terms refer is offered below (Ch. 4, §41). The translation chosen for this volume reflects common usage in English translations medieval astronomical texts, and aim at literalness more than explanatory force.

2.1.6 §40 aux/oppositum augis

While the Latin term 'aux' can be rendered as 'apogee' in English, Grosseteste's terminology has no separate term for 'perigee'. Instead, he simply uses 'oppositum augis', or 'what is opposite the apogee'. The translation here follows Grosseteste's mode of expression closely, giving priority to fidelity to the text over ease of reference for modern readers.

2.1.7 §40 circulus egresse cuspidis

The literal rendering of this Latin phrase chosen for the present translation, 'the circle of the displaced cusp', requires some explanation; circles, by definition, do not have cusps. This phrase was adopted into Latin through translations of Islamicate astronomy (see Ch. 7, §4.2 below), where the Latin word 'cuspis' or 'cusp' translated the Arabic مركز (markaz). This Arabic term means 'centre' or 'cusp'; not in the sense of a geometrical feature of a shape but in reference to the place where a flag is planted into the ground to mark the centre of an area. The root of the term is ر ك ز • (r-k-z), the literal meaning of which is 'to stick something into the ground'. For an Arabic readership, therefore, the term would connote the marker of a centre, or possibly even the place where the needle point of a pair of compasses was planted as a circle was drawn. The Latin 'cuspis' would not import similar connotations. The literal and

92 THE SCIENTIFIC WORKS OF ROBERT GROSSETESTE

initially puzzling translated term chosen for the present translation therefore stays closer to how the original Latin readers might have experienced Grosseteste's text at this point.

3 Synopsis

Grosseteste begins *On the Sphere* (DS §1) by declaring that his purpose is to describe the shape of the 'world machine', the position and shapes of its parts, and their movements, using language which looks back to his work *On the Liberal Arts* and placing the present treatise within the remit of astronomy as there described. Its title is explained when Grosseteste goes on to say that the 'world machine' (meaning the universe as then conceived and understood in terms of its structure and operation) is spherical, and the first part of the treatise duly focuses on the geometry of the nested spheres of which it was thought to comprise.

The argument proceeds from first principles and using careful geometric demonstrations and implied diagrams to show how the outermost zone (the quintessence) is necessarily spherical at both its outer and its inner boundaries, and is 'circularly mobile' holding the planets and fixed stars, leaving no void between it and the inner spheres which share its nature and shape (DS §§2–6). This is a demonstration of why (*propter quid*) the machine is formed as it is, derived by logical deduction, following Aristotelian scientific methodology, and only once that is established does Grosseteste turn (following the same methodology rather than that of other sources available to him) to arguments concerning how the machine operates in practice (*quia*) based inductively on observation (not necessarily, or at all, his own, rather of his authorities, principally Ptolemy).

That the sphericity can be perceived through observation is presented in §§7–8: from the differently limited view of the heavens at different latitudes (showing north-south curvature); and from the different timings of day and night, and eclipses, at different longitudes (showing east-west curvature). A further argument (DS §9) is then offered for the sphericity of the heavens based on the apparent

MANUSCRIPTS, TRANSLATION, HISTORIOGRAPHY 93

movement of the stars in circles of increasing size, but at constant magnitude, around the pole.

Grosseteste's attention now turns (DS §§10–19) following the statement of purpose at the head of the treatise, from the shape of the world machine to the (relative) position and shape of its parts. The North and South Poles are defined as the extremities of the axis around which the heavens revolve, and great circles are introduced which circumscribe the heavens from pole to pole (twice, each at right angles to the other) and around the equinoctial at right angles to both and equidistant between the poles (DS §§10–12). From this Grosseteste proceeds to define another great circle, the ecliptic, which is oblique to the equinoctial and which, with parallel smaller circles to each side of it, forms the path of the zodiac (DS §§13–14). This is followed by geometrical demonstrations of the shape and site of the circles of the tropics (DS §15) and the Arctic and Antarctic Circles (DS §16), establishing five parallel circles which are linked to the five zones referenced in writers such as Virgil and Ovid, who are quoted. §§17–18 describe its oscillating spiral path of the sun following the ecliptic rather than the equinoctial, while §19 defines the horizon.

The progression of Grosseteste's treatment now moves on to consider movement over time, not indicated in his initial statement of purpose for this treatise but given in the related section of *On the Liberal Arts* (see Ch. 7), so that consideration of the horizon leads to a demonstration of day and night lengths, shadows and the times when the sun is directly overhead (DS §§20–8), ending with treatment of the 'days' and 'nights' of the areas within the Arctic and Antarctic Circles (DS §§29–30). Staying with the subject of time, §§31–7 give an extended geometrical demonstration of the way in which the signs of the zodiac are seen to require different lengths of time to rise according to how oblique or not their angle of rising is, to demonstrate which the reader is invited to imagine an appropriate diagram. This (DS §38) is one cause of the variation in day lengths, and Grosseteste goes again beyond his sources in adding a second (DS §§38–43), the eccentricity of the orbit of the sun, that is, the displacement of the centre of the sun's orbit from the centre of the earth.

94   THE SCIENTIFIC WORKS OF ROBERT GROSSETESTE

Movement over time is also the theme of §§44–7 in the sense of the effect of the motions of the heavens on climate. §44 argues that the southern polar region cannot be inhabited, because of the combined effect of the summer sun being closer to the earth in southern regions (because of its eccentric orbit) and striking it at a less oblique angle. The northern region is more temperate and (DS §45) can be divided into seven climes on the basis of day lengths as measured on a sundial. Great circles (DS §§46–7) are again used to demarcate these.

Grosseteste then turns in §§48–53 to the movement of the fixed stars in addition to that which they have from east to west. He clarifies that they are called fixed not because they have no movements but because their relative position within constellations is constant (DS §48). He cites Ptolemy's description of their having a slow eastwards precession around the zodiac (DS §49) but uses Thābit's trepidation hypothesis (DS §§50–3) to propose an alternative model in which the fixed zodiac in the firmament has a moving projection in the sphere of the fixed stars, which oscillates following the ecliptic (DS §§51–3). He finishes this section with the almost throwaway remark that the apogees of the planets follow the same movement (implying familiarity with Ptolemy's theory of their eccentric paths).

On the Sphere next describes the movement over time of the moon. §54 sets up a geometric model of its motion along a path tilted from the ecliptic and eccentric to the earth's centre, with the further complication that it orbits a point on this path (making an epicycle) rather than following it directly (DS §55). A complex geometrical demonstration of the resultant path of the moon, especially in its relationship to the position of the sun, follows in §56 (see Ch. 4), and this allows Grosseteste in §§57–9 to specify why and when lunar eclipses occur. The final paragraphs of the treatise (DS §§60–3) then consider solar eclipses, taking into account the further complexity of the divergence between the actual position of the moon and its apparent position to a terrestrial observer, which is cited as an explanation of why solar eclipses are less frequent than lunar ones. The treatise ends at this point without a formal conclusion but having covered the ground required by its introduction.

Roberti Grosseteste

De sphera

De sphera

Edited by Cecilia Panti

1. Intentio nostra in hoc tractatu est describere figuram machine mundane et situm et figuras corporum eam constituentium et motus corporum superiorum et figuras circulorum suorum. Quia igitur huius mundi machina spherica est dicendum est in primis quid sit sphera.

2. Est autem sphera transitus semicirculi diametro eius fixa quousque ad locum suum unde incepit redeat. Si igitur semicirculus ACB circumvolvatur super AB diametrum fixam, manifestum est quod motu suo describet corpus a cuius medio, scilicet O, omnes linee exeuntes ad eius circumferentiam sunt equales. Et erit, corpus illud, cuiusmodi corpus dicimus esse spheram. Tale autem corpus est tota mundi machina.

3. Ymaginemur iterum super O centrum DFE semicirculum describi. Manifestum est ergo quod superficies inter ACB semicircumferentiam et DFE semicircumferentiam, si circumvolvatur super AB diametrum, motu suo describet corpus cuius ultima superficies et intima erunt spherice, et corpus illud totum interius et exterius sphericum, nihil habens extra se, omnia corpora continens intra se. Consimilis figure et situs corpus huius mundi est; unum quod quintam essentiam nominant philosophi, sive ethera sive corpus celi, et preter elementares proprietates circulariter mobile, in quo 7 planete cum stellis fixis continentur.

4. Posito iterum O centro et OG spatio occupato, describatur semicirculus GHI. Superficies igitur contenta inter DEF semicircumferentiam et GHI semicircumferentiam corpus interius et exterius sphericum describet, contiguum exterius quinte essentie et intra se continens reliqua corpora. Huius figure et situs est corpus ignis. Superficies iterum inter GHI semicircumferentiam et KLM contenta circumvolutione sua describet corpus, cuius figure et situs optinet aer similitudinem. Item superficies inter KLM et NRP semicircumferentiam contenta circumrotatione sua describet corpus, cuius corporis figure et situs similitudinem optinet aqua. Circuitio iterum NRP semicirculi describet corpus sphericum, in medio predictorum corporum contentum, cuius figure et situs similitudinem optinet terra. Verumtamen ut animalia terrena habitaculum et receptaculum haberent,

On the Sphere

Translated by Sigbjørn Olsen Sønnesyn

§1 Our purpose in this treatise is to describe the shape of the world machine and the [relative] position and shapes of its constituent bodies, and the movements of higher bodies and the shapes of their orbits. Therefore, since the machine of this world is spherical, we should state at the beginning what a sphere is.

§2 Now, a sphere is the passage of a semicircle with its diameter fixed until it returns to the place from which it started. Therefore, if a semicircle ACB is turned around over its fixed diameter AB, it is obvious that it describes by its movement a body from the middle of which—that is, O—all lines going out to its circumference are equal. And this body will be the sort of body we call a sphere. Now, the whole machine of the world is such a body.

§3 Let us again imagine that a semicircle, DFE, is described over the centre O. It is then obvious that the area between the half-circle ACB and the half-circle DFE, if it [that is, DFE] is turned around over the diameter AB, will describe by its movement a body, the outer and inner surfaces of which will be spherical, and that whole body will be spherical internally and externally, having nothing outside of itself, containing all bodies inside itself. The body of this world is of this shape and site; [that is] the one [body] that the philosophers name the fifth essence, or ether, or the body of the heavens, and which, contrary to elementary properties, [is] circularly mobile, and in which the seven planets are contained together with the fixed stars.

§4 The centre O again being posited, and a span OG being taken up, let a semicircle GHI be described. The area contained between the DEF semicircle and the GHI semicircle will then describe a body internally and externally spherical, externally contiguous to the fifth essence and containing within itself all other bodies. Of this shape and position is the body of fire. Again, the area contained between the semicircles GHI and KLM will describe in its rotation a body of such shape and position as that which air possesses. Likewise, the area contained between KLM and the semicircle NRP in its rotation will describe a body of such shape and position as the body that water possesses. The rotation of the NRP

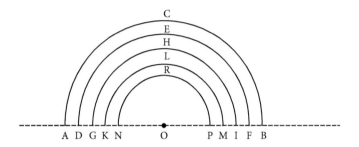

aqua in concavitatem terre recessit et apparuit superficies terre separata. Estque terra cum aquis in se contentis sicut sphera terre solum.

5. Quod autem corpora predicta spherica sint et rationibus naturalibus et experimentis astronomicis ostenditur. Quia namque a natura rei est forma et unumquodque predictorum corporum naturalium nature unius est, cuius scilicet quelibet pars participat cum toto in nomine et diffinitione, necesse fuit ut unumquodque haberet uniformem figuram, cuius quelibet pars esset toti consimilis. Talis autem nulla est preter sphericam. Preterea, quia omne ponderosum tendit ad centrum, et locus profundior est qui est circa centrum, necesse est duo corpora ponderosa sphericam habere figuram; et similiter de duobus levibus, quia locus elevatior est qui magis distat a centro, et omne leve ad magis elevatum tendit.

6. De quinta essentia ostendit philosophus quod ipsa est spherica quia necesse est motus rectos, qui sunt elementorum gravium et levium, reduci ad motum circularem, qui est de necessitate quinte essentie. Sed si movetur circulariter de necessitate spherica est, quia si esset angulosa de necessitate esset locus vacuus.

7. Experimento etiam scitur quod terra est rotunda. Si enim esset plana, cum visus recte procedat, visus omnium super superficiem terre existentium ad eundem locum in celo terminaretur. Sed notum est experimento quod illi qui sunt in terra Indie super Arim civitatem vident polum septemtrionalem, et ipse est finitor visus eorum. Et quanto homines magis recedunt ab illa civitate versus septemtrionem, tanto magis elevatur eis polus, et finitur visus eorum sub polo. Illud autem non posset accidere nisi hac via terra esset rotunda.

8. Quod autem sit rotunda versus oriens et occidens patet per hoc quod prius est dies hiis qui magis accedunt orienti et tardius hiis qui sunt propinquiores occidenti, et similiter nox. Et hoc scitur per eclipses

MANUSCRIPTS, TRANSLATION, HISTORIOGRAPHY 99

semicircle will again describe a spherical body, contained in the centre of the abovementioned bodies, of the shape and position such as the one earth possesses. Nevertheless, so that terrestrial creatures should have a home and shelter, water retreated to the hollow of the earth, and there appeared a separate surface of the earth. And the earth together with the waters contained within it is, as it were, the sphere of earth alone.

§5 Both natural reasoning and astronomical observations show that the abovementioned bodies are spherical. For since the form [of a thing] comes from the nature of a thing, and each of the abovementioned natural bodies is of a single nature, any part of which shares the same name and definition as the whole, it was necessary that each [body] should have a uniform shape, any part of which should be like the whole. There is no such [shape] except the spherical. Moreover, because every heavy object tends towards the centre, and the deeper place is the place around the centre, it is necessary that the two heavy bodies [earth and water] have a spherical shape; and similarly the two light bodies [fire and air], because the higher place is the one the furthest from the centre, and everything light tends towards what is higher.

§6 Concerning the fifth essence the Philosopher shows that it is spherical because it is necessary that rectilinear movements, which belong to heavy and light elements, are brought back to circular movement, which by necessity belongs to the fifth essence. However, if it is moved in a circle, it is by necessity spherical, because if it were angular there would be by necessity an empty place.

§7 It is also known from observation that the earth is round. For if it were flat (since vision proceeds in a straight line) the limit of what everyone living on the surface of the earth could see would be set at the same place in the heavens. However, it is known from observation that those who are in the land of India in the city of Arim see the northern pole [the Pole Star], and this is the limit of what they can see. And the more human beings depart from that city towards the north, the higher is the pole raised for them, and the limit of what they can see is below the pole. However, this could not happen unless the earth were to be round in this direction.

§8 However, that it is round towards the east and west is clear from this: that it is day earlier for those who are nearer to the east and later for those who are closer to the west, and the same for night. And this is known by the eclipses of the moon. For the same eclipse was seen in Arim in the evening

100 THE SCIENTIFIC WORKS OF ROBERT GROSSETESTE

lunares. Eadem enim eclipsis visa fuit apud Arim in vespere, que fuit in media nocte eorum qui fuerunt in oriente, et non apparuit eis qui fuerunt in occidente. Similiter alia eclipsis que fuit in media nocte eorum qui fuerunt apud Arim, fuit in vespere occidentalium, et in mane orientalium.

9. Quod autem celum sit sphericum patet per apparentiam nobis in visu. Videmus enim stellam unam in celo non motam et omnes reliquas moveri circulariter circa ipsam, et stellas ei propinquiores breviores circulos describere et remotiores maiores. Apparet etiam nobis unaqueque stella in ortu suo et in medio celi et in occasu suo eiusdem magnitudinis. Iste autem dispositiones non possunt esse nisi in spherico et spherice moto circa diametrum immobilem. Scimus itaque quod quinta essentia circulariter mota est circa diametrum fixam.

10. Diameter fixa axis vocatur, ebraice quidem *magal'*, et extremitates axis poli appellantur, quorum unus qui semper nobis apparet articus appellatur ab *arthos* grece, quod est ursa latine, eo quod prope ipsum est tam maior ursa quam minor. Polus ei oppositus antarticus dicitur, quasi contra articum polum positus. Super hos duos polos, ut diximus, circumvolvitur celum cum omnibus stellis et planetis, que sunt in eo motu equali et uniformi per diem et noctem semel, cuius motus causa efficiens est anima mundi.

11. Ymaginemur itaque circulum magnum per duos polos circumductum et alium per eosdem polos circumductum, secantem priorem orthogonaliter. Hii duo circuli vocantur coluri a *colon* quod est membrum et *uros* quod est bos silvester, eo quod apparens nobis in visu de circulis illis caude bovis assimilatur.

12. Ymaginemur iterum circulum magnum circumductum distantem ab utroque polo secundum latus quadrati. Hic circulus utrumque predictorum secabit orthogonaliter, et hic vocatur equinoctialis, eo quod quando sol circumrotatione firmamenti describit illum circulum, equalis est dies nocti in omni regione.

13. Ab equinoctiali itaque circulo accipiantur 24 gradus, sive 23 gradus cum 33 minutis, versus polum articum in uno predictorum colurorum, et a puncto equinoctialis priori opposito sumantur in eodem coluro totidem gradus et totidem minuta versus polum antarticum, et circumducatur circulus magnus per terminos predictorum

MANUSCRIPTS, TRANSLATION, HISTORIOGRAPHY 101

that happened in the middle of the night for those in the east and that was not visible to those who were in the west. Similarly, another eclipse that was in the middle of the night for those who were in Arim, was in the evening for the westerners and in the morning for the easterners.

§9 Moreover, that the heavens are spherical is clear by how they appear to us visually. For we see a single star unmoved in the heavens and all the others being moved in a circle around this; and closer to it [the unmoved star], [we see] stars describing shorter circles and those further away longer. Each star, indeed, appears to us to be of the same magnitude in its rising, in the middle of the heavens, and in its setting. These features cannot exist except in something spherical and in something moved spherically around an immobile diameter. We know, therefore, that the fifth essence is moved in a circle around a fixed diameter.

§10 A fixed diameter is called an axis, or in Hebrew *magal'*, and the extremities of the axis are called poles, of which the one that is always visible to us is called the Arctic, from the Greek *arthos*, which is *ursa* [bear] in Latin, because it is close to both Ursa Major [Great Bear] and Ursa Minor [Little Bear]. The pole opposite to it is called the Antarctic, that is, placed opposite to the Arctic. Upon these two poles, as we have said, the heavens revolve with all the stars and planets, which are in this same continuously equal and uniform movement, once per day and night, and the efficient cause of this movement is the world soul.

§11 Let us therefore imagine a great circle drawn around through the two poles, and another drawn around through the same two poles, cutting the first at right angles. These two circles are called 'colures' from *colon*, which is 'member', and *uros*, which is 'an untamed ox', because what appears to us visually of these circles is like the tail of an ox.

§12 Let us again imagine a great circle drawn around at a distance from each of the poles according to the side of a square. This circle will cut each of the abovementioned [circles] at right angles, and this is called the 'equinoctial', because, when the sun through a rotation of the firmament describes that circle, the day is equal to night in all regions.

§13 Therefore, from the equinoctial circle let 24 degrees, or rather 23 degrees and 33 minutes, be counted towards the Arctic pole in one of the aforementioned colures, and from a point of the equinoctial opposite to the first let there be taken in that same colure as many degrees and as many minutes towards the Antarctic pole; and let a great circle be drawn around

102 THE SCIENTIFIC WORKS OF ROBERT GROSSETESTE

graduum et minutorum hinc inde sumptos, qui circulus de necessitate transibit per duo puncta, ubi equinoctialis reliquum secat colurum. Hic circulus magnus vocatur linea ecliptica, sive cingulus signorum. Et si huic circulo circumducantur duo circuli equedistantes hinc inde, quorum uterque distat ab eo per 6 gradus, superficies circularis habens latitudinem 12 graduum inter eosdem circulos contenta zodiacus vocatur a *zoas*, quod est animal, eo quod eius partes ymaginibus sunt insignite nominibus animalium nuncupatis.

14. Hic enim circulus in 12 partes dividitur et vocatur unaqueque duodecima signum unum. Quodlibet iterum signum in 30 gradus dividitur. Et erunt in toto circulo 360 gradus. Quilibet iterum gradus in 60 minuta dividitur. Initio itaque sumpto in zodiaco ubi eum secat equinoctialis circulus, a qua sectione, si fiat processus contra motum firmamenti procedendo in partem septemtrionalem, principium primi signi invenitur. Prima enim duodecima vocatur aries, secunda taurus, tertia gemini, quarta cancer, quinta leo, sexta virgo, septima libra, octava scorpius, nona sagittarius, decima capricornus, undecima aquarius, duodecima pisces.

15. Principium cancri est in puncto cinguli signorum qui magis appropinquat polo artico; et motu ipsius puncti describitur circulus equedistans equinoctiali, qui vocatur paralellus, eo quod est equedistans, et tropicus estivalis, quia sol accedens ad ipsum in estate convertit motum suum versus austrum. *Tropos* namque idem est quod conversio. Circumvolutione iterum capricorni describitur circulus equedistans equinoctiali equalis priori equedistanti, qui tropicus hyemalis appellatur, eo quod sol ad eum accedens in hyeme, cum ibi pervenerit convertit motum suum ad septemtrionem.

16. Si igitur ymaginemur lineam rectam orthogonaliter penetrantem superficiem cinguli signorum per centrum eius, linea illa erit axis zodiaci, cuius extremitates, scilicet poli, erunt in coluro qui transit per puncta tropica, tantum remota utrimque a polis mundi quanta est declinatio punctorum tropicorum ab equinoctiali. Circumrotatione itaque polorum zodiaci describuntur duo circuli equedistantes ab equinoctiali eiusdem magnitudinis. Et ille qui est prope polum articum, vocatur paralellus articus sive septemtrionalis, et reliquus ei oppositus paralellus antarticus sive

MANUSCRIPTS, TRANSLATION, HISTORIOGRAPHY   103

through the end points of the abovementioned degrees and minutes taken on both sides, which circle by necessity will pass through the two points through which the equinoctial will cut the other colure. This great circle is called the line of the 'ecliptic', or the girdle of signs. And if two circles, equidistant on both sides, are drawn around this circle, each of which is distant from the first by six degrees, the circular area of a latitude of 12 degrees contained between both circles is called the zodiac after *zoas*, which is 'animal', because its parts are marked with images designated by names of animals.

§14 This circle, then, is divided into 12 parts, and each of these twelve is called a single sign. Every sign is again divided into 30 degrees. And in the whole circle there will be 360 degrees. Every degree is then divided into 60 minutes. Starting, then, from the point of the zodiac where the equinoctial circle cuts it, if one proceeds from this cut in the opposite direction to the movement of the firmament and towards the north, the beginning of the first sign is found. And the first is called Aries, the second Taurus, the third Gemini, the fourth Cancer, the fifth Leo, the sixth Virgo, the seventh Libra, the eighth Scorpio, the ninth Sagittarius, the tenth Capricorn, the eleventh Aquarius, the twelfth Pisces.

§15 The beginning of Cancer is at the point of the girdle of signs which is closest to the Arctic pole; and by the movement of this point is described a circle equidistant from the equinoctial, and this is called the 'parallel', because it is equidistant, and the 'summer tropic', because when the sun reaches it in summer it turns its movement towards the south. For *tropos* means the same as 'turning'. Again, by the rotation of Capricorn a circle is described, equidistant from the equinoctial and equal to the prior equidistant, and this is called the 'winter tropic' because the sun, approaching it in winter, turns its movement towards the north when it reaches it.

§16 If, then, we imagine a straight line passing through the area enclosed by the girdle of signs at a right angle through its centre, that line will be the axis of the zodiac, the extremes of which—that is, its poles—will be in the colure that passes through the tropic points, as far removed on both sides from the poles of the world as the declination of the tropical points are from the equinoctial. By the revolution of the poles of the zodiac two circles of the same magnitude are described at equal distance from the equinoctial. And the one that is closer to the Arctic pole is called the Arctic or the northern parallel, and the other opposite to it is called the Antarctic or southern parallel. These five parallels are the five zones about which

104 THE SCIENTIFIC WORKS OF ROBERT GROSSETESTE

australis. Hii 5 paralelli sunt 5 zone, de quibus Vergilius: *quinque tenent celum zone*, <et Ovidius:> *totidemque plage tellure premuntur.*

17. Ymaginemur itaque circulum sub cingulo signorum recte dispositum nusquam a cingulo signorum declinantem. In circulo sic disposito currit corpus solis, ita quod centrum corporis solis semper est in circumferentia predicti circuli et movetur in hoc circulo motu proprio contra firmamentum, ita quod in 365 diebus et quarta diei fere percurrit circulum illum. Motu tamen firmamenti circumfertur ab oriente in occidens et ab occidente in oriens per diem et noctem semel. Si itaque sol esset immobilis quoad motum proprium, circumvolutione sua in occidens describeret paralellum equedistantem equinoctiali, aut ipsum equinoctialem, si esset in principio arietis vel libre. Sed quia mobilis est, dum circumfertur motu celi, iam recessit a puncto in quo fuit in principio illius revolutionis, unde circumrotatione firmamenti spiram unam cotidie describit, que spira quasi paralellus est, et propter insensibilem differentiam paralellum nominamus quandoque.

18. Manifestum est igitur quod quotiens revolvitur firmamentum dum transit sol a principio cancri usque in principium capricorni, tot spiras vel paralellos motu firmamenti describit. Dum vero revertitur sol a capricorno usque ad cancrum per eosdem paralellos, iterum circumfertur.

19. Orizon vero est circulus qui dividit medietatem celi visam a medietate non visa, et interpretatur orizon finitor visus. Radius enim visualis est sicut linea recta contingens terram. Et si ponatur linea terram contingens super aliquem punctum terre protensa usque ad firmamentum, et circumrotetur linea in eodem puncto contingens terram, ipsa circumrotata faciet circulum dividentem celum in duo equalia, cum magnitudo terre sit insensibilis respectu celi. Talis circulus est orizon a radio visuali descriptus, unde quot sunt loca super terram et circumferentiam, tot possibile est esse orizontes.

20. Ex situ orizontum et paralellorum predictorum facile est videre quid accidit in omni situ terre de equalitate et inequalitate dierum ac noctium. Videndum est in primis quid accidit hiis quorum cenith capitis est in equinoctiali circulo. Voco autem cenith capitis extremitatem linee

MANUSCRIPTS, TRANSLATION, HISTORIOGRAPHY   105

Virgil [said] 'five zones hold the heavens', <and Ovid> 'and the same number of regions are stamped upon the earth'.

§17 Let us then imagine a circle below the girdle of signs, disposed in a straight manner and never declining from the girdle of signs. In the circle disposed in this way runs the body of the sun, so that the centre of the body of the sun is always on the circumference of the aforementioned circle and is moved in this circle by its own movement against the firmament, so that it completes that circle in about 365 days and a quarter of a day. By the movement of the firmament, however, it is carried around from east to west and from west to east once every day and night. Therefore, if the sun were immobile with respect to its own movement, it would describe by its rotation towards the west a parallel equidistant from the equinoctial, or the equinoctial itself, if it were in the starting point of Aries or Libra. However, because it is mobile, while it is carried around by the movement of the heavens it recedes at the same time from the point where it was at the beginning of its revolution, from which by the circular movement of the firmament it daily describes a single [turn of] the spiral. This spiral is close to a parallel, and because of the imperceptible difference we sometimes call it a parallel.

§18 It is therefore manifest, that the sun describes as many spirals or parallels through the movement of the firmament as there are revolutions of the firmament during the course of the sun's passing from the beginning of Cancer to the beginning of Capricorn. And while the sun returns from Capricorn to Cancer along the same parallels, it is once again carried around.

§19 The horizon is the circle that divides the visible half of the heavens from the hidden half, and 'horizon' is understood as the limit of what can be seen. For the visual ray is like a straight line touching the earth. And if a line is posited, touching the earth upon a point on the earth, and extended all the way to the firmament, and if this line, touching the earth, is rotated around that same point the line once rotated makes a circle dividing the heavens into two equal [parts], since the magnitude of the earth is imperceptible with respect to the heavens. The horizon is such a circle described by the visual ray, so that as many horizons are possible as there are places on the earth and the circumference [that is, the firmament].

§20 From the position of the horizons and parallels mentioned above it is easy to see what happens in every position on earth in terms of equality and inequality of days and nights. We should first see what happens to those whose zenith [*lit.* 'zenith-of-the-head'] is on the equinoctial circle. And what

recte directe a centro terre per capud hominis usque ad firmamentum. Eorum ergo orizon, quorum cenith est in equinoctiali circulo punctus aliquis, de necessitate transit per utrumque polum mundi, cum semper a cenith capitis sit quarta circuli usque ad orizonta.

21. Cum igitur duo poli sint immobiles, poli sic semper erunt in confinio visus eorum. Et cum omnis circulus descriptus super polos mundi secetur a predicto orizonte orthogonaliter et per equalia, et omnis punctus in celo alius a polis mundi motu celi describit circulum supra polos mundi, manifestum est quod omnis punctus in celo alius a polis illis habet ortum et occasum, et est per medietatem unius revolutionis supra orizonta eorum et per tantum spatium sub orizonte.

22. Oriturque omnis punctus et omnis stella secundum angulos rectos. Ex hoc patet quod quilibet dies eorum est equalis sue nocti, et quilibet dies cuilibet diei et cuilibet nocti. Cum enim quilibet paralellus descriptus a sole per unam celi revolutionem secetur a predicto orizonte orthogonaliter et per equalia, et motus celi semper est uniformis et quelibet revolutio equalis alii—dumque describit sol medietatem para-lelli, scilicet supra orizonta, est dies, et dum describit medietatem sub orizonte est nox—manifestum est quod omnis dies est equalis sue nocti, et quilibet dies cuilibet diei et cuilibet nocti.

23. Preterea existentibus sub equinoctiali circulo contingit quod sol bis in anno transit per cenith capitis eorum, scilicet quando sol est in principio arietis et iterum quando est in principio libre. Tunc enim motu celi describit equinoctialem circulum et umbra eorum ante mer-idiem tendit directe versus occidens, et umbra post meridiem directe versus oriens, et umbra meridiana rei erecte nulla. Istud patet per hoc quod umbra semper fertur in oppositum lucidi. Dum vero sol describit signa septemtrionalia, oritur eis sol inter oriens et septemtrionem, et ascendit cotidie inter ipsos et septemtrionem, estque in meridie recte inter cenith capitis eorum et septemtrionem, unde umbre meridiane directe flectuntur ad austrum. Dum vero sol describit signa australia, oritur eis sol cotidie inter oriens et austrum et ascendit et descendit

I call the zenith is the extremity of a straight line going directly from the centre of the earth, through the head of a man, and up to the firmament. The horizon, then, of the people whose zenith is any point on the equinoctial circle, necessarily passes through both poles of the world, since there is always a quarter of a circle between the zenith and the horizon.

§21 Since, therefore, the two poles are immobile, accordingly the poles will be always within the confines of what they can see [standing on the equator]. And since every circle described over the poles of the world will be cut by the aforementioned horizon at a right angle and into equal parts, and any point in the heavens except the poles of the world will describe, by the movement of the heavens, a circle over the poles of the world, it is manifest that any point in the heavens except those poles has a rising and a setting. For half the duration of a single revolution [any point] is above their horizon and for the same duration below the horizon.

§22 Moreover, every point and every star rise at right angles. From this it is clear that any of their [the inhabitants on the equator] days is equal to its night, and that any day [is equal] to any day and any night. For any parallel described by the sun through one revolution of the heavens will be cut by the horizon mentioned above at a right angle and [divided] into equal parts, and the movement of the heavens is always uniform and any revolution equal to another. Moreover, when the sun describes the half of the parallel that is over the horizon, it is day, and when it describes the half below the horizon, it is night. It is therefore manifest that every day is equal to its night, and any day to any day and any night.

§23 Furthermore, for those living under the equinoctial circle it happens that the sun passes through their zenith twice a year, that is when the sun is at the beginning of Aries and again when it is at the beginning of Libra. For then by the movement of the heavens the sun describes the equinoctial circle and before midday their [the inhabitants] shadow points directly towards the west, and after midday their shadow directly towards the east, and at midday a vertical object casts no shadow. This is clear through the fact that that shadow is always directed away from the light source. For while the sun describes the northern signs, the sun rises for them [the inhabitants on the equator] between east and north, and ascends every day between them and the north, and is at midday right between their zenith and the north, hence the shadows at midday are turned directly towards the south. And while the sun describes the

108 THE SCIENTIFIC WORKS OF ROBERT GROSSETESTE

inter ipsos et austrum, et umbre meridiane flectuntur directe versus septemtrionem.

24. Omnibus vero hiis quorum cenith est inter equinoctialem circulum et tropicum estivum accidit similiter quod sol bis in anno transit super cenith capitis eorum, et umbre meridiane nulle. Et dum sol ascendit partem zodiaci inter tropicum estivalem et paralellum transeuntem super cenith capitis eorum, transit sol in meridie inter cenith capitis eorum et septemtrionem, et flectuntur umbre meridiane ad austrum. Et econtrario est quando sol ascendit partem zodiaci inter tropicum hyemalem et paralellum transeuntem per cenith capitis eorum.

25. Eis vero qui sunt sub capite cancri, accidit quod semel in anno, scilicet quando sol est in capite cancri, transit supra cenith capitis eorum in meridie.

26. In omni vero loco inter septemtrionalem circulum et equinoctialem est dies maior nocte dum sol est in signis septemtrionalibus, et econtrario dum sol est in signis australibus. Et hoc patet quia in omni tali loco elevatur polus septemtrionalis supra orizonta quantum cenith capitis eorum distat ab equinoctiali. Orizon vero, qui est circulus magnus, secat equinoctialem per equalia, et omnem paralellum quem secat inter equinoctialem et polum septemtrionalem, secat sic quod magis medietate relinquitur supra orizonta et minus medietate sub orizonte. Et cuiuslibet paralelli sic divisi per orizonta remotioris ab equinoctiali circulo pars relicta super orizonta maior est respectu sui circuli quam sit pars paralelli propinquioris equinoctiali relicta super orizonta respectu sui circuli.

27. Cum igitur quelibet revolutio firmamenti sit equalis alii et in qualibet revolutione firmamenti describat sol paralellum unum, motu firmamenti semper uniformi, manifestum est quod omnis dies, dum sol est in signis septemtrionalibus, in omni loco versus septemtrionem ab equinoctiali linea, est maior nocte sua, quia paralelli pars relicta super orizonta, descripta a sole revolutione firmamenti, est maior parte eiusdem paralelli sub orizonte relicta. Et quanto sol magis accedit ad caput cancri, tanto facit dies estivos maiores, quia pars paralelli descripti a sole

MANUSCRIPTS, TRANSLATION, HISTORIOGRAPHY   109

southern signs, the sun rises for them [the inhabitants] every day between the east and the south and ascends and descends between them and the south, and the shadows at midday are turned directly towards the north.

§24 For all of the people, however, whose zenith is between the equinoctial circle and the summer tropic, it happens similarly that the sun passes over their zenith twice a year, and there are no shadows at midday. And while the sun ascends the part of the Zodiac between the summer tropic and the parallel passing above their zenith, the sun passes at midday between their zenith and the north, and their shadows at midday are turned towards the south. And the opposite happens when the sun ascends the part of the Zodiac between the winter tropic and the parallel passing through their zenith.

§25 And for those people, moreover, who are under the head of Cancer, it happens once every year, that is, when the sun is in the head of Cancer, that it passes over their zenith at midday.

§26 By contrast, in every place between the northern circle and the equinoctial, the day is longer than the night when the sun is in the northern signs, and vice versa when the sun is in the southern signs. And this is clear because in every such place the northern pole [the Pole Star] is elevated above the horizon by as much as their zenith is distant from the equinoctial. For the horizon, which is a great circle, cuts the equinoctial into equal parts, and every parallel that it cuts between the equinoctial and the northern pole it cuts in such a way that more than half remains over the horizon and less than half below the horizon. And in the case of any parallel divided in this way by the horizon, and more remote from the equinoctial circle, the part [of that parallel] remaining over the horizon is greater with respect to its circle than is the part of a parallel closer to the equinoctial remaining over the horizon with respect to its circle.

§27 Therefore, since any revolution of the firmament is equal to any other, and the sun will describe a single parallel in any revolution of the firmament, the movement of the firmament being always uniform, it is manifest that every day is longer than its night in every place north of the equinoctial line while the sun is in the signs of the north, because the part of the parallel remaining over the horizon and described by the sun through the revolution of the firmament is greater than the part of that same parallel remaining under the horizon. And the more the sun approaches the head of Cancer, the longer it makes the days of summer,

propinquioris capiti cancri relicta supra orizonta est maior respectu sui circuli quam sit pars remotioris paralelli a capite cancri supra orizonta relicta respectu sui circuli. Et econverso se habet de diebus et noctibus dum sol est in signis australibus. Quorum etiam cenith capitum magis distat ab equinoctiali habent dies estivos longiores hiis qui magis accedunt equinoctiali. Declinatio namque orizontis eorum maior est, et ab orizonte magis declivi magis relinquitur de quolibet paralello inter equinoctialem et polum elevatum, quam ab orizonte minus declivi.

28. Quod autem accidit de diversitate dierum et noctium, scilicet in omni loco inter equinoctialem circulum et septemtrionalem, dum sol est in signis septemtrionalibus, idem accidit in loco tantum remoto versus austrum dum sol est in signis australibus. Et quod accidit nobis dum sol est in signis australibus, idem accidit in locis versus austrum dum sol est in signis septemtrionalibus. Hoc patet totum si ymaginetur polus australis elevatus et septemtrionalis depressus. In omni itaque loco inter circulum equinoctialem et septemtrionalem dividitur una revolutio in diem et noctem, quia orizon cuiuslibet talis loci semper secat zodiacum et duos tropicos paralellos et omnes eis interpositos.

29. In omni vero loco sub circolo septemtrionali accidit totam unam revolutionem diem esse, scilicet cum sol est in capite cancri, et totam aliam revolutionem noctem esse, scilicet cum sol est in capite capricorni. Cum autem circulus septemtrionalis sit a polo zodiaci descriptus, accidit semel in qualibet revolutione polum zodiaci esse cenith capitis hiis qui sunt sub circulo illo, et tunc zodiacus et eorum orizon sunt simul loco, et est totus tropicus estivalis supra orizonta et totus tropicus hyemalis sub orizonte, sed quilibet paralellus inter hos secatur ab orizonte. Sole enim existente in capite cancri et describente tropicum estivalem erit tota illa revolutio dies, quia aut sol est in confinio visus aut supra orizonta. Sole vero existente in capite capricorni, dum revolutione firmamenti describit tropicum hyemalem, erit sol sub orizonte, et ita erit una revolutio nox. Sole vero existente in signis intermediis, erit quelibet revolutio divisa in diem et noctem, quia quilibet paralellus intermedius ab orizonte intersecatur.

MANUSCRIPTS, TRANSLATION, HISTORIOGRAPHY 111

because the part remaining over the horizon of the parallel which the sun describes closer to the head of Cancer is greater in terms of its circle than is the part remaining over the horizon of the parallel more remote from the head of Cancer in terms of its circle. And the converse is the case concerning days and nights while the sun is in the southern signs. Those whose zenith is further away from the equinoctial have longer summer days than those who are closer to the equinoctial. For the declination of their horizon is greater, and from a more declined horizon more remains of any parallel between the equinoctial and the raised pole than from a horizon less declined.

§28 As regards the difference between days and nights, that is to say, in any place between the equinoctial circle and the northern, while the sun is in the northern signs, the same thing happens in the place equally removed towards the south when the sun is in the southern signs. And the same happens for us while the sun is in the southern signs, as happens in places to the south when the sun is in northern signs. This is entirely clear if the southern pole is imagined as raised and the northern pole as lowered. In every place, then, between the equinoctial circle and the northern [circle] a single revolution is divided into day and night, because the horizon of every such place always cuts the zodiac and the two tropic parallels and all [parallels] placed between them.

§29 Indeed, in every place beneath the northern circle it happens that an entire revolution is day, that is, when the sun is in the head of Cancer, and another entire revolution is night, that is, when the sun is in the head of Capricorn. However, since the northern circle is described by the pole of the zodiac, it happens once in every revolution that the pole of the zodiac is the zenith for those who are under that circle, and then the zodiac and their horizon are in the same place, and the entire summer tropic is over the horizon and the entire winter tropic is under the horizon, but each parallel between these is cut by the horizon. When, therefore, the sun is present in the head of Cancer and describing the summer tropic, that entire revolution will be day because the sun is within the confines of what can be seen, or over the horizon. When the sun is in the head of Capricorn, during which time by the revolution of the firmament it describes the winter tropic, the sun will be under the horizon, and in this way a single revolution will be night. But while the sun is in the intermediate signs, every revolution will be divided into day and night, because every intermediate parallel will be intersected by the horizon.

112 THE SCIENTIFIC WORKS OF ROBERT GROSSETESTE

30. In omni vero alio loco est declinatio orizontis ab equinoctiali minor declinatione zodiaci. Unde paralelli plures de hiis qui sunt a tropico versus equinoctialem ex parte poli elevati semper sunt super orizonta et ex opposito semper sub orizonte. Unde, dum revolutione firmamenti describit sol paralellos apparentes, semper est dies. Et dum describit ex opposito totidem paralellos occultos, semper est nox. Et quanto maior est accessus cenith ad polum, cum tanto minor fit declinatio orizontis ab equinoctiali, tanto plures de paralellis quos describit sol sunt toti apparentes et totidem toti occulti. Unde plures revolutiones sunt dies unus et totidem una nox. Sub polo vero est orizon cum equinoctiali semper simul loco, et propter hoc una medietas celi semper est apparens, et alia medietas semper occulta. Et una medietas anni, scilicet dum sol est in signis septemtrionalibus, est dies et alia medietas anni est nox; unde totus annus est unus dies cum sua nocte.

31. Hiis prelibatis considerandum est quid accidat de ortu et occasu signorum tam in sphera recta quam obliqua. Sciendum igitur quod tam in sphera recta quam obliqua ascendit equinoctialis circulus semper uniformiter, scilicet in temporibus equalibus equales partes ascendunt. Motus enim celi uniformis est, et angulus quem facit equinoctialis cum orizonte aliquo non diversificatur in aliquibus horis. Arcus vero de equinoctiali circulo qui ascendit cum aliqua parte zodiaci dicitur ascensio eiusdem partis.

32. Partes igitur zodiaci equales habentes ascensiones in temporibus equalibus oriuntur, et que in temporibus equalibus oriuntur, equales habent ascensiones. Partes vero zodiaci equales non de necessitate equales habent ascensiones, quia quanto aliqua pars zodiaci rectius oritur, tanto maius tempus ponit in ortu suo, et quanto obliquius oritur tanto minus tempus ponit in ortu suo. Hoc patet sensui et etiam ymaginationi, si ymaginentur circuli magni descripti super utrosque polos mundi transeuntes in zodiaco per sectiones signorum: resecabunt enim zodiacum in 12 partes equales, et equinoctialem in totidem inequales. Et pars in equinoctiali que respondet parti zodiaci resecte ad angulos magis acutos minor est parte equinoctialis respondente parti zodiaci resecte ad angulos minus acutos.

MANUSCRIPTS, TRANSLATION, HISTORIOGRAPHY   113

§30 In every other place the declination of the horizon from the equinoctial is smaller than the declination of the zodiac. Therefore, on the side with the elevated pole, more of the parallels that are from the tropic towards the equinoctial are always over the horizon and on the other side always under the horizon. Therefore, while the sun describes visible parallels by the revolution of the firmament, it is always day. And on the contrary, while it describes the same number of hidden parallels, it is always night. And the more the zenith approaches the pole, since a correspondingly smaller declination of the horizon from the equinoctial is made, the greater number of the parallels described by the sun are entirely visible and the same number entirely hidden. Therefore, more revolutions are a single day and the same number a single night. Under the pole, by contrast, the horizon is always in the same place as the equinoctial, and for this reason one half of the heavens is always visible, and the other half always hidden. And for half of the year, that is, when the sun is in the northern signs, it is day, and the other half of the year it is night; and therefore the whole year is a single day with its night.

§31 Having had a taste of these issues we should consider what happens concerning the rising and setting of signs, in the right sphere as well as in the oblique. It should be known, therefore, that in the right sphere as well as in the oblique the equinoctial circle always ascends in a uniform way, that is, equal parts ascend for equal periods of time. For the movement of the heavens is uniform, and the angle that the equinoctial makes with any horizon is not changed at any hour. Indeed, the arc of the equinoctial circle that ascends with any part of the zodiac is called the ascension of that same part.

§32 The parts of the zodiac, therefore, that have equal ascensions take equal amounts of time to rise, and those that take equal time to rise have equal ascensions. However, equal parts of the zodiac do not necessarily have equal ascensions, because the more vertically a part of the zodiac rises, the longer the time it requires in its rising, and the more obliquely it rises, the shorter the time it requires in its rising. This is evident to the senses and also to imagination, if great circles described over both poles of the world are imagined, passing through the zodiac by way of the divisions between the signs, they will divide the zodiac into 12 equal parts, and the equinoctial into the same number of unequal [parts]. And the part of the equinoctial corresponding to the part of the zodiac divided at more acute angles will be smaller than the part of the equinoctial corresponding to the part of the zodiac divided at less acute angles.

33. In sphera utique recta quelibet medietas zodiaci cuilibet medietati equinoctialis equales habet ascensiones. Quelibet enim medietas zodiaci cum medietate equinoctialis ascendit; quelibet etiam quartarum, que sunt inter puncta tropica et equinoctialia, oritur cum quarta equinoctialis circuli. Quodlibet etiam signum signo sibi opposito equalem habet ascensionem. Ascensio namque cuiuslibet signi equalis est occasui sibi opposito, et equalis est occasus et ortus cuiuslibet eiusdem signi in sphera recta.

34. Puncta autem tropica in sphera recta recte oriuntur. Puncta autem equinoctialia in eadem sphera maxime oblique oriuntur. Quanto igitur aliquod signum propinquius est puncto tropico in sphera recta tanto tardius oritur et maiorem habet ascensionem. Quanto vero propinquius est puncto equinoctiali, tanto citius oritur et minorem habet ascensionem. Et quelibet duo signa eque propinqua puncto tropico, equales habent ascensiones. Similiter quelibet duo eque propinqua altrinsecus puncto equinoctiali.

35. In sphera vero obliqua omnis medietas zodiaci inchoata in aliquo puncto signorum septemtrionalium maiorem habet ascensionem quam medietas sibi opposita. Oritur namque quelibet medietas in die cum arcu de equinoctiali simili arcui paralelli descripti a principio eiusdem medietatis, arcui dico existenti supra orizonta. Arcus autem supra orizonta obliquum existens ex parte septemtrionali ab equinoctiali maior est medietate sua. Et arcus cuiuslibet paralelli ex parte australi ab equinoctiali existens supra orizonta est minor sua medietate, sicut dictum est supra.

36. Quelibet vero medietas zodiaci oritur cum arcu de equinoctiali simili arcui existenti supra orizonta de paralello descripto a principio eiusdem medietatis. Medietas igitur utrimque incohata equedistanter ab utroque tropico puncto equales habet ascensiones. Et quanto propinquius incohatur tropico estivo, tanto maiorem habet ascensionem signo sibi opposito, et medietas que est a capite cancri usque ad caput capricorni maximam, et omne signum huius medietatis habet ascensionem maiorem signo sibi opposito. Ascensiones autem quorumlibet duorum signorum sibi oppositorum coniuncte in qualibet sphera obliqua equantur ascensionibus eorundem in sphera recta coniunctis, suntque quorumlibet signorum oppositorum 30 gradus coniuncte, id est due hore equinoctiales. Hora enim equinoctialis est ascensio 15 graduum de circulo equinoctiali.

MANUSCRIPTS, TRANSLATION, HISTORIOGRAPHY 115

§33 Assuredly, in a right sphere any given half of the zodiac will have ascensions equal to any given half of the equinoctial. For any given half of the zodiac ascends with half of the equinoctial; and any one of the quarters that are between the tropic and equinoctial points rises with a quarter of the equinoctial circle. Indeed, any given sign has an ascension equal to that of the sign opposite to it. For the ascension of any given sign is equal to the setting opposite to it, and the setting and rising of any same sign is equal, in a right sphere.

§34 Tropic points, moreover, rise vertically in the right sphere, but equinoctial points in the same sphere rise very obliquely. Therefore, in the right sphere, the closer a sign is to a tropic point the slower it rises and the greater is its ascension. However, the closer it is to an equinoctial point, the faster it rises and the smaller is its ascension. And any two signs equally close to a tropical point have equal ascensions. It is the same for any two points equally close on opposite sides to an equinoctial point.

§35 In an oblique sphere, then, every half of the zodiac starting from any point of the northern signs has greater ascension than the half to which it is opposed. For any half rises in the day with an arc of the equinoctial similar to the arc of the parallel described from the starting point of that same half—to the arc, I mean, being present over the horizon. In contrast, the arc present over the oblique horizon is greater than its half on the northern side from the equinoctial. And the arc of any given parallel on its southern side from the equinoctial being present over the horizon is smaller than its half, as stated above.

§36 Any half of the zodiac rises with an arc of the equinoctial similar to the arc present above the horizon of the parallel described from the starting point of the same half. The half, therefore, begun on both sides equidistant from both tropic points has equal ascensions. And the closer it begins to the summer tropic, the greater ascension it has than the sign to which it is opposed, and the half that is from the head of Cancer to the head of Capricorn has the greatest, and every sign of this half has a greater ascension than the sign to which it is opposed. Now, in any oblique sphere the ascensions of any two opposite signs, added together, are equal to the ascensions of the same [two signs], added together, in the right sphere, and the ascensions of any opposite signs, added together, are 30 degrees—that is, two equinoctial hours. For one equinoctial hour is an ascension of 15 degrees from the equinoctial circle.

116   THE SCIENTIFIC WORKS OF ROBERT GROSSETESTE

37. Hiis vero quorum cenith est sub circulo descripto a polo zodiaci, oriuntur 6 signa que sunt a capite cancri usque ad caput capricorni subito. Cum enim polus zodiaci est cenith capitis eorum, orizon et zodiacus simul sunt et statim post intersecant se per equalia.

38. Postquam de ortu et occasu signorum diximus, quorum directio et obliquitas est una causa inequalitatis dierum naturalium ad invicem, restat subiungere de alia causa inequalitatis que provenit ex eo quod sol est ecentricus, ut duabus causis inequalitatis coniunctis tota pateat inequalitatis ratio dierum naturalium ad invicem.

39. Ymaginemur lineam ductam a 18° gradu geminorum per centrum terre usque in gradum sagittarii oppositum. Et a centro terre computentur in eadem linea duo gradus et dimidius, de diametro circuli solis, versus geminos. Et ubi finitur talis computatio ponatur centrum et describatur circulus super centrum illud secundum eandem quantitatem, que est semidiametri circuli solis, in superficie cinguli signorum. Erit igitur ille circulus recte dispositus sub ecliptica nusquam ab ea declinans. Et is est circulus solis in cuius circumferentia fertur centrum corporis solaris, et movetur corpus solis in hoc circulo motu proprio ab occidente in oriens motu uniformi et equali, ita quod centrum corporis eius semper est in circumferentia huius circuli.

40. Punctus autem per quem transit predicta linea directa a geminis in sagittarium ex parte geminorum est maxime accedens ad firmamentum et maxime remotus a terra inter omnia puncta eiusdem circumferentie. Punctus vero oppositus ex parte sagitarii est maxime recedens a firmamento et maxime accedens ad terram. Et maxime elevatus a terra vocatur aux vel longitudo longior, et punctus oppositus vocatur oppositum augis vel longitudo propior. Et circulus solis vocatur ecentricus solis, eo quod centrum eius egressum est a centro terre. Et eadem ratione vocatur circulus egresse cuspidis, eo quod cuspis, id est centrum eius, egressa est a centro terre. Sol igitur, cum uniformiter moveatur sub hoc circulo, inuniformiter movetur in celo.

MANUSCRIPTS, TRANSLATION, HISTORIOGRAPHY 117

§37 For those people, then, whose zenith is beneath the circle described by a pole of the zodiac, the six signs from the head of Cancer to the head of Capricorn rise at once. For when the pole of the zodiac is their zenith, the horizon and the zodiac coincide, and immediately afterwards divide each other into equal parts.

§38 Since we have spoken of the rising and setting of signs, whose rightness and obliquity is one cause of why natural days are unequal to one another, it remains to add [something] concerning another cause of the inequality that comes from the fact that the sun is eccentric, so that from the combination of these two causes of the inequality the whole reason for why natural days are unequal to one another may become evident.

§39 Let us imagine a line drawn from the 18th degree of Gemini through the centre of the earth to the degree of Sagittarius to which it is opposite. And from the centre of the earth on the same line let two and a half degrees be calculated from the diameter of the circle of the sun towards Gemini. And let a centre be posited where this calculation concludes, and a circle is described over that centre according to the same quantity as the radius of the circle of the sun, on the surface of the girdle of signs. Therefore, that circle will be disposed in a straight manner under the ecliptic, never falling away from it. And this is the circle of the sun in whose circumference the centre of the solar body is carried, and the solar body is moved in this circle from the west to the east through its own movement with a uniform and equal movement, in such a way that the centre of this body is always in the circumference of this circle.

§40 The point, however, through which the line mentioned above drawn from Gemini to Sagittarius passes [the circle of the sun] on the side of Gemini, is the closest to the firmament and the furthest removed from the earth among all the points of this same circumference. The opposite point, on the side of Sagittarius, is the furthest away from the firmament and the closest to the earth. And that [point] most elevated from the earth is called the 'apogee' or the further longitude, while the opposite point is called 'opposed to the apogee' [perigee] or the closer longitude. And the circle of the sun is called the eccentric of the sun, because its centre is displaced from the centre of the earth. And for the same reason it is called the circle of the displaced cusp, because the cusp, that is, its centre, is displaced from the centre of the earth. The sun, therefore, since it is moved uniformly under this circle, is moved non-uniformly in the heavens.

118   THE SCIENTIFIC WORKS OF ROBERT GROSSETESTE

41. Motus ergo solis inuniformis in celo est una causa inequalitatis dierum naturalium. Cum enim dies naturalis sit una revolutio firmamenti et insuper ascensio eius quod describit sol in celo interim motu suo proprio—et durante una revolutione plus aut minus describit sol quam sequenti revolutione—manifestum est quod quantum est de ista causa erunt dies naturales inequales.

42. Preterea, cum partes zodiaci obliquius orientes minores habeant ascensiones quam equales partes rectius orientes, etiam si sol uniformiter moveretur in celo, contingeret inequalitas dierum ex hac parte, quia illud quod describit sol in una revolutione minorem vel maiorem haberet ascensionem quam quod describeret in sequenti revolutione.

43. Si igitur motus solis procedit augmentando in firmamento et partes sequentes rectius oriuntur in zodiaco erit duplex causa coniuncta maioritatis dierum naturalium; et dicuntur huiusmodi dies maiores. Quando vero procedit motus solis diminuendo in firmamento et partes sequentes obliquius oriuntur, coniuncta est duplex causa minoritatis dierum naturalium; et dicuntur huiusmodi dies minores. Quando vero tantum addit una causa quantum reliqua diminuit, dicuntur dies mediocres.

44. Ex eo autem quod sol est predicto modo ecentricus, accidit quod regio ultra equinoctialem circulum non potest inhabitari. Sole enim existente in opposito augis per 5 gradus est terre propinquior quam quando est in auge. Cum enim, sole existente in signis australibus, multum appropinquat sol terre et sic recte super loca australia, duplicatur causa caliditatis in eorum estate. Cum vero est in signis septemtrionalibus, recedit sol a cenith locorum australium et elongatur a terra, unde duplex est causa frigiditatis in eorum habitatione. Cum vero accedit sol ad cenith capitum nostrorum elongatur a terra, et cum recedit a nostro cenith appropinquat terre. Et ideo est regio septemtrionalis temperata.

45. Regio igitur septemtrionalis dividitur in 7 climata. Et dicitur clima tantum spatium terre per quantum sensibiliter variatur horologium. Idem namque dies estivus aliquantus est in una regione et sensibiliter est minor in regione propinquiori austro. Spatium igitur tantum per

MANUSCRIPTS, TRANSLATION, HISTORIOGRAPHY   119

§41 The non-uniform movement of the sun in the heavens, therefore, is one cause of the inequality of natural days. For since the natural day is a single revolution of the firmament added to the ascension of that [path] which the sun describes in the heavens in the same time by its own movement—and during a single revolution the sun will describe more or less than during the subsequent revolution—it is manifest that, to the extent that it is from this cause, the natural days will be unequal.

§42 Moreover, since the parts of the zodiac rising more obliquely have shorter ascensions than equal parts rising more rightly, there would be inequality of days for this reason even if the sun were moved uniformly in the heavens, because what the sun describes in a single revolution would have shorter or longer ascension than what it describes in the subsequent revolution.

§43 If therefore the sun's movement proceeds by increasing in the firmament and the following parts rise more rightly in the zodiac, there will be a conjoined twofold cause for the natural days being greater; and days of this sort will be called greater days. However, when the movement of the sun proceeds by diminishing in the firmament, and the following parts rise more obliquely, there is a twofold conjoined cause for natural days being lesser, and they will in this way be called lesser days. When, then, one cause adds as much as the other lessens, they are called median days.

§44 Since, then, the sun is eccentric in the way mentioned above, it follows that the region beyond the equinoctial circle cannot be inhabited. For when the sun is at the point opposite the apogee it is five degrees closer to earth than when it is at the apogee. For when the sun, being present in the southern signs, draws much closer to the earth and therefore right above southern locations, then the cause of heat is doubled in their summer. Now, when the sun is in the northern signs, the sun draws away from the zenith of southern locations and is further away from the earth, from which there is a twofold cause of coldness in their habitations. Now, when the sun reaches our zenith it becomes distant from the earth, and when it draws away from our zenith it draws closer to earth. And therefore the northern region is temperate.

§45 Now, the northern region is divided into seven climes. And what we call a clime is the measure of space on the earth by which the sundial varies perceptibly. For the same day of summer is of a certain length in one region and is perceptibly shorter in the region nearer to the south. The measure of space, therefore, by which the day begins to

120 THE SCIENTIFIC WORKS OF ROBERT GROSSETESTE

quantum incipit sic idem dies sensibiliter variari dicitur clima. Nec est idem horologium in principio et fine huius spatii observatum. Hore enim eiusdem diei sensibiliter variantur qualiter et horologium.

46. Distinctiones igitur horum climatum sic possunt ymaginari. Intelligatur circulus magnus cingens corpus terre sub utroque polo et alius circulus magnus cingens corpus terre sub equinoctiali circulo: secundum situm horum duorum circulorum cingunt duo maria totam terram. Et illud quod cingit terram sub polis Amphitrites, reliquum vero Occeanus vocatur. Hec duo maria dividunt terram in 4 partes, quarum una sola inhabitatur.

47. Angulus vero sectionis duorum marium ex parte orientis quarte inhabitate dicitur simpliciter oriens, et angulus oppositus dicitur occidens. Si igitur fiat dimensio ab Occeano versus septemtrionem secundum spatium prescriptum, et per finem illius dimensionis ducatur linea in superficie terre equedistans Occeano utrinque terminata in Amphitrite, spatium terre contentum inter lineam sic descriptam et Occeanum est unum clima. Hoc etiam modo sumpta dimensione a fine primi climatis versus septemtrionem secundum spatium prescriptum, et a fine illius dimensionis ducatur linea utrinque in Amphitrite terminata equedistanter linee terminanti primum clima, spatium contentum inter has duas lineas erit secundum clima. Et ad huius similitudinem signantur sequentia climata.

48. Post hoc videndum est de motu stellarum fixarum quem habent preter motum ab oriente in occidens, qui est communis omnibus corporibus celestibus. Verumtamen, ex eo quod dicuntur fixe, videtur quod non habent motum aliquem preter quam predictum. Sed sciendum est quod non dicuntur stelle fixe quia non habent motum alium, sed quoniam figura et ymago quam constituunt aliquot ex eis que dicuntur stelle fixe, semper retinentur ab eis. Verbi gratia, si tres stelle triangulum faciant, semper retinent eandem figuram. A fixione igitur figurarum quas faciunt, fixe dicuntur.

49. Ptolomeus igitur in libro *Almagesti* posuit quod omnes stelle fixe et omnes auges planetarum moventur super polos zodiaci, et quod unaqueque describit in 100 annis unum gradum de circulo in quo situm est corpus stelle, descripto super polos zodiaci contra motum firmamenti. Et accidit per hanc viam Ptolomei quod aux solis et stelle

MANUSCRIPTS, TRANSLATION, HISTORIOGRAPHY   121

vary perceptibly is called a clime. And neither is the sundial observed to be the same at the beginning and end of this space. For the hours of the same day vary perceptibly, and so does the sundial.

§46 The distinctions, then, of these climes may be imagined in the following way: let us consider a great circle girdling the body of the earth beneath each pole, and another great circle girdling the body of the earth beneath the equinoctial circle: according to the position of these two circles two seas girdle the whole earth. And the one that girdles the earth beneath the poles is called Amphitrites, the other Occeanus. These two seas divide the earth into four parts, of which only one is inhabited.

§47 The corner that cuts the two seas on the eastern side of the inhabited quarter is simply called the east, and the opposite corner is called the west. If, then, a measure is made from Occeanus towards the north according to the space described previously, and through the end of this measuring a line is drawn on the surface of the earth equidistant from Occeanus on both sides and terminating in Amphitrites, the space of the earth contained between the line so drawn and Occeanus is a single clime. In this same way the measure taken from the end of the first clime towards the north according to the space mentioned above, and from the end of that measure is drawn a line terminating in Amphitrites on both sides equidistant to the line terminating the first clime, the space contained between these two lines will be the second clime. And in this same way the following climes are marked out.

§48 After this, we should examine the movement that the fixed stars have in addition to the movement from east to west that is common to all celestial bodies. Nevertheless, because they are called fixed, it seems that they do not have any movement beside that mentioned above. However, it should be known that they are not called fixed stars because they have no other movement, but because the shape and image are always retained by groups constituted of stars that are called fixed. For instance, if three stars make a triangle, they always retain that same shape. From the fixity of the shapes that they make, then, they are called fixed.

§49 Ptolemy, then, in the book *The Almagest* posited that all fixed stars and all apogees of planets are moved around the poles of the zodiac, and that each one describes in 100 years one degree of the circle described around the poles of the zodiac contrary to the movement of the firmament, on which the body of the star is positioned. And this approach of Ptolemy means that the apogee of the sun and the stars that

122 THE SCIENTIFIC WORKS OF ROBERT GROSSETESTE

que sunt in signis septemtrionalibus pervenirent in signa australia, fieretque regio habitata inhabitabilis. Quod patet per rationem superius dictam, qua ostenditur per solis ecentricitatem quod regio inter equinoctialem et paralellum australem est inhabitabilis.

50. Thebit vero, qui operatus est super operationes Ptolomei, invenit per certa experimenta motum stellarum fixarum esse alium. Ad ymaginandum igitur motum stellarum fixarum quem invenit Thebit, ymaginemur in celo zodiacum ex 12 signis constantem, sicut predictum est, et divisum in 4 partes per duo puncta equinoctialia et duo solstitialia. Incipientque aries et libra a punctis equinoctialibus, et cancer et capricornus a punctis solstitialibus. Diciturque zodiacus iste zodiacus fixus. Eruntque 12 signa 12 spatia solum firmamenti. Sub firmamento autem est sphera stellarum fixarum. Ymaginemur iterum in sphera stellarum fixarum zodiacum alium numero a predicto, constantem ex 12 ymaginibus ex stellis fixis compositis. Et hic circulus magis proprie dicitur zodiacus a *zoas*, quod est animal, propter ymagines animalium ex quibus constat.

51. Principio igitur arietis zodiaci fixi centro posito et super ipsum circulo descripto, occupatis 8 gradibus et 37 minutis et super caput libre fixe huic circulo equali circulo descripto, ymaginemur caput arietis et libre ymaginum circumferri in duabus predictis circumferentiis duorum predictorum circulorum. Moventur igitur caput arietis et libre zodiaci ymaginum in duabus predictis circumferentiis: quando sunt in parte septemtrionali, caput arietis cum motu firmamenti caput vero libre contra motum firmamenti, et quando sunt in parte australi movetur caput arietis contra motum firmamenti, caput vero libre cum motu firmamenti, in 12 annis unum gradum et 2 minuta fere describendo. Quoniam igitur hii duo zodiaci sunt ita siti quod unus sub alio est, cum caput arietis mobilis erit in 19° minuto 5^i gradus arietis fixi, erit caput libre mobilis in consimili loco libre fixe. Et cum caput arietis mobilis erit in 42° minuto 26^i gradus piscium fixorum, erit caput libre mobilis in consimili loco virginis fixe.

MANUSCRIPTS, TRANSLATION, HISTORIOGRAPHY   123

are in the northern signs would end up in the southern signs, and this inhabited region would become uninhabitable. This is evident from the reason stated above, by which the eccentricity of the sun shows that the region between the equinoctial and the southern parallel is uninhabitable.

§50 Now, Thebit, who worked through Ptolemy's results, found through accurate observations that the movement of the fixed stars was different. To imagine the movement of the fixed stars that Thebit found, let us imagine in the heavens the zodiac consisting of 12 signs, as stated above, and divided into four parts by the two equinoctial points and the two solstice points. Let Aries and Libra begin from the equinoctial points, and Cancer and Capricorn from the solstice points. This zodiac is called the fixed zodiac. And the 12 signs will simply be 12 spaces in the firmament. Now, under the firmament is the sphere of fixed stars. Let us again imagine in the sphere of fixed stars another zodiac numerically different from the aforementioned, consisting of 12 images composed of fixed stars. And this circle is more properly called a zodiac from *zoas*, or animal, because of the images of animals of which it consists.

§51 A centre, therefore, being placed at the beginning of Aries of the fixed zodiac, and a circle being described over this centre, with 8 degrees 37 minutes being occupied, and over the head of fixed Libra a circle equal to this circle being described: let us imagine the heads of the images of Aries and Libra being carried around in the two circumferences mentioned above of the two circles mentioned above. The heads, therefore, of the images of Aries and Libra in the zodiac are moved in the two circumferences mentioned above. When they are in the northern part, the head of Aries is moved in the direction of the movement of the firmament while the head of Libra [is moved] contrary to the movement of the firmament, and when they are in the southern part the head of Aries is moved contrary to the movement of the firmament, while the head of Libra along with the movement of the firmament, describing in 12 years approximately one degree and two minutes. Since, therefore, these two zodiacs are situated in such a way that one is below the other, when the head of mobile Aries will be in the 19th minute of the 5th degree of fixed Aries, the head of mobile Libra will be in a similar position with respect to fixed Libra. And when the head of mobile Aries will be in the 42nd minute of the 26th degree of fixed Pisces, the head of mobile Libra will be in a similar place of the fixed Virgo.

124 THE SCIENTIFIC WORKS OF ROBERT GROSSETESTE

52. Caput vero cancri et capricorni ymaginum adheret in ecliptica progrediendo et regrediendo in ea. Cum enim caput arietis mobilis a predicto minuto piscium recedit, recedit ab ecliptica, et similiter caput libre mobilis in partem oppositam, nec veniat caput arietis ad eclipticam donec veniat ad predictum minutum arietis fixi. Verumtamen caput cancri mobilis fuit in consimili loco in ecliptica in geminis, cum caput arietis mobilis fuit in minuto predicto in piscibus, et cum caput arietis ascendit versus predictum minutum 5^i gradus arietis fixi in circumferentia circuli predicti, caput cancri semper progreditur in ecliptica donec veniat in 19^i minutum 5^i gradus cancri, capite arietis perveniente in consimile minutum arietis fixi.

53. Descendente igitur capite arietis iterum versus pisces in circumferentia predicta, retrocedit caput cancri in ecliptica eodem spatio quo progressum est. Et iste motus, quem sic ymaginati sumus in predictis circulis duobus, est motus totius sphere stellarum fixarum et augium omnium planetarum.

54. Cursus vero lune est sub zodiaco. Verumtamen non est cursus eius directe sub ecliptica sicut est cursus solis, sed circulus lune secat eclipticam in duobus punctis oppositis, declinans ab eius circumferentia 5 gradibus. Est autem circulus lune ecentricus sicut circulus solis. In circumferentia vero ecentrici est centrum circuli brevis quem devehit ecentricus. Sunt autem circulus brevis et ecentricus in superficie una. Centrum vero corporis lune est semper in circumferentia circuli brevis. Ecentricus itaque lune circumvolvitur super diametrum terre ab oriente in occidens motu continuo et uniformi; describit itaque centrum ecentrici motu predicto circulum circa centrum terre et centrum ecentrici semper est in circumferentia illius circuli.

55. Centrum vero circuli brevis movetur econtrario ab occidente in oriens ita quod, si ducatur linea a centro terre per centrum circuli brevis in firmamentum, terminus linee motu equali movetur. Et is motus linee vocatur motus lune medius in celo.

MANUSCRIPTS, TRANSLATION, HISTORIOGRAPHY   125

§52 The heads of the images of Cancer and Capricorn adhere to the ecliptic, going back and forth on it. For when the head of mobile Aries withdraws from the minute of Pisces mentioned above, it draws away from the ecliptic, and similarly the head of mobile Libra towards the opposite part, nor does the head of Aries arrive at the ecliptic until it arrives at the above-mentioned minute of fixed Aries. Nevertheless, the head of mobile Cancer was in the corresponding place in the ecliptic in Gemini when the head of mobile Aries was in the minute in Pisces mentioned above, and when the head of Aries ascends towards the above-mentioned minute of the 5th degree of fixed Aries in the circumference of the circle mentioned above, the head of Cancer always proceeds along the ecliptic until it arrives at the 19th minute of the 5th degree of Cancer, with the head of Aries arriving in the corresponding minute of fixed Aries.

§53 While the head of Aries, then, descends towards Pisces along the circumference mentioned above, the head of Cancer goes back along the ecliptic in the same space through which it went forth. And this movement, which we have imagined in this way in the two aforementioned circles, is the movement of the entire sphere of the fixed stars and apogees of all the planets.

§54 The course of the moon, however, is beneath the zodiac. Nevertheless, its course is not directly beneath the ecliptic, as is the course of the sun, but the circle of the moon cuts the ecliptic at two opposite points, declining from its circumference by 5 degrees. Moreover, the circle of the moon is eccentric as is the circle of the sun. Now, on the circumference of the eccentric is the centre of a little circle that the eccentric carries around. Moreover, the little circle and the eccentric are on one plane. Now, the centre of the body of the moon is always on the circumference of the little circle. Furthermore, the eccentric of the moon is rotated about the diameter of the earth from east to west with a continuous, uniform movement; and so by this movement the centre of the eccentric will describe a circle around the centre of the earth, and the centre of the eccentric is always on the circumference of that circle.

§55 Now, the centre of the little circle is moved conversely from the west to the east in such a way that if a line is drawn from the centre of the earth through the centre of the little circle to the firmament, the end of the line would be moved with an equal movement. And this movement of the line is called the mean movement of the moon in the heavens.

126 THE SCIENTIFIC WORKS OF ROBERT GROSSETESTE

56. Et quotiens iste motus lune medius est in eodem puncto cum medio motu solis, centrum epicicli est in auge ecentrici lune. Separanturque statim aux ecentrici lune et centrum epicicli, et relinquitur motus medius solis in medio inter augem ecentrici et centrum epicicli equaliter distans ab utroque. Evenitque de necessitate quod cum medius motus lune opponitur medio motui solis, occurrit centrum epicicli augi ecentrici, separanturque ibi iterum. Et occurrunt sibi invicem cum iterum coniungitur medius motus lune medio motui solis. Ex quo patet quod centrum epicicli bis in uno mense describit ecentricum. Estque motus epicicli velocior motu ecentrici quantum est medius motus solis. Aliter enim non semper equaliter distarent a medio motu solis medius motus lune et aux ecentrici. Luna vero movetur in circumferentia sui epicicli, ita quod in superiori parte sui epicicli movetur ab oriente, scilicet cum firmamento, in occidens, et in parte inferiori ab occidente in oriens.

57. Ecentricus lune, ut predictum est, secat eclipticam in duobus punctis oppositis. Vocaturque punctus per quem transit luna a parte australi ecliptice in partem septemtrionalem caput draconis, et punctus oppositus cauda draconis. Coniunctio enim duorum circulorum, eo quod facit figuram tortuosam, vocatur draco lune. Luna igitur existente in capite vel cauda vel prope caput vel caudam draconis, et sole in opposito ipsius, erit eclipsis lune. Si enim removeatur ab altero nodorum plus 12 gradibus, non patietur eclipsim. Accidit enim eclipsis lune per hoc quod ipsa transit per umbram terre que proicitur semper in oppositum solis. Cum enim sol sit corpus luminosum et terra corpus umbrosum et radii recti sint et sol sit maior terra, necesse est ut sol proiciat umbram figure piramidalis, et conus umbre terminetur e directo puncti opposti soli in ecliptica. Cum ergo sol semper est sub ecliptica, sic conus umbre terre semper est sub ecliptica.

MANUSCRIPTS, TRANSLATION, HISTORIOGRAPHY   127

§56 And whenever this mean movement of the moon is on the same point as the mean movement of the sun, the centre of the epicycle is on the apogee of the eccentric of the moon. And the apogee of the eccentric of the moon and the centre of the epicycle become steadily separated, and the mean movement of the sun is left in the middle between the apogee of the eccentric and the centre of the epicycle, equidistant from both. And it happens by necessity that when the mean movement of the moon is opposed to the mean movement of the sun, the centre of the epicycle meets the apogee of the eccentric, and then they are again separated there. And they come to meet each other when once again the mean movement of the moon is joined to the mean movement of the sun. From this it is clear that the centre of the epicycle describes twice a month the eccentric. And the movement of the epicycle is faster than the movement of the eccentric to the same extent as is the mean movement of the sun. For otherwise the mean movement of the moon and the apogee of the eccentric would not always be equidistant to the mean movement of the sun. The moon, however, is moved along the circumference of its epicycle, so that in the upper part of its epicycle it is moved from the east towards the west, that is, with the firmament, and in the lower part from the west to the east.

§57 The eccentric of the moon, as has been said, cuts the ecliptic at two points opposite to each other. And the point through which the moon passes from the southern part of the ecliptic to the northern is called the head of the dragon, and the point opposite is called the tail of the dragon. For the conjunction of the two circles is called the dragon of the moon, because it makes a winding shape. When the moon, therefore, is present in the head or the tail, or close to the head or the tail, of the dragon, with the sun in opposition to it, there will be an eclipse of the moon. However, if it [the moon] is removed from either of the nodes by more than 12 degrees, it will not undergo an eclipse. For a lunar eclipse happens because the moon travels through the shadow of the earth, which is always projected opposite the sun. For since the sun is a luminous body, and the earth is a shadowy body, and rays are straight, and the sun is larger than the earth, it is necessary that the sun projects a shadow in the shape of a cone, and that the apex of the shadow terminates directly on the point on the ecliptic opposite the sun. So, since the sun is always beneath the ecliptic, so the apex of the shadow of the earth is always beneath the ecliptic.

128 THE SCIENTIFIC WORKS OF ROBERT GROSSETESTE

58. Corpus vero lune corpus umbrosum est, et non habet lumen nisi a sole. Unde pars illa quam respicit sol semper est illuminata, reliqua vero umbrosa. Cum ergo currat luna inferior sole, cum sol et luna coniunguntur, pars lune que terram respicit tota est umbrosa. Cum ergo paulatim recedit a sole, incipit pars illa paulatim illuminari, quia radii solis paulatim eam attingunt. Et quanto plus a sole recedit luna, tanto pars eius terram respiciens plus est illuminata. Cumque venit luna in oppositum solis, tota medietas terram respiciens est illuminata, et tunc dicitur *panselenos*, quasi plena lumine. Deinde, sicut paulatim accedit versus solem, sic paulatim lumen decrescit in parte terre opposita.

59. Luna igitur plena existente in altero nodorum vel prope, cum tunc sit sub ecliptica vel prope eclipticam, necesse est ut transeat per umbram terre, et patietur defectum luminis aut tota aut secundum partem. Si vero multum removeatur ab altero nodorum in plenilunio, cum tunc removeatur ab ecliptica—et conus umbre terre fertur sub ecliptica—corpus lune non attinget umbram, sed a latere umbre pertransibit. Et ita non patietur luna defectum.

60. Notandum quod crebriores sunt defectus lune quam solis propter diversitatem aspectus lune. Est autem diversitas aspectus lune arcus circuli magni transeuntis per cenith capitis interceptus inter verum locum lune et locum in visu apparentem. Verus locus lune est terminus linee ducte a centro terre per centrum corporis eius in firmamentum. Cum igitur terra ad lune circulum sensibilem habeat magnitudinem, linea recta ducta ab oculo videntis non existentis sub luna per centrum corporis lune in firmamentum, secat predictam lineam in centro corporis lune, et terminatur alibi quam predicta linea, et locus ubi terminatur vocatur locus lune apparens.

61. Arcus vero inter hec duo loca comprehensus vocatur diversitas aspectus lune. Hec tamen diversitas aspectus lune divisa est in diversitatem aspectus lune in longitudine et diversitatem aspectus lune in latitudine. Et neutrum eorum est diversitas aspectus lune quam supra

MANUSCRIPTS, TRANSLATION, HISTORIOGRAPHY 129

§58 Now, the body of the moon is a shadowy body, and it does not have light except from the sun. For this reason, that part that faces the sun is always illuminated, and the other always shaded. Since, therefore, the moon runs below the sun, when the sun and the moon are in conjunction, the part of the moon that faces the earth is entirely in shadow. Therefore, as it little by little recedes from the sun, that part begins little by little to be illuminated, because the rays of the sun little by little reach it. And the further the moon recedes from the sun, the more its part facing the earth is illuminated. And when the moon arrives opposite the sun, the entire half facing the earth is illuminated, and it is then called *panselenos*, as though full of light. Afterwards, as it approaches the sun little by little, so little by little the light decreases in the part opposite the earth.

§59 Therefore, since, when the full moon is in one of the nodes, it will then be beneath the ecliptic or close to the ecliptic, it must pass through the earth's shadow and undergo a total or partial lack of light. Now, if it [the moon] is very remote in full moon from either of the nodes, then, since it is then remote from the ecliptic—and the apex of the shadow of the earth is carried beneath the ecliptic—the body of the moon will not reach the shadow, but will pass from the side of the shadow. And therefore the moon will not undergo an eclipse.

§60 It should be noted that eclipses of the moon are more frequent than those of the sun because of the diversity of aspect of the moon. For the diversity of the aspect of the moon is the arc of a great circle passing through the zenith cut off between the true place of the moon and its position as apparent to sight. The true place of the moon is the end of a line drawn from the centre of the earth through the centre of its [the moon's] body to the firmament. Therefore, since the earth has a perceptible magnitude with respect to the circle of the moon, a straight line drawn from the eye of an observer, who is not present beneath the moon, through the centre of the body of the moon to the firmament will cut through the line mentioned above in the centre of the body of the moon, and ends at another place than the line mentioned above [ends]. And the place where it ends is called the apparent place of the moon.

§61 Now, the arc contained between these two places is called the diversity of the aspect of the moon. However, this diversity of the aspect of the moon is divided into the diversity of the aspect of the moon in longitude and the diversity of the aspect of the moon in latitude. And neither of these is

130 THE SCIENTIFIC WORKS OF ROBERT GROSSETESTE

diximus. Has duas diversitates aspectus lune sic ymaginabimur. Ymaginetur circulus ductus per verum locum lune eque distans ab ecliptica, si non sit verus locus lune in ecliptica. Si vero sit in ecliptica, ipsa erit circulus quem querimus. Et ducatur similiter alius circulus equedistans priori per locum lune apparentem. Deinde per polos orbis signorum transeant duo magni circuli, quorum alter transeat per verum locum lune et alter per locum lune apparentem. Intersecabunt se hii 4 circuli ita quod constituent quadrangulum ex 4 arcubus existentibus inter sectiones.

62. Arcus igitur circuli equedistanter ducti ecliptice per locum lune apparentem dicitur diversitas aspectus lune in longitudine. Arcus vero circuli transeuntis per polos orbis signorum interceptus inter duos circulos ductos equedistanter ecliptice dicitur diversitas aspectus lune in latitudine. Diversitas autem prima quam diximus est sicut diagonalis huius quem prediximus quadranguli. Cum vero ita est quod ambo circuli ducti equedistanter linee ecliptice sunt simul loco, diversitas aspectus in latitudine nulla est. Cum vero accidit quod reliqui duo circuli sunt simul loco, diversitas aspectus in longitudine nulla est.

63. Ex hiis igitur patet quod quamvis coniungantur in puncto capitis vel caude, vel prope caput vel caudam draconis, luna existente australi in regionibus septemtrionalibus non erit eclipsis solis, eo quod locus lune apparens erit tunc ex parte australi respectu ecliptice. Sed ad hoc quod esset eclipsis solis oporteret quod locus lune apparens esset simul cum loco solis, vel quod distantia inter ipsa esset minor quantitate duorum semidiametrorum, scilicet solis et lune. Cum igitur coniunguntur sol et luna, luna existente septemtrionali, accidit eclipsis in regionibus septemtrionalibus. Verumtamen non est necesse quod in omnibus, sed in hiis locis in quibus locus lune apparens in ecliptica vel prope eclipticam minus distat ab ea quam quantitate duorum semidiametrorum solis et lune.

MANUSCRIPTS, TRANSLATION, HISTORIOGRAPHY 131

the diversity of the aspect of the moon of which we spoke above. We will imagine these two diversities of the aspect of the moon in the following way. Let a circle be imagined, drawn through the true place of the moon at an equal distance from the ecliptic, if the true place of the moon is not on the ecliptic. If, however, it is on the ecliptic, [the ecliptic] itself will be the circle we seek. And similarly let another circle be drawn, equidistant to the former, through the apparent place of the moon. Then let two great circles pass through the poles of the orb of signs, of which one passes through the true place of the moon and the other through the apparent place of the moon. These four circles will intersect each other so as to constitute a quadrangle of four arcs being present between the sections.

§62 The arc, therefore, of the circle drawn at an equal distance to the ecliptic through the apparent place of the moon is called the diversity of the aspect of the moon in longitude, whereas the arc of the circle passing through the poles of the orb of signs bounded between the two circles drawn equidistantly to the ecliptic is called the diversity of the aspect of the moon in latitude. However, the first diversity we spoke of is, as it were, the diagonal of this quadrangle we mentioned above. Now, when both circles drawn equidistantly to the line of the ecliptic are together in the same place, there is no diversity of the aspect of the moon in latitude. Indeed, when the other two circles are together in the same place, there is no diversity of the aspect of the moon in longitude.

§63 From these points, therefore, it is clear that when the moon is present in the south there will not be an eclipse of the sun in the northern regions, however much they may be conjoined in the point of the head or of the tail, or close to the head or to the tail, of the dragon, because the apparent place of the moon will then be on the southern side with respect to the ecliptic. Nevertheless, for there to be an eclipse of the sun, the apparent place of the moon would have to be together with the place of the sun, or the distance between them would have to be less than the quantity of the two radii, that is, of the sun and of the moon. When, therefore, the sun and the moon are in conjunction, the moon being present in the north, there will be an eclipse in the northern regions. Nonetheless, this is not necessary in all of these [regions], but in these places in which the apparent place of the moon on the ecliptic or close to the ecliptic is distant from it by less than the quantity of the two semidiameters of the sun and the moon.

4

Commentary on the Model
of the World Machine

What is presented below is a technical commentary to elucidate and explain the mathematical, geometrical, and astronomical content of each section of *On the Sphere*.

§1. Grosseteste's opens his treatise with the statement that, since the world machine is spherical, it is important to understand the properties of a sphere.

§2. Such a sphere is generated by rotating a semicircle out of the plane of the diagram about its diameter AB (Fig. 4.1(a)). This rotation generates a sphere in the same way that spinning a coin appears to the eye as a sphere on the table (Fig. 4.1(b)). A circle or semicircle is the locus of points equidistant from the centre of the circle, or equivalently the centre point of the semicircle base. When the body is rotated, all points on the resulting sphere must be equidistant from the centre. This feature is important in the later discussion of the evidence for the heavens being spherical. The world machine is such a spherical body.

§3. Here, it is demonstrated that similar rotation of a second semicircle of a different diameter DF (Fig. 4.2(a)) will generate another sphere enclosed by the original (Fig. 4.2(b)). The ensuing rotation of the area between the first and second semicircles generates a body which necessarily has inner and outer surfaces that are spherical. The statement that there is nothing outside of this body, and that all bodies are contained within it, sets up the Aristotelian universe of nested spheres. The material contained in the first body of the world machine is the fifth essence or ether, and the circularly mobile spherical shell contains the seven planets and the fixed stars.

COMMENTARY ON THE MODEL OF THE WORLD MACHINE

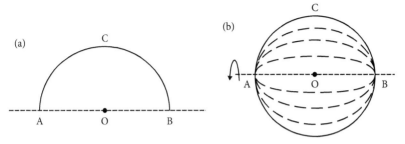

Fig. 4.1. (a) A semicircle drawn about point O on diameter AB. (b) The sphere generated by rotating the semicircle about the diameter AB.

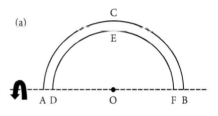

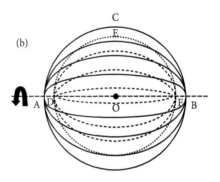

Fig. 4.2. (a) Second semicircle, about the same origin O, of smaller diameter DF. (b) Creation of a second sphere, contained within the first, by rotation of the semicircle around DF.

§4. Paragraph 4 extends the exposition. By drawing a further semicircle on an even smaller diameter GI (Fig. 4.3) and rotating it to form a third concentric sphere GHI, a second spherical shell is generated. This, continuous to the inside of the spherical shell of the ether, is the body of fire. A further semicircle of diameter KM, when rotated generates another sphere KLM. This has a third spherical shell between it and the sphere GHI, which is the body of air. A final semicircle with diameter NP, when rotated generates a sphere NRP. This sphere is the sphere of

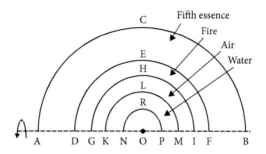

Fig. 4.3. Cross-section of nested spheres containing the five elements.

earth and the shell between it and the sphere KLM is the body of water. This conforms to the principles laid out by Aristotle which state that the natural place of earth is within the sphere of water.[1] However, this raises the question why, if the earth is assumed to be in its natural place, it is not enclosed by water but is in contact with the sphere of air. In order to allow terrestrial creatures to have a home and shelter on dry land, it is supposed that water retreated to cavities or basins within the earth. In this way, the earth and the water are contained within it as a single sphere.

§5. It is noted that both reasoning and observation show that the bodies containing the five elements are spherical. Here, the reasoning is based on what, in modern mathematics, is called 'self-similarity'. Self-similar objects have component parts that are exactly or approximately similar to the object as a whole. The most common examples are fractal systems, such as ferns or coastlines, in which fine detail replicates the overall shape of the object.[2] In *On the Sphere*, it is argued that, as the form comes from the nature and each of the bodies of the fifth essence, fire and air is of a single nature, each part within these bodies must be like the whole. Therefore, each body must have a uniform shape, and the only finite three-dimensional body that satisfies this requirement is the sphere.

Further, within the spheres of the changeable elements (fire, air, water/earth), all heavy bodies move towards the centre, that is, they fall along lines which radiate from the centre. For all heavy bodies to move towards

---

[1] Aristotle, *De caelo*, IV.4–5.
[2] B. Mandelbrot, 'How Long is the Coast of Britain? Statistical Self-Similarity and Fractional Dimension', *Science*, 156 (1967), 636–8.

COMMENTARY ON THE MODEL OF THE WORLD MACHINE 135

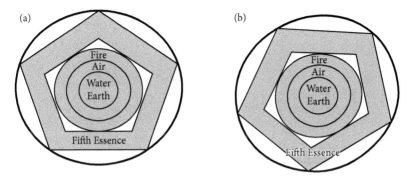

Fig. 4.4. (a) A model universe with a non-spherical body. (b) For circular motion of and in a non-spherical body, there must be unfilled space between bodies.

a place equidistant from the centre, the body containing earth and water, onto which they fall, must have a spherical shape. Similarly, all light bodies tend towards a higher place and for this to be the same for all bodies, air and fire must be contained in spheres.

§6. The argument that the body of the quintessence must be spherical is based on space-filling. It has been shown in §5 that the bodies of the elements are nested spheres. Viewed in two dimensions, these are nested circles (Fig. 4.4(a)). There are no gaps between them. Since linear motion is normal in the bodies of earth, water, air, and fire, then motion anywhere within these (shaded) volumes can be accommodated. However, in the body of the fifth element, according to Aristotle (invoked explicitly), only spherical motion is possible and necessary. If the body of the fifth element were angular, for motion to be circular, there would of necessity be space between the spheres and the body of the fifth element (Fig. 4.4(b)). It must therefore be spherical.

§7. A single piece of observational evidence for the sphericity of the earth is presented in this paragraph. If the earth is flat, the limit of what people could see in the heavens would be the same for all places on the earth. The validity of the argument is seen in (Fig. 4.5(a)) where each person on a flat earth sees the heavens as an identical hemisphere, which is displaced when viewed from different positions. However, it is known that the Pole Star is at the horizon for people living at Arim (see Ch. 5, §4.3) on the equator (Fig. 4.5 (b)) and that this is not the case for people

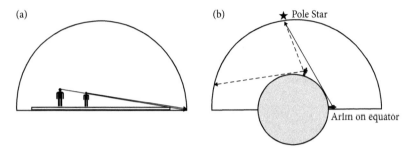

Fig. 4.5. (a) The horizon is the same for persons on a flat earth. (b) The horizon differs, depending on the latitude of a person situated on the earth, if the earth is round in a north-south direction.

living further north. Therefore, the earth must be round in a north-south direction.[3] The Pole Star appears progressively higher in the sky as the viewer proceeds northwards and at the North Pole, the Pole Star reaches a limit, always being vertically upwards.

§8. Additional arguments demonstrate curvature in the east-west direction. As evident from Fig. 4.5(a), if the earth were flat, sunrise would occur at the same time for people in the two positions. However, as people further east experience sunrise earlier than those in the west, there must be curvature on the earth's surface. Furthermore, because the start of an eclipse of the moon only occurs at a very specific instant, namely when the moon enters the shadow of the earth, if the eclipse is seen at different times of day in different positions on the earth, it implies that the time of day must be different in different parts of the earth. As the time of day is linked to sunrise and sunset, this requires that the earth must be round in the east-west direction for this to be possible. This is demonstrated by analysis of the times of lunar eclipses. For example, an eclipse of the moon seen at midnight at Arim would occur in the morning for those who find themselves east of Arim and in the evening for those who are westwards. In the former case, sunrise and hence the time of day of the start of the eclipse, would occur before that in Arim. That leads to the timing of the eclipse later than that at Arim. Further west, the timing

---

[3] Grosseteste is correct in noting that this argument only demonstrates that there is north-south curvature. It, in itself, does not prove that the earth is spherical.

COMMENTARY ON THE MODEL OF THE WORLD MACHINE 137

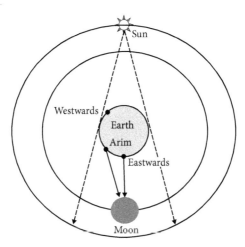

Fig. 4.6. The variation in the times of a lunar eclipse demonstrates that the earth is round in an east-west direction.

would be earlier, because sunrise does not occur until later. Similarly, for a midnight eclipse east of Arim, the timing at Arim would be later, and further west it is possible that the eclipse might not be visible at all. Strictly, the invisibility provides a slightly different reason, in that because of the spherical shape of the earth, the horizon for those that far west cuts off the line of sight to the moon (Fig. 4.6). Grosseteste is here giving theoretical examples based on the computus, astronomical tables, and possibly other astronomical literature (See Ch. 5, §§2 and 4.3). A practical observation of an eclipse occurring at a different time in different places was reported by Walcher of Malvern in 1091.[4]

§9. The key to understanding this paragraph is to recognize that all stars appear to move in circles around the Pole Star and that the further a star is from the Pole Star, the larger is the diameter of the circle (Fig. 4.7). The diurnal circular motion implies symmetry about the fixed rotation axis with one end at the centre of the earth and the other at the Pole Star. The increase in diameter away from the Pole Star implies that the surface of the body of the heavens increases with distance from the latter point.

---

[4] Walcher of Malvern, *De Dracone*, in C. Philipp E. Nothaft (ed. and trans.), *Walcher of Malvern,* De lunationibus *and* De Dracone: *Study, Edition, Translation, and Commentary* (Turnhout: Brepols, 2017), 31–6, 114–17, 240–5.

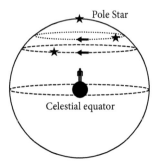

Fig. 4.7. Diurnal circular motion of the fixed stars about the Pole star.

Although this symmetry is not sufficient to prove the spherical nature of the heavens, the observation that stars appear of constant magnitude when rising, setting, or in the middle of the heavens does provide this proof. The constant magnitude indicates that stars always remain at a constant distance from the earth. If stars were at greater and shorter distances from the earth as the heavens rotate they would appear dimmer and brighter depending on where they were in the sky. The only body which can have the necessary symmetry is a sphere. Movement of bodies of the fifth essence, a motion which is necessarily circular according to Aristotle (DS §6), must have a constant diameter and lie on a sphere.

§10. The following paragraphs sequentially set up the model of the universe which is shown below as Fig. 4.8. Although there are an infinite number of diameters to a sphere; the first move is to select one such diameter, which is defined as an axis, about which the heavens rotate. The extremities of this diameter are called poles, the one which is visible in the northern hemisphere being called the Arctic because of its proximity to the Great Bear and Little Bear stellar constellations. There is naturally a diametrically opposite point, called the Antarctic. It is about these two celestial points that the spherical heavens containing the stars and planets rotate diurnally at a constant rate.

§11. A 'great circle' is defined as a circle on the surface of a sphere which lies in a plane passing through the centre of the sphere. To define an origin in the direction of rotation, a great circle is drawn through the celestial poles on the surface of the sphere (Fig. 4.8). This is called a *colure* and at this point in the exposition the first one is selected arbitrarily. As the ecliptic is subsequently defined from it, the choice

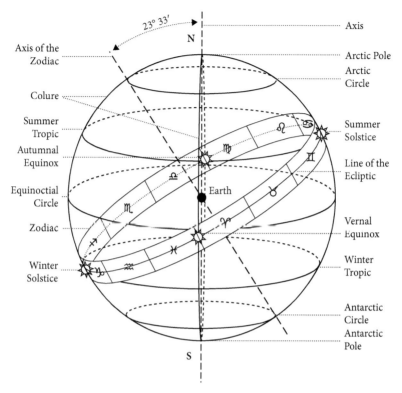

Fig. 4.8. The principal circles on the celestial sphere as described in the text.

determines the relative orientation of the whole model. The second of the *colures* is at right angles to the first and therefore, once the first of the *colures* has been defined, the second one is fixed in its orientation. The association of these intersecting great circles with the tail of an ox arises because the circles appear incomplete to an observer.[5]

---

[5] The etymology for the *coluri* is given variously in the sources on which Grosseteste seems to rely at this point. William of Conches states in his *Dragmaticon*, III. 7. 12, ed. I. Ronca and A. Badia, CCCM, 152 (Turnhout: Brepols, 1997): 'These circles are called by the common name of "colures", as if "colon uri", that is, a limb of an untamed ox. For "colon" means "member", "urus" "ox of the wilderness". Its members, and particularly its tail, are incomplete. And so, these circles, because they are incomplete, are called "colures"—not because they are incomplete in themselves, but [because they are incomplete] from our perspective, since the part of the other pole through which they cut [sc. the southern hemisphere] is never seen by us [Isti circuli communi nomine dicuntur coluri, quasi 'colon uri', idest membrum bouis siluestris: colon enim est 'membrum', urus 'bos siluestris'. Huius membra, et maxime cauda, sunt imperfecta. Vnde isti

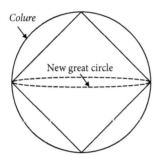

Fig. 4.9. A new *colure* drawn 'according to the sides of a square'.

§12. The phrase 'according to the sides of a square' is not immediately clear but probably refers to how the position of the new great circle corresponds to the corners of the inscribed square (Fig. 4.9). The sides of a square are equal in length and so the great circle should be drawn on the celestial sphere such that any point on it lies at equal distances from the two poles. There is only one great circle that satisfies this requirement and it also circumscribes the plane that is perpendicular to the axis of the two poles (Fig. 4.8). Such a circle cuts the *colures*, drawn previously, at right angles. When the sun is in a position to describe this great circle, as the heavens revolve around the earth once each day about the axis defined by the Arctic and Antarctic poles, there are equal amounts of day- and night-time. Accordingly, the great circle is known as the equinoctial circle (in the heavens) (Fig. 4.8). This equality of day and night-time is true for all positions on the earth's surface.

§13. The ecliptic or the girdle of signs, around which the stars move in an annual rotation, is defined from a point on one of the *colures* which is inclined at 23°33' from the equinoctial towards the Arctic pole. The position of this point A is defined by an angular, not linear position, and this angle is the one made by the line connecting A on the celestial

---

circuli, propter sui imperfectionem, dicuntur coluri: non quia in se sint imperfecti, sed quantum ad nos, quia pars alterius poli, quam intersecant, numquam a nobis uidetur].' Sacrobosco's definition is as follows, *De sphera* 90: 'The colure is so called from "colon", which means 'member', and 'uros', which means 'ox of the wilderness', since, in the way the tail of the wild ox stands erect (the is, its member), it makes a semicircle and not a complete [circle], so the colure also always appears incomplete to us, since only half of it is visible [Dicitur autem colurus a colon, quod est membrum, et uros, quod est bos silvester, quoniam quemadmodum cauda bovis silvestris erecta, que est eius membrum, facit semicirculum et non perfectum, ita colurus semper apparet nobis imperfectus, quoniam tantum una est eius medietas apparens].'

COMMENTARY ON THE MODEL OF THE WORLD MACHINE 141

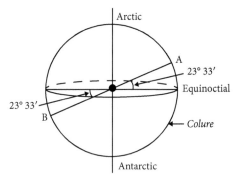

Fig. 4.10. Position of the ecliptic on the celestial sphere with respect to the equinoctial.

sphere to the centre of the earth with the diameter of the equinoctial great circle (Fig. 4.10). A similar point B is found 23°33' towards the Antarctic pole on the part of the same *colure* which is on the far side of the heavens.

A great circle is drawn through these points A and B which is defined as the ecliptic. Two parallel circles are then drawn, one on each side of the ecliptic, 6° of arc from the ecliptic on each side (Fig. 4.8). These are not great circles but are defined by being parallel to the ecliptic. The area on the celestial sphere enclosed by these two little circles is named the zodiac. This zodiac band contains certain constellations (mostly) named after animals.

§14. Reference to Fig. 4.8 shows the division of the zodiac into twelve equal divisions, called *signs*. These are then named starting with Aries, and it is stressed that they are all equal in size, measured in degrees of the ecliptic. This is unlike the constellations of the same names that vary in size. The curved dividing lines on the surface of the celestial sphere are perpendicular to the two little circles. Division and labelling of the signs start from the point where the ecliptic cuts the equinoctial; labels are assigned in the opposite direction to that of the diurnal rotation of the heavens and towards the north. Once set, this convention is maintained henceforth.

§15. The ecliptic is tilted with respect to the equinoctial. As a result, there is a single point on it, at the beginning of Cancer, 90° around the

142 THE SCIENTIFIC WORKS OF ROBERT GROSSETESTE

ecliptic from the starting point, which is closest to the Arctic pole (Fig. 4.8). Through this point a little circle is now drawn that is everywhere parallel to the equinoctial. An identically sized circle is drawn in the southern hemisphere, passing through the start of Capricorn. These circles are stated to be the northern and southern limits of the sun's annual motion.

§16. Since the summer solstice point, on the *colure* drawn in the plane of the page (Fig. 4.8), is at an angular distance of 23°33' from the equinoctial, the axis of the ecliptic makes the same angle with respect to the north-south celestial axis about which the heavens rotate. This is evident in Fig. 4.8, allowing for the small distortion of the visual perspective in the diagram. The point of intersection of the axis of the zodiac with the celestial sphere is the north pole of the zodiac. There is an equivalent south pole.

Although not stated explicitly, the poles of the zodiac are rotated about the celestial north-south axis, drawing two circles on the celestial sphere around the celestial poles. The northern one is the Arctic parallel, 23°33' from the north pole, while the southern one is the Antarctic parallel. There are now five circles drawn, all perpendicular to the north-south celestial axis: the Arctic and Antarctic parallels, the summer and winter tropics and the equinoctial. These five circles have their equivalents on the earth, marking the boundaries of climatic zones as identified by ancient authors.

§17. Having set up the features on the celestial sphere, Grosseteste turns his attention to the motion of the sun. The sun moves on a circle which is below the ecliptic, that is, it does not move on the sphere of the fixed stars, but on a circle with a smaller diameter.[6] The circle is parallel to the circle of the ecliptic and in the same plane (Fig. 4.11). If the sun were in a constant position with respect to the fixed stars, it would be carried round the earth once per day on the dashed circle of Fig. 4.11, a circle parallel to the equinoctial. Such a circle would be below the equinoctial, the dot-dash line in Fig. 4.11, if the sun were at the points of Aries or Libra.

However, the sun is carried around the circle below the ecliptic once every year (365¼ days). This means that, except at either of the solstices,

---

[6] Note that in Fig. 4.8 the sun's position has been indicated as if on the celestial sphere. This is in order to avoid the diagram becoming too complex and difficult to interpret.

COMMENTARY ON THE MODEL OF THE WORLD MACHINE 143

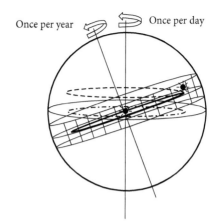

Fig. 4.11. The diurnal motion of the sun, assuming it is in a fixed position with respect to the fixed stars for two possible positions on the ecliptic.

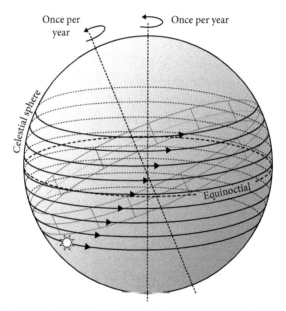

Fig. 4.12. The combination of the sun's diurnal motion and annual motion on the ecliptic results in a tight spiral motion. In practice there would be 182 spiral turns, one per day, between the extremes of winter and summer solstices. The reduced number of nine turns are shown here for clarity.

the celestial latitude of the sun differs slightly from one day to the next (see on §38 below), and as a result exhibits a spiral motion (Fig. 4.12). The sun's spiral motion (approximately ninety circuits to move 23°), is tightly wound and, as it is perceived as approximating to a circular motion, it is sometimes called a parallel.

§18. This paragraph reinforces the concept of the spiral motion, noting that there are as many turns of the spiral as there are days. It also introduces the idea that, through the motion on the circle below the ecliptic, the sun moves from Cancer to Capricorn, the direction of its north-south motion being reversed on reaching each of the tropic points.

§19. A shift of focus enters at this point in the text, and the next ten paragraphs are concerned with the view of the observer (with the orientation of the diagrams here changed accordingly). Since the observer is small compared with the diameter of the celestial sphere, from the surface of the earth, the observer sees what is almost a hemisphere. A visual ray tangential to the surface of the earth, corresponds to the limiting ray from the eye of an observer small compared with the diameter of the earth (Fig. 4.13). Extension of this line to the celestial sphere accounts for all that can be seen between the observer and the celestial sphere. Under the approximation that the diameter of the earth is very small compared with that of the celestial sphere, rotation through 360° of that tangent line about the point where the observer is located creates a circle which divides the celestial sphere into two very nearly

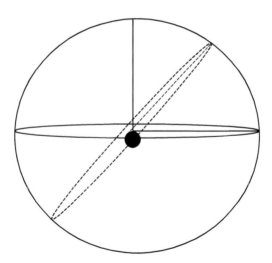

Fig. 4.13. The horizon and the visible celestial hemisphere viewed from two different positions on the earth.

COMMENTARY ON THE MODEL OF THE WORLD MACHINE 145

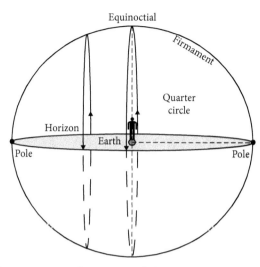

Fig. 4.14. Horizon as seen by a person below the equinoctial circle. Circles centred on and perpendicular to the polar axis are at right angles to the horizon.

equal parts. This is the horizon which is different for all positions of the observer on the surface of the earth, as illustrated by the equivalent dashed horizon construction in Fig. 4.13.

§20. For those for whom the equinoctial circle is directly overhead, the horizon passes through the celestial poles (Fig. 4.14).

§21. The phrase 'every circle described over the [immobile] poles' refers to any circle, great or little, on the celestial sphere which has the perpendicular to the plane of the circle passing through the poles. As evident in Fig. 4.14, such circles are perpendicular to the horizon of those under the equinoctial, the intersection of the two planes therefore always being at right angles. The plane of the horizon divides the celestial circle into two equal parts. In this way, as the heavens rotate diurnally, for any such circle any point spends an equal time above and below the horizon. The poles, which do not move, are always visible.

§22. As a point in the heavens traces out such a circle daily, it follows that, for those under the equinoctial, the rising and setting of any star is perpendicular to the horizon. Furthermore, because the motion of the sun approximates to diurnal movement on such a circle, there are equal

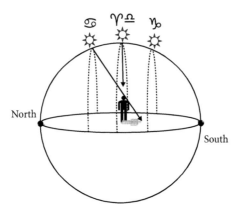

Fig. 4.15. Shadow of a person on the equator when the sun is in the various zodiac signs.

periods of daylight and night. This is true whatever approximate circle is described by the sun. As the motion of the heavens is the same throughout the year, under the equinoctial there are always equal periods of day and night (Fig. 4.14). It is worth noting, perhaps, that this is not precisely correct, as the declination of the ecliptic means that, even at the equator, days are not exactly equal during the year. There are longer days nearer the equinoxes and shorter ones at the solstices, when the component of the solar motion parallel to the equinoctial is a maximum.

§23. When the motion of the sun, on its circle below the ecliptic, carries it to the start of the 30° of either Aries or Libra (that is, the vernal and autumnal equinoxes), the diurnal motion of the heavens carries the sun along the equinoctial. Then the sun rises due east and the morning shadow lies due west and vice versa in the evening. At midday the sun is directly overhead and there is no shadow (Fig. 4.15). When the daily rotation of the sun describes a circle north of the equinoctial (in the northern hemisphere, spring and summer), the midday shadow of someone on the equator points south (Fig. 4.15). The northerly limit is when the sun is at the beginning of Cancer. The reverse occurs when the sun describes a circle south of the equinoctial.

§24. For people north of the above equatorial position but south of the summer tropic (tropic of Cancer) line, in order to retain the orientation of the observer vertically up the page, the diagram of Fig. 4.15

COMMENTARY ON THE MODEL OF THE WORLD MACHINE 147

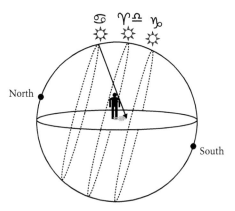

Fig. 4.16. Shadow of a person north of the equator, but south of the tropic of Cancer, points south at the solstice.

must be rotated by an angle corresponding to the number of degrees north. The circle of the sun's daily rotation is then at an angle to the horizontal, but, of course, still perpendicular to the north-south polar axis (Fig. 4.16). For those living between the equator and tropic of Cancer, on two days of the year, either side of the summer solstice, the sun will be directly over their latitude. It will be, therefore, directly overhead at midday. Between those two days and the summer solstice, or on the solstice itself, the sun will be north of their latitude, so their midday shadow points south (Fig. 4.16).

§25. For those beneath the summer tropic, that is, living at a terrestrial latitude of 23½°N, there is only one day in the year when the sun is directly overhead at midday, because this is the northern limit of the sun's north-south motion. This corresponds to the sun being at the beginning of the sign of Cancer, namely, the summer solstice (Fig. 4.8).

§26. For all latitudes north of the equator, because all the circles of the sun's motion are at right angles to the polar axis, these circles are always tilted to the right in equivalent diagrams to Fig. 4.16. In such rotated diagrams the elevation of the north celestial pole is always equal to the angle between the zenith and the equinoctial. The horizon and equinoctial are both great circles and hence the horizon cuts the equinoctial into two equal parts. For the left hand solar parallel in Fig. 4.16, because of the tilt and the shape of the sphere, the plane of the horizon here cuts the

148 THE SCIENTIFIC WORKS OF ROBERT GROSSETESTE

circle into two unequal segments. As the sun is in the northern signs, that is, it is north of the equinoctial towards the tropic of Cancer, the larger segment of the sun's daily circle is that above the horizon. Accordingly, days are longer than nights in summer at all places between the equator and the Arctic Circle. The more remote is any circle (parallel) of the sun's motion from the equinoctial, the larger is the fraction of the circle that remains above the horizon. Although not explicitly included in the text, an equivalent argument for the sun in the southern signs shows that winter nights are longer than days, as stated at the beginning of the paragraph.

§27. The discussion on the length of days continues into this paragraph with the notion that because all days are equivalent and the motion of the heavens uniform with respect to time, every day is longer than night in the northern hemisphere while the sun is in the northern signs, between the equinoctial and the summer tropic. The closer the sun is to the head of Cancer, the longer the days will be for those in the northern hemisphere, because the further the sun travels north, the more of its daily circle will be above the observer's horizon. The limit of the longest day is when the sun reaches the position of Cancer and follows the line of the summer tropic because the fraction of the circle above the horizon increases the further north is the sun's circle (parallel). As the argument could have been made equivalently for the sun in the southern signs, the reverse occurs when the sun is south of the equator, with winter nights always being longer than winter days.

The further north on the earth, the longer are the summer days. This can be seen by considering the limit of the sun following the summer tropic, as this corresponds to the longest day. Moving further north tilts the north-south polar axis by a greater amount and the tilt of the sun's circles (parallels) as shown in Fig. 4.16 increases. When this happens, says Grosseteste, the fraction of the circle (parallel) above the horizon increases and, on progressing further north, the summer days become longer.

§28. The same variations in length of days that can be observed in the northern hemisphere when the sun is north of the equator, will be observed in the southern hemisphere when it is south of the equator. This can be shown by 'imagining the south pole raised and the north pole lowered', that is, rotating Fig. 4.16 by 180° in the plane of the page, such that the north and south poles are interchanged. With this reversal of labelling, the above arguments can be repeated, with north everywhere replaced by south. As a result of the symmetry, the length of the day at points equidistant north and south from the equinoctial circle is the same when the sun is in the equivalent signs south and north.

In all places between the equator and the arctic circle, each day will have day and night (sunrise and sunset), because wherever the sun is between the tropics, its daily circle will be cut by the observer's horizon.

§29. For someone living at the Arctic Circle, once in every daily rotation of the heavens about the north-south polar axis, the zenith coincides with the pole of the zodiac. Fig. 4.17 is drawn such that this situation occurs when the pole of the zodiac is in the plane of the page. At this instant, the plane of the horizon, shaded grey, coincides with the plane of the ecliptic or zodiac. Rotation about the north-south polar axis results in the axis of the poles of the zodiac moving out of the plane of the page and therefore the plane of the ecliptic moving away from the fixed grey plane of the horizon. From the viewpoint of the observer in Fig. 4.18, it is evident that when the sun reaches the northern limit of its range in Cancer the circle it follows just fails to cut the horizon when

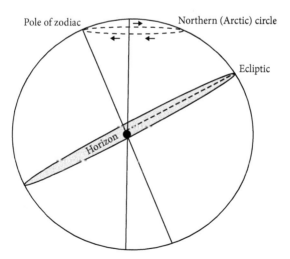

Fig. 4.17. Coincidence of the planes of the horizon and ecliptic.

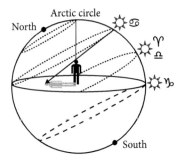

Fig. 4.18. The formation of the 'Midnight Sun' for someone below the Arctic circle when the sun is in the sign of Cancer.

the sun is due north. The result is the 'Midnight Sun'; on this day, the sun does not dip below the horizon and there is no night. Similarly, from the same figure it is evident that when the sun is at its southernmost limit at the head of Capricorn, that circle too does not cut the horizon and therefore it is always night. Fig. 4.18 illustrates that when the sun is between the two extreme positions of Cancer and Capricorn, an observer at the Arctic Circle will always experience some day and some night because every circle followed by the sun cuts the horizon somewhere.

§30. For 'every other place' north of the Arctic Circle, any movement further north towards what Grosseteste refers to as the elevated pole, the more turns of the sun's annual spiral occur completely above the horizon; these summer days in the Arctic are always light. However, because of the symmetry of the sphere, when the sun is in the opposite signs, there will be correspondingly more turns of the spiral that do not appear above the horizon and are completely dark. By the same token on proceeding northwards, more Arctic winter 'days' are completely dark. In the limit of being beneath the north celestial pole, the horizon aligns exactly with the celestial equator and then for the summer half of the year, the diurnal circles followed by the sun never cut the horizon, always remaining above it. On the other hand, in the winter half of the year, the circles (parallels) always remain below the horizon. For half the year the sun never rises and for the other half it never sets.

§31. The right sphere here refers to the situation of an observer at the equator (Fig. 4.19). There, as explained in §22, the axis of daily rotation is on the horizon and the stars rise at right angles to the horizon. For an observer anywhere else on earth, the stars will rise at oblique angles. In

COMMENTARY ON THE MODEL OF THE WORLD MACHINE 151

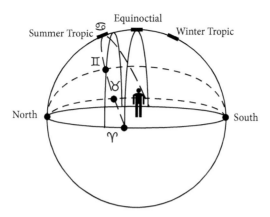

Fig. 4.19. The right sphere of an observer situated at the equator. Here, the stars rise at right angles.

any location on earth, the equinoctial will rise in a uniform way, equal arcs in equal amounts of time, maintaining the same angle to the horizon, because it is perpendicular to the axis of heavenly rotation. Equal increments of angle always ascend in equal amounts of time. The final sentence of the paragraph defines the term '[right] ascension' as the arc of the celestial equator that rises in the same time as the part of the zodiac one is looking at. Since the equator rises uniformly, the ascension is a measure of the time any star takes to rise. Modern astronomers use Right Ascension, the projection onto the equinoctial plane, as a coordinate equivalent to a celestial longitude.

§32. As the equinoctial rises uniformly, parts of the zodiac with equal ascensions take equal amounts of time to rise. However, different parts of the zodiac rise at different angles; those making circles at a more oblique angle with the vertical to the horizon take a shorter time to rise than those which rise more vertically (Fig. 4.20). This is shown by imagining great circles drawn through the celestial poles which cut the zodiac in equal increments of angle. Fig. 4.19 illustrates this for one-quarter of the sphere with great circles cutting the beginning of Aries, Taurus, Gemini, and Cancer. These great circles intercept the equinoctial, but because of the projection, the distances on the equinoctial are not now equal. The distance from Aries to Taurus, the more acute segment, has a shorter distance on the equinoctial than the less acute segment from Gemini to Cancer.

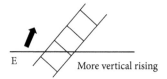

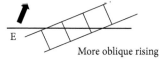

Fig. 4.20. More vertical and more oblique rising.

§33. Here, there is an appeal to the symmetry of the sphere followed by the deduction that the above argument applies equally to rising and setting of the signs.

§34. In this paragraph 'tropic points' refers to the heads of Cancer and Capricorn on the ecliptic, since the sun, on its annual path around the ecliptic, passes those points when it is over the tropics. 'Equinoctial points' refers to the heads of Aries and Libra: days and nights are of equal length when the sun is at those points. All twelve signs rise and set each day, but during each day the angle of the ecliptic to the observer's horizon changes. At the equator (the right horizon), when the tropic points of the ecliptic are rising, the ecliptic will be perpendicular to the horizon. When the equinoctial points are rising, the ecliptic will be at an oblique angle to the horizon. This is perhaps easiest to visualize if one imagines an observer standing at the equator facing east, with northern hemisphere constellations such as Cancer on their left (north) and southern hemisphere constellations on their right (south), and the ecliptic as a line joining the signs. When an equinoctial point rises, directly in the east, the ecliptic will be sharply angled from north to south, as the constellations (and the seasonal path of the sun) pass from northern to southern hemisphere (Fig. 4.21(a)). When a tropic point rises (corresponding to the sun's northernmost or southernmost limit at the solstices), the ecliptic is vertical (Fig. 4.21(b)) as it prepares to turn back in the other direction (a globe showing the ecliptic will have it parallel to the equator at those points).

§35. While the previous paragraph demonstrated that at the equator, tropic points rise vertically and equinoctial points rise obliquely it has

COMMENTARY ON THE MODEL OF THE WORLD MACHINE 153

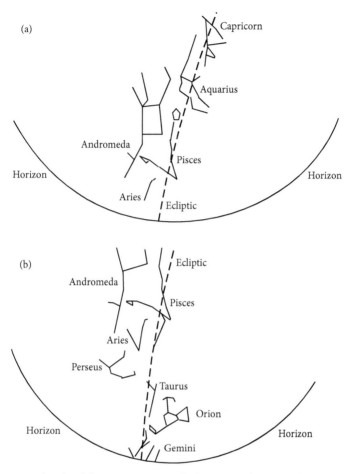

Fig. 4.21. Sketch of the projection on the heavens of some zodiac and near-ecliptic constellations. (a) The rising of an equinoctial point. (b) The rising of a tropic point.

been shown earlier in the treatise (DS §32) that the more vertically a part of the ecliptic rises, the greater its ascension, that is, the longer it will take to rise. Now it is demonstrated that, for an observer with an oblique horizon, that is, not at the equator, one half of the ecliptic starting from a northern hemisphere sign (for example, Cancer) will take longer to rise than the half-ecliptic starting from its opposite in the southern hemisphere (for example, Capricorn). On any given day, half the zodiac will

154   THE SCIENTIFIC WORKS OF ROBERT GROSSETESTE

rise with (that is, in the same time as) a 180° arc of the equinoctial corresponding to the parallel at that point (as, for example, the tropic of Cancer corresponds to the head of Cancer). This takes longer for people in the northern hemisphere, just as the days are longer in the northern hemisphere when the sun is under the tropic of Cancer, because the horizon cuts the equinoctial unequally (see DS §26).

§36. Half the zodiac rises with an arc of the equinoctial corresponding to the parallel at that point. The points of the zodiac 'equidistant from both tropic points' are the equinoctial points, that is the heads of Aries and Libra, also known as the vernal and autumnal equinoxes. Just as, whatever the observer's horizon, day and night are of equal length at the equinoxes, when the sun traces a parallel over the earth's equator, so at the oblique horizon, the half of the zodiac beginning with the head of Aries will rise in the same time as its opposite half. If a 180° arc of the zodiac is chosen, then the closer the starting point of that arc is to the head of Cancer, the greater the difference in rising time compared with the opposite 180° arc. However, because of the symmetry mentioned in §33, they will always average out, and taken together will add up to the same as if they were rising in the right sphere. Since 360° of the equinoctial (or ecliptic) rise in twenty-four hours, one hour corresponds to 15° of rising-time, and therefore one 30° sign rises in two hours.

§37. Although no location on earth will always be beneath the ecliptic pole, at the Arctic Circle, the circle traced daily by the ecliptic pole around the celestial pole will pass through the zenith. Directly beneath the pole of the zodiac (that is the ecliptic, not the celestial pole), the horizon and the plane of the zodiac coincide. The zodiac will rise horizontally, then, for any observer directly under the ecliptic pole.

§38. While one cause of the variation in the length of the natural day is the variation in obliquity in the rising and setting of signs discussed in the previous paragraphs, there is a second cause which arises because the sun's annual rotation is not centred on the earth and therefore the sun passes through the signs at different speeds.

§39. The centre of the sun's orbit is displaced from the centre of the earth in the direction of Gemini (Fig. 4.22) in the plane of the ecliptic, with which the sun's circle is perfectly aligned. The centre of the sun's orbit lies on the line drawn from the 18th degree of Gemini, through the

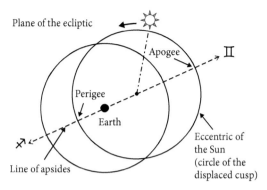

Fig. 4.22. The eccentric of the sun, its circular orbit not being centred on the earth.

earth, to the opposite side of the ecliptic. Since the sun's circle of annual rotation is not centred on the earth, the line of apsides, which is the line connecting the nearest (perigee) and furthest (apogee) points of the sun with respect to the earth, is the only diameter of that circle which passes through the earth (Fig. 4.22). The value of 2½ 'degrees' displacement of the centre of the sun's circle with respect to the centre of the earth is the solar eccentricity as a fraction of the full radius of 60.

§40. Since the sun's circle is displaced from earth, its orbit will take it closer to and further from earth. The closest point of approach is 'perigee', at Sagittarius 18°, and the furthest point is 'apogee', at Gemini 18°. Perigee and apogee are defined both in terms of the distances to the firmament and to the earth, the latter being more pertinent to the subsequent discussion on climate. Uniform motion about the eccentric circle results in a non-uniform motion of the sun with respect to the heavens and therefore the motion is non-uniform with respect to an observer on the surface of the earth.

§41. That non-uniform motion results in the natural day, defined by the 360° rotation of the heavens each day (which is uniform) together with the sun's own motion against the heavens, being non-uniform through the year.

§42. The two effects, namely the eccentricity of the sun's orbit and the obliquity of the rising of the signs of the zodiac, independently give rise to the variations in the natural day.

156 THE SCIENTIFIC WORKS OF ROBERT GROSSETESTE

§43. When the two effects combine in a way to maximize the variation in the natural day, the period over which this enhancement occurs is named the 'greater days'. The 'lesser days' refer to the time period when the eccentricity and obliquity combine to minimize the length of the natural day. The term 'median days' is used for days of length between the two extremes.

§44. The difference between the closest and furthest approach of the sun to the earth is twice the 2½° displacement of the centre of the sun's orbit from the centre of the earth, namely 5°. From the viewpoint of an observer on the earth, the sun is at its closest distance to the earth when it is in the sign of Sagittarius that is during the summer of the southern hemisphere (Figs. 4.22 and 4.8). The effect of the sun being simultaneously at its zenith and at its perigee is increased heating in summer. (Note that the cause of the heating effect is doubled, not the resultant heating.) On the other hand, in the winter of the southern hemisphere, the sun is at its furthest from the earth and at its lowest angle and therefore winters are intemperately cold. Therefore, the southern hemisphere, which is beyond the equinoctial circle, is uninhabitable. However, because the sun is further from the earth in the summer of the northern hemisphere, the two effects oppose one another and the greater distance of the earth when the sun is at its zenith tempers the climate of the northern hemisphere. Winters are also more temperate, since, when the sun is at its lowest in the sky, it is at its closest approach to earth. Such an effect does exist, but quantitatively the impact of the sun's obliquity far outweighs the change in light flux due to the changing distance.

§45. To proceed northwards over the earth's surface is to experience an increase in the length of summer days and the angular range over which the sundial shadow can be observed. Grosseteste's criterion for perceptibility of difference is not made clear, but division of the northern hemisphere into seven zones allows for the difference in range to be clearly distinguished. As the hour was defined as one-twelfth of the length of the day, the length of the hour is unequal in places of different latitude.

§46. The great circles described here are shown in Fig. 4.23, the great circle below the equinoctial circle being the equator.

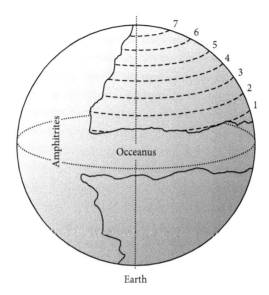

Fig. 4.23. The seven climes in the inhabited part of the earth and the two great oceans.

§47. Circles parallel to the equator define each of the seven climes in the inhabited part of the earth. Neither of the two southern lands are inhabited, as is the case for one of the two northern lands (Fig. 4.23). The construction starts at the northern shore of Occeanus, the first parallel being at the distance where the length of day is perceptibly shorter. As noted, this distance is not specified in terms of distance over the surface of the earth, but it is one-twenty-eighth part of the earth's circumference, if the width of Occeanus is neglected.

§48. The next six paragraphs relate to the observation, noted by Ptolemy in the *Almagest*, that the positions of the fixed stars with respect to the solsticial and equinoctial points are slowly changing with time (see also Ch. 8). Here, the term 'fixed stars' is specified as meaning that the relative positions of the stars with respect to one another remain constant.

§49. Ptolemy's explanation of the eastwards drift of the stars is in terms of precession of the equinoxes. This precession carries the sun's apogee and perigee around the ecliptic in a constant unidirectional motion, at a speed of 1° in approximately a hundred years. Since the

158   THE SCIENTIFIC WORKS OF ROBERT GROSSETESTE

perigee would eventually enter the northern hemisphere, instead of the effects of the sun's angle and closest position adding in summer in the southern hemisphere of the earth, making it uninhabitable, it would at some future time do so in the northern hemisphere. Were this to have been the case the temperate part of the world, that suitable for habitation, would become uninhabitable in approximately 180 centuries.

§50. As a counter to Ptolemy's proposal the trepidation hypothesis ascribed to Thebit (Thābit ibn Qurra) is presented. In §14, Grosseteste emphasized that the twelve zodiac signs, all of equal (30°) size, are distinct from the constellations of the same names (which are unequal in size). The trepidation model posits that the signs and constellations are on two separate zodiacs, in two separate spheres, able to move with respect to one another. In effect, this is a virtual zodiac in the firmament. Below it, in the sphere of the fixed stars, is another zodiac which is formed by the images conjured up by the configuration of the fixed stars.

§51. As indicated above, the 'fixed zodiac' here refers to the signs; 'images' refers to the constellations. The two circles around the head of Aries and Libra are drawn in the fixed zodiac of the firmament and the heads of Aries and Libra in the zodiac of the constellations are displaced from the position in Aries and Libra in the fixed zodiac onto the two circles. The two little circles rotate in opposite directions, causing the zodiacal band of constellations to change plane with respect to the plane of the ecliptic signs. Sliding the two spheres with respect to one another results in the head of Aries in the constellations being displaced northwards, when the head of Libra in constellations is displaced southwards (Fig. 4.24). There is counter-rotation of the two equinoctial points in the constellations on the two circles in the firmament of just over 5' of arc per year, resulting in an apparent oscillation of the position of the constellations with respect to the firmament.

§52. The oscillation is such that solstice points in the star system remain at all times on the ecliptic in the firmament system, the heads of Cancer and Capricorn just moving back and forth along the fixed ecliptic in an east-west direction parallel to the equinoctial.

§53. This motion applies both to the sphere of the fixed stars and to the apogees, namely the points furthest from the earth, of the planets. The implication of this requirement, and his noting in §49 that Ptolemy

COMMENTARY ON THE MODEL OF THE WORLD MACHINE 159

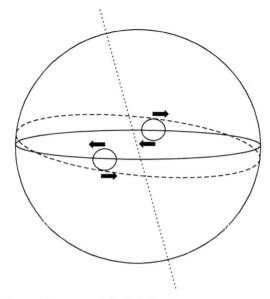

Fig. 4.24. The trepidation model of Thābit ibn Qurra.

posited that the apogees of the planets moved around the poles, is that Grosseteste was familiar with the model of all the planets moving on eccentric circles whose centres were displaced from the centre of the earth (See Ch. 5, §4.2).

§54. The remainder of the treatise concerns the motion of the moon. First, Grosseteste notes that this motion is below the zodiac. This refers both to distance and angular position. The zodiac is a band corresponding to an angular width of 12° and the moon moves on a circle that lies within this range. However, that is not on the ecliptic, which lies exactly along the centre of the zodiac, but tilted at 5° with respect to it. The 10° range contained by this circle lies within, or beneath, then, the 12° band of the zodiac.

The centres of the circles of both the sun and the moon are displaced from the centre of the earth, that is, they are eccentric. As shown in Fig. 4.25, there is a small diameter circle in the same plane as the eccentric circle and whose centre is always on the eccentric circle. This little circle travels around the eccentric circle and the moon always lies on the circumference of the small circle. In addition, the centre of the

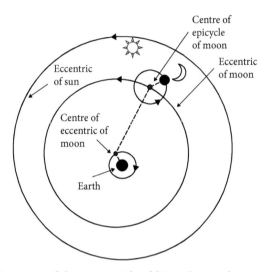

Fig. 4.25. Eccentric of the moon with additional epicycles.

eccentric circle is not only displaced from the centre of the earth, but also moves in a circle around the centre of the earth in a motion from east to west (distinct from the precession of apogees referred to in previous paragraphs) (see Ch. 9, §2.4).

§55. The centre of the little circle moves from west to east, uniformly with respect to an observer on earth. The uniform rotation of this epicycle centre is defined as the mean motion of the moon (Fig. 4.26(a)).

§56. Despite not being immediately obvious, there is enough information in this paragraph to enable the reader to deduce the rates of rotation of the mean position of the moon and the centre of the eccentric, although no actual numbers are provided. Grosseteste first specifies that when the mean motion of the moon and sun coincide, the centre of the little circle (epicycle) coincides with the apogee of the lunar eccentric. This is equivalent to stating that the apogee of the eccentric and the centre of the epicycle are at the same point and that the mean position of the sun lies on the line drawn from the centre of the earth through the centre of the moon's epicycle to the firmament (Fig. 4.26(b)).

The respective opposite motions of the apogee of the eccentric and the centre of the epicycle result in them becoming separated in such a way that the mean position of the sun always remains equidistant from each.

COMMENTARY ON THE MODEL OF THE WORLD MACHINE 161

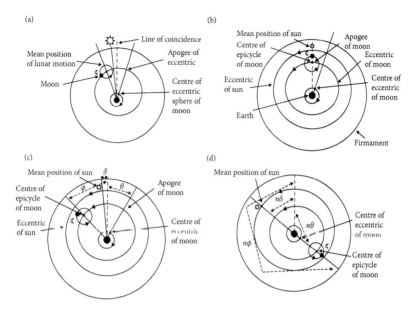

Fig. 4.26. (a) Mean position of moon. (b) Position of coincidence of mean solar and lunar positions. (c) Displacement after one day, corresponding to an angular increment δ. (d) Position of opposition of sun and moon.

In one day, let the angular change of the mean position of the epicycle be ϕ, the angular change of the position of the apogee be θ, and the angular change of the sun's mean position be δ. Then, as seen in Fig. 4.26 (c), for the sun to be always equidistant from the apogee and centre of epicycle,

$$\phi - \delta = \theta + \delta \quad \text{or equivalently,} \quad \phi - \theta = 2\delta. \qquad \text{Equation 4.1}$$

Grosseteste proceeds to note that there will be a position where the mean position of the moon is opposite to the mean position of the sun, and here the apogee of the moon coincides with the centre of the epicycle. Let this occur in a certain number of days, n. Then, as seen in Fig. 4.26(d), the centre of the epicycle will have travelled an angle of $n\phi$, the centre of the eccentric $n\theta$ and the sun $n\delta$. For alignment to occur,

$$n\phi - n\delta = 180° \qquad \text{Equation 4.2}$$

162 THE SCIENTIFIC WORKS OF ROBERT GROSSETESTE

and

$$n\theta + n\delta = 180° \qquad \text{Equation 4.3}$$

adding equations 4.2 and 4.3, yields

$$n(\phi + \theta) = 360° \qquad \text{Equation 4.4}$$

Now, the sun traverses the 360° of the zodiac in 365¼ days, so it advances approximately 1° per day. Thus, $\delta \approx 1°$. Further, the moon travels around the earth once per month, there being 29.5 (approximately 30) days between new moons. Therefore $2n \approx 30$ days, or $n \approx 15$ days.

$$\phi + \theta \approx 24° \qquad \text{Equation 4.5}$$

From equation 4.1,

$$\phi + \theta \approx 2° \qquad \text{Equation 4.6}$$

Adding equations 4.5 and 4.6, gives $2\phi \approx 26°$, that is a movement of approximately 13° per day for the epicycle centre and subtracting equations 4.5 and 4.6 gives $2\theta \approx 22°$, that is a movement of approximately 11° per day for the eccentric centre. The sun's motion of approximately 1° per day results in it remaining equidistant from each, and this of course remains the case if the above calculation is made with higher precision.

The statement that 'the centre of the epicycle describes twice a month the eccentric' refers to the relative variation of the centre of the epicycle and the centre of the eccentric. As noted in the above discussion, the moon travels round the earth once, not twice per month. The time to orbit is twenty-eight days, which differs from the time between new moons due to the apparent motion of the sun and centre of the eccentric. As seen by comparing Figs. 4.26(c) and 4.26(d), the time between the centre of the epicycle coinciding again with the apogee of the eccentric is fifteen days, the time taken to go from alignment to opposition with respect to the sun. With respect to the reference frame of the centre of the eccentric, the epicycle centre goes from the apogee to perigee and back to apogee twice per month; it describes the eccentric twice per month.

COMMENTARY ON THE MODEL OF THE WORLD MACHINE 163

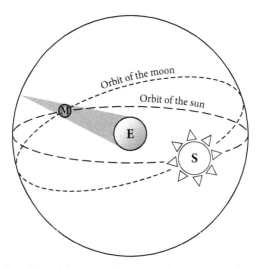

Fig. 4.27. Inclination of the moon's orbit with respect to that of the sun.

The rotation rate of the moon around the epicycle is not given; the last sentence in this paragraph specifies the direction of rotation.

§57. As a result of the inclination of the lunar orbit by 5° with respect to that of the sun, Fig. 4.27, there are two points (nodes) where the two orbits intersect. Only when the sun and moon are close to these positions in opposition can an eclipse of the moon occur. As the sun is larger than the earth, the shadow of the earth forms a circular cone, the axis of which always lies beneath the ecliptic (since the sun never deviates from the ecliptic). Grosseteste states that a lunar eclipse cannot occur when the moon is further than 12° from a node,

The value of this angle can be established in the following manner. As the umbra of the earth is about 2.7 moon diameters at the moon's orbit, in order for the moon to cut the umbra and there to be an eclipse, the centre of the moon must be less than $(2.7/2 + \frac{1}{2})$ moon diameters away from the ecliptic. As the moon's orbit is inclined at 5° to the ecliptic (§54) and the diameter of the moon subtends an angle of 0.5°, the critical angle β away from a node at which an eclipse will fail to take place is given by

$$5 \sin \beta = 0.5 \left(\frac{2.7}{2} + \frac{1}{2} \right) \qquad \text{Equation 4.7}$$

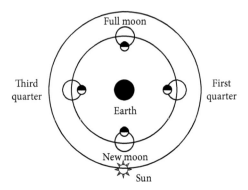

Fig. 4.28. The phases of the moon.

Thus $\beta = 11°$ and the agreement with the 12° quoted in the text may be regarded as satisfactory.

§58. Moonlight is light reflected from the sun, and the side of the moon facing the sun is always illuminated. From a viewpoint on the earth (Fig. 4.28), as the moon orbits the earth, the lunar hemisphere which is illuminated goes from being fully visible to invisible as the moon progresses through its phases. The moon is full when the moon is in opposition to the sun; the new moon occurs when the moon is in conjunction with the sun.

§59. Since the moon is full when it is in opposition to the sun, the earth is between the two. If this happens when the moon is at one of its nodes where its path crosses the ecliptic, it will be fully shadowed by the earth, that is, eclipsed (Fig. 4.27). Where the full moon occurs and the moon is relatively far from the ecliptic, an eclipse will not occur as the earth's shadow always lies in the plane of the ecliptic.

§60. The final paragraphs are designed to explain why eclipses of the sun are less frequent than eclipses of the moon. Although the discussion focuses on the position of the moon, its concern is with the issue of the moon blocking the sun and therefore with the relative diameters of the earth and the moon. Since the earth's diameter is significant in relation to the size of the moon's orbit, there is a difference in the direction of two lines, one drawn from the centre of the earth through the moon's centre and one drawn through the moon's centre, from an observer on the surface of the earth, for whom the moon is not directly overhead.

COMMENTARY ON THE MODEL OF THE WORLD MACHINE   165

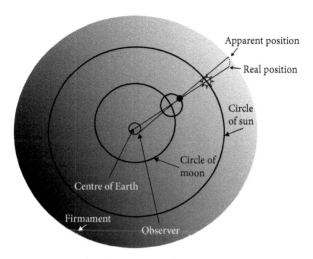

Fig. 4.29. Apparent and real positions of the moon on the firmament.

The former (Fig. 4.29) passes through the position of the centre of the sun when at the position of eclipse, whereas the latter does not. Extension of these two lines to the sphere of the firmament gives the true and apparent position of the moon. The following paragraph deals with the diversity of the aspect of the moon, this being defined as the arc of a great circle passing through these two points.

§61. The diversity of aspect of the moon is resolved into two components, the diversity of aspect in longitude and the diversity of aspect in latitude. The text contains a description of the construction of Fig. 4.30, in which these two components can be seen geometrically. The key geometric constructions are four circles on the firmament. Two are parallel to the ecliptic. The first passes through the true position of the moon and if the moon lies on the ecliptic, this first circle coincides with it. As drawn in Fig. 4.30, it is displaced. The second circle passes through the apparent position of the moon on the firmament, while the other two are great circles which pass through the poles of the zodiac, one passing through the true position of the moon on the firmament, the other through the apparent position of the moon. The area shaded in Fig. 4.30 is the quadrangle to which Grosseteste refers.

§62. The definition of the diversity of aspect in longitude and latitude are the distances AT' and TT' in Fig. 4.30. When either pair of circles

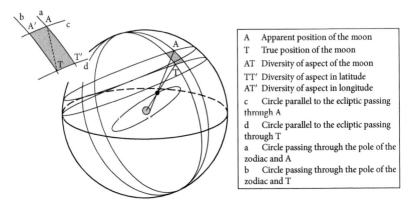

Fig. 4.30. Resolution of the diversity of aspects of the moon into latitudinal and longitudinal components.

coincide, there is no diversity of aspect in longitude or latitude as appropriate.

§63. Grosseteste then states that because of the diversity of aspect, it is not possible for alignment of a northern hemisphere observer, the moon and the sun to occur when the moon is in the south as the apparent place of the moon is to the south of the ecliptic. For there to be a partial eclipse, the misalignment must be less than the sum of the two apparent radii of the sun and moon. Under these conditions, eclipses can be seen in the northern hemisphere when the moon is in the north.

5

Intellectual Inheritances

Absence and Presence

One of the crucial elements in understanding Robert Grosseteste's *On the Sphere* is how selective were his choices across a wide range of possible models and sources. As noted in the Introduction to this volume the identification of the treatise as a 'textbook' is not helpful. *On the Sphere* is not, in any way, a comprehensive account of astronomy and its implications. Grosseteste does not, for example, discuss, except in passing, the motions of the planets apart from those of the sun and moon (Ch. 4). In order to gauge the selectivity of *On the Sphere* it is important to establish the scale of the intellectual inheritances on which he would draw for his treatise, and from which he would make his choices of content. Placing *On the Sphere* in its wider intellectual context allows its particular characteristics to be drawn in sharper relief.

What follows is a detailed discussion of three such inheritances: Islamicate astronomy, astrology, and the Latin astronomical tradition. The latter includes consideration of the traditions and tools of observational astronomy. While Grosseteste separated observation and scientific reasoning in *On the Sphere* and presented the treatise firmly in concord with the latter, he was certainly no advocate for the former's futility, and may be expected, reasonably enough, to have been familiar with its practice (see Ch. 6, §4.3). All of these areas have relevance to the themes and questions of Grosseteste's treatise and reveal not only the depth of learning to which he very probably had access but also the broader conceptual frameworks from which his particular thought emerged.

Islamicate astronomy and its complex legacy within medieval Europe holds a prominent place in this discussion. Grosseteste was part of the generation of European scholars who had access to ever larger parts of this wide and deep inheritance through Latin translations

168 THE SCIENTIFIC WORKS OF ROBERT GROSSETESTE

from the mid-1120s onwards. These new texts revealed far more advanced methods of astronomical calculation and more accurate observations. Islamicate thinkers also showed various ways of appropriating the Ptolemaic tradition. Since there are many traditions that present divergent knowledge, an overview of some of these debates provides cardinal points around which Grosseteste's particular use and description of astronomy can be mapped. One of the results of the influx of Islamicate thinking about astronomy into medieval Europe throughout the twelfth century was the equal, if not greater, interest displayed in its ancient and medieval partner, astrology. In this sense one of the principal ways in which twelfth- and early-thirteenth-century Europe engaged with astronomy was by way of astrology. That Grosseteste was no stranger to this tradition is shown in his first treatise, *On the Liberal Arts*, which demonstrates his familiarity with astrology.[1] It was a subject supported and practiced in Hereford, the city, shire, and diocese to which he can be connected from about 1195 until the mid-1220s (See Ch. 1, §3.3; Ch. 2, §1).[2] Astrology provides another model for thinking about the heavens and the planets, and one which it is important to rehearse for the wider dimensions of Grosseteste's astronomical thinking.

The final inheritances considered here are those of observational astronomy as practised in Grosseteste's day, and traditions of astronomical reasoning inherited from classical Rome, early medieval Europe, and Grosseteste's near-contemporary John of Sacrobosco. These turn on the question of how to reconcile the tension between cosmological and mathematical astronomy and introduce a tension between observation and calculation. This is particularly the case with regard to planetary motion and how to account for the fact that planets appear to move in a non-constant motion. Retrograde motion was the subject of considerable and long-standing discussion in this respect. Not only would Grosseteste have had sufficient sources from the classical Roman and early medieval corpus to have been aware of this debate, it is also clearly articulated in Sacrobosco's *On the Sphere*, which Grosseteste almost certainly knew (see Ch. 1, §1). Examination of this debate in other Latin sources underlines once more the need to establish the inheritances around which

---

[1] *Knowing and Speaking*, 166–95.   [2] *Knowing and Speaking*, 26–35.

INTELLECTUAL INHERITANCES   169

understanding of Grosseteste's particular interests should be placed. Why Grosseteste did not mention retrograde motion when Sacrobosco, at corresponding points in his treatise, did, allows for consideration of the former's rather different purposes in writing. Another possible inheritance for *On the Sphere* within the Latin tradition are the writings of Petrus Alfonsi with particular reference to the demonstration of the sphericity of the world, eclipses, and the location of the Indian city of Arim.

1 Islamicate Astronomy

Astronomy in the context of the Islamicate world was an influential scientific paradigm until the sixteenth century CE, and contributed novel mathematized notions that were supported by a large body of empirical observational data, while also critically revising and expanding the inherited legacies from the ancient Greek, Byzantine, Babylonian, and Indian astronomical systems.[3] Arabic astronomy focused its analytical and critical studies primarily on the Ptolemaic astronomical model and Aristotelian cosmological doctrines.[4] Such pathways in scientific inquiry generated various calculations of the motions of the sun, the moon, the planets, and the fixed stars, along with sightings of comets, and the modelling of the positions and configurations of the heavenly spheres. These advances were made through geometrical methods and observational instruments, and had practical applications in aiding travellers on land and sea, and in determining the direction of the *Qibla* towards the Kaaba in Mecca (south-west Arabian Peninsula), the timekeeping of the five daily ritual prayers, and the setting of lunar calendars for festivities and the fasting month of Ramadan.[5]

Arithmetic, algebra, geometry, optics, and mechanics all played an important role in this science. The reliance on spherical trigonometry

---

[3] For example, the use of the sine function instead of the arc and chord as in Greek trigonometry displays influences from the *Zīj al-Sind hind* of Indian provenance and the *Zīj al-Shāh* from the Sassanid Persian sources.

[4] George Saliba, *Islamic Science and the Making of the European Renaissance* (Cambridge, MA: MIT Press, 2007).

[5] David A. King, *In Synchrony with the Heavens: Studies in Astronomical Timekeeping and Instrumentation in Medieval Islamic Civilization* (Leiden: Brill, 2005).

170　THE SCIENTIFIC WORKS OF ROBERT GROSSETESTE

was fundamental in terms of perfecting the use of sines and cosines, along with the analysis of their secant and cosecant reciprocals in studying spherical polygons. Such investigations were set around the *Spherica* of Menelaus of Alexandria, and they expanded the penchant of mathematizing astronomy beyond Ptolemy's *Almagest* (Μαθηματικὴ Σύνταξις; Μεγίστη; *al-Majistī*). These resulted in further developments in celestial kinematics and its theories based on new research in spherical geometry and trigonometry as aided by empirical observational data.[6] Mathematical astronomy benefited in this regard from designing and using instruments like astrolabes, armillary spheres, ephemerides tables, dust-boards, sundials, equatoria, mechanical-calendars, wall-quadrants, and data-tabulations.

Tensions arose between those favouring natural philosophy and those engaged in cosmology, and mathematized physics, with the former dominant over the latter.[7] The Aristotelian notion of the perfected order of the cosmos, set alongside the astronomical structure associated with Ptolemy, gave rise to problems as to how to understand the earth, moon, sun, and planets. The phenomenology of immediate visual perception, as mediated via lived experiences, in addition to religious outlooks on physical reality with an embedded earthly anthropocentric perspective, all reinforced the complex factors that underpin the notion of the universe as having a geocentric order. In the absence of an overall heliocentric model, the return to a philosophical interpretation of the results of mathematics and observation required further centuries of theoretical refinements and recensions, as aided by the practical progress in material culture in the design and use of instruments. This eventually allowed the geocentric model to be overthrown and replaced with a heliocentric system by way of a deconstructive revolution in astronomy

---

[6] Dominique Raynaud, 'Abū al-Wafā Latinus? A Study of Method', *Historia Mathematica*, 39 (2012), 34–83; Roshdi Rashed, *Ibn al-Haytham, New Astronomy and Spherical Geometry: A History of Arabic Sciences and Mathematics*, Vol. IV, trans. Judith Field (London: Routledge, 2014).

[7] An aspect of the tension between natural philosophy and the mathematizing of physics is discussed in Nader El-Bizri, 'In Defence of the Sovereignty of Philosophy: al-Baghdādī's Critique of Ibn al-Haytham's Geometrisation of Place', *Arabic Sciences and Philosophy*, 17 (2007), 57–80.

INTELLECTUAL INHERITANCES 171

the effect of which was an epistemological rupturing of the classical paradigm and its replacement.

The Ptolemaic model was the subject of mathematical corrections and empirical doubts in Arabic astronomical circles from the early ninth century CE. A genre of sceptical treatises (*shukūk*) concerning the Ptolemaic system and its deconstruction developed accordingly.[8] For example, the *Shukūk 'alā Baṭlāmyūs* (*Aporias Concerning Ptolemy*) of Ibn al-Haytham (Alhazen; *c*.965–1040 CE) set out to readjust the geocentric model.[9] This criticism shifted in some quarters into scepticism towards the cosmological system as a whole, and speculations even emerged regarding the possibility of assuming the rotation of the earth around its axis, as suggested in the *Qānūn al-Mas'ūdī* (Masudic Canon) of al-Bīrūnī (973–1048 CE), albeit without doing away with the geocentric model.[10] The decisive position over this matter was given over to cosmologists and natural philosophers rather than to mathematical astronomers.[11] Such studies posited the Ptolemaic model in abstract terms as being geometrical rather than physical.

---

[8] A. I. Sabra, 'An Eleventh-Century Refutation of Ptolemy's Planetary Theory', in E. Hilftein, P. Czartoryski, and F. D. Grande (eds.), *Science and History* (Wrocław-Warszawa: Studia Copernicana, 1978), 117–31; George Saliba, 'Ibn Sīnā and Abū 'Ubayd al-Jūzjānī: The Problem of the Ptolemaic Equant', *Journal for the History of Arabic Science*, 4 (1980), 376–403; George Saliba, 'The Role of the *Almagest* Commentaries in Medieval Arabic Astronomy: A Preliminary Survey of Ṭūsī's Redaction of Ptolemy's *Almagest*', *Archives Internationales d'Histoire des Sciences*, 37 (1987), 3–20; Julio Samsó, 'Ibn al-Haytham and Jābir b. Aflaḥ's Criticism of Ptolemy's Determination of the Parameters of Mercury', *Suhayl*, 2 (2001), 99–225; Miquel Forcada, 'Ibn Bājja's *Discourse on Cosmology* (*Kalām fī al-hay'a*) and the "Revolt" Against Ptolemy', *Zeitschrift für Geschichte der Arabisch-Islamischen Wissenschaft*, 20–1 (2012–14), 64–167.

[9] This geometrical technique was based on the construction of epicycles, and, as a geometric equivalent, it predated the 'Fourier series' that devised the asymptotic approximation and representation of periodic functions by trigonometric series (see Ch. 8). Jean-Baptiste Joseph Fourier (d. 1830 CE) introduced the series in 1807 in the context of his harmonic analysis and research on heat as set in his *Mémoire sur la propagation de la chaleur dans les corps solides* (Paris: Bernard, 1808), 112–16 (*Œuvres complètes*, 'Tome 2', esp. pp. 217–19); Jacques Peyrière, *Convolution, séries et intégrales de Fourier* (Paris: Ellipses, 2012). For epicycles and 'Fourier series', see Ch. 9 of this volume.

[10] Edward S. Kennedy, 'Al-Bīrūnī's Masudic Canon', *al-Abḥāth*, 24 (1971), 59–81; Seyyed Hossein Nasr, *An Introduction to Islamic Cosmological Doctrines* (Albany: SUNY Press, 1993), esp. 135–6. Donald R. Hill, 'Al-Bīrūnī's Mechanical Calendar', *Annals of Science*, 42 (1985), 139–63; J. L. Berggren, 'Al-Bīrūnī on Plane Maps of the Sphere', *Journal for the History of Arabic Science*, 6 (1982), 47–112.

[11] Bernard R. Goldstein, 'Theory and Observation in Medieval Astronomy', *Isis*, 63 (1972), 39–47.

172 THE SCIENTIFIC WORKS OF ROBERT GROSSETESTE

The approach taken by Ibn al-Haytham in his *Aporias Concerning Ptolemy* was particularly critical of the Ptolemaic astronomical models as presented in the *Almagest, Planetary Hypotheses,* and *Optics.* In this he drew on, and developed, the work of astronomers in the Islamicate tradition from the ninth century onwards.[12] For instance, Ibn al-Haytham argued that some of the mathematical devices that Ptolemy introduced into astronomy, especially the equant, failed to satisfy the physical requirement of uniform circular motion, and pointed to the absurdity of relating actual physical motions to imaginary mathematical points, lines, and circles.[13] Ptolemy's model, Ibn al-Haytham noted, was based on imagining the motions of the planets along some circles in the heavens, which does not bring about a description of the actual physical movement. Ibn al-Haytham rather aimed at describing a novel planetary model based on spherical geometry, infinitesimal geometry, and trigonometry. The overall schemata of an ordered geocentric universe were retained alongside the assumption that the celestial motions were uniformly circular, which itself necessitated the inclusion of the epicycle and the eccentric, while doing away with the equant. Other recensions of the Ptolemaic model were introduced for example by the Andalusian Avempace (Ibn Bājja; 1085–1138 CE) in terms of deploying concentric circles without epicycles, or the 'Ṭūsī-couple' device that appealed later to the Copernican tradition.[14]

Scientific studies in astronomy and cosmology opened the way also for occult practices in astrology and the tracking of planetary motions for the projection of their purported impacts on terrestrial life. This effort was guided by the mathematical capacity to calculate the projected future positions of celestial bodies and to come up with a model that

---

[12] Robert R. Newton, *The Crime of Claudius Ptolemy* (Baltimore: Johns Hopkins University Press, 1977); Bernard R. Goldstein, 'Casting Doubt on Ptolemy', *Science*, 199 (1978), 872; Ibn al-Haytham, *On the Configuration of the World*, ed. and trans. Y. Tzvi Langermann (London: Routledge, 2016).

[13] Roshdi Rashed, 'The Celestial Kinematics of Ibn al-Haytham', *Arabic Sciences and Philosophy*, 17 (2007), 7–55.

[14] Jamil Ragep, *Naṣīr al-Dīn al-Ṭūsī's Memoir on Astronomy (al-Tadhkira fī 'ilm al-hay'a)*, ed. trans. (Dordrecht: Springer Verlag, 1993); Jamil Ragep, 'Copernicus and his Islamic Predecessors: Some Historical Remarks', *History of Science*, 45 (2007), 65–81; A. I. Sabra, 'Configuring the Universe: Aporetic, Problem Solving, and Kinematic Modeling as Themes of Arabic Astronomy', *Perspectives on Science*, 6 (1998), 288–330; George Saliba, 'The First Non-Ptolemaic Astronomy at the Maraghah School', *Isis*, 70 (1979), 571–6.

INTELLECTUAL INHERITANCES   173

aligns such coordinates in trigonometry and spherical geometry with earthly seasons and specific calendrical calculations. Zodiac iconographic constellations evoked the proclaimed occult powers of the planets, stars, and comets. This reflected a fascination with celestial phenomena such as shooting stars, the passing of meteors, bolides, and comets, which were not conceived as being within the ethereal realm and its uniform circular motions. Such phenomena may not therefore have the same impact on the sub-lunary world of generation and corruption as that of the planetary spheres. Comets, shooting stars, and meteors may have influences on earth that are meteorological, as is the case with thunder, rain, lightning, storms, in terms of terrestrial weather, floods, or droughts.[15] On the whole, many scientists, philosophers, and theologians in the premodern Arabic/Islamic milieu considered astrology as a pseudo-science that exploited the calculations of astronomy; its superstitions deserving of rejection on scientific and religious grounds.[16]

These are modes of learning which would prove popular in Latin Europe especially when fuelled by the translation of key texts. The Greco-Arabic astronomical inheritance was the foundation of Grosseteste's conceptual framework for *On the Sphere*. That framework was complex, however, as much in what he left out as in what he included. This is particularly the case for traditions of Latin astronomy and cosmology.

2 Saturn, Jupiter, Mars, Venus, and Mercury: Grosseteste and the Toledan Tables

An obvious absence in *On the Sphere*, as noted above, is any detailed mention of the planets or, indeed of their spheres. Grosseteste, like all

---

[15] Arlstotle, *Meteorologica*, 1.4, 1.6, 1.10–13.
[16] Yahya J. Michot, 'Ibn Taymiyya on Astrology: Annotated Translation of Three *Fatwas*', *Journal of Islamic Studies*, 11 (2000), 147–208; George Saliba, 'The Role of the Astrologer in Medieval Islamic Society', *Bulletin d'Études Orientales*, 44 (1992), 45–67; Ignaz Goldziher, 'The Attitude of Orthodox Islam Toward the "Ancient Sciences"', in *Studies on Islam*, trans. and ed. Merlin L. Swartz (Oxford: Oxford University Press, 1981), 185–215; Ibn Sīnā, *Risāla fī ibṭāl aḥkām al-nujūm* (Avicenna's Epistle on Falsifying Astrology), in Hilmi Zia Ülken (ed.), *Opuscules d'Avicenne*, ed. Hilmi Zia Ülken (Istanbul: Ibrahim Horoz Basimevi, 1953), 49–67; August Ferdinand Mehren, 'Vues d'Avicenne sur l'astrologie et sur le rapport de la responsabilité humaine avec le destin', *Le Muséon*, III/3 (1884), 1–38.

174 THE SCIENTIFIC WORKS OF ROBERT GROSSETESTE

other authorities, is clear that the universe is a sphere, and also writes about the inner and outer surfaces of the smaller spheres; but he never states definitely that the spheres are solid. That the exact make-up of the spheres was a tricky subject is suggested by a lack of clarity on the inter-relationships involved.[17] Unlike Sacrobosco he sets out a single statement:

> The body of this world is of this shape and site; [that is] the one [body] that the philosophers name the fifth essence, or ether, or the body of the heavens, and which, contrary to elementary properties, [is] circu-larly mobile, and in which the seven planets are contained together with the fixed stars.   (DS §3)

While this supports the view that there was a fundamental division between the quintessence and the elemental region, the statement does not go into clear detail. These things in combination create the impres-sion that *On the Sphere* represents an attempt to sum up the state of the argument in relation to the fast-moving, and potentially controversial, subject matter of astronomy, whilst being very careful about its related fields of cosmology and astrology.

Grosseteste's relative silence on the planets may relate, at least in part, to the problem of the bad match between Aristotelian philosophy and Ptolemaic astronomy. Setting out a full account of the movements of all seven planets would entail covering those which periodically stood still and changed the direction of their motion (that is to say, went retro-grade). It is worth noting again here that these issues were particularly hard to reconcile with the Aristotelian model of the universe, although valiant attempts to do just this had been made by those working on the cutting edge of Islamicate astronomy and mathematics (§1). Such the-ories were part of the great body of work in these fields which was translated into Latin during the twelfth century and was still at a rela-tively early stage of assimilation in the first decades of the thirteenth.[18]

---

[17] See the discussion in E. Grant, 'Celestial Orbs in the Latin Middle Ages', *Isis*, 78 (1987), 152–73; Richard C. Dales, 'The De-Animation of the Heavens in the Middle Ages', *Journal of the History of Ideas*, 41 (1980), 531–50.
[18] For a survey, see J. Chabás, 'Aspects of Arabic Influence on Astronomical Tables in Medieval Europe', *Suhayl*, 13 (2014), 23–40.

INTELLECTUAL INHERITANCES 175

Central to this body of knowledge were the complex sets of astronomical tables which provided updated versions of Ptolemy's models and calculations of the movements of the planets, which will be discussed below. These were crucial for calculation of the positions of the planets on any given date, knowledge which was fundamental for any practical application of astronomical theory. For it was not only the sun and the moon whose positions in relation to the earth and to one another affected matters such as climate, weather, and the abundance or scarcity of food and natural resources.

Roman writers, including Pliny, had passed on a great deal of knowledge on the movements of the planets and the interaction between the heavens and human endeavours. Parts of this had been absorbed into the computus of the Christian Church and its broader intellectual hinterland. In this way, for instance, Bede wrote in his influential treatises that a 'transit' of Saturn brings cold weather.[19] However, direct observation of Saturn and calculation based on textual authority were likely to produce different results in medieval Europe until at least the mid-twelfth century. Experts in computus had access to texts which gave information on the average speeds of each planet (usually omitting to mention retrogradation). These could be used to produce estimates of planetary positions for chosen dates, by consulting a world chronology (to establish the exact time elapsed since the creation) and related texts on the zodiac position of each planet when it was created. The calculations involved would be laborious, given the long periods of time involved; and an accurate prediction was so unlikely that this method was more likely to produce confusion than certainty. The fact that such attempts were made suggests that this knowledge was valued.[20] It is precisely this whole field of study which Grosseteste omitted from *On the Sphere*.

---

[19] Bede, *De natura rerum*, ed. C. Jones, CCSL 123A (Turnhout: Brepols, 1975), *c*. 11.

[20] For discussion, see A. Lawrence-Mathers, *Medieval Meteorology: Forecasting the Weather from Aristotle to the Almanac* (Cambridge: Cambridge University Press, 2019), 40–65. On the calculation of planetary conjunction without the aid of tables, see David Juste, 'Neither Observation nor Astronomical Tables: An Alternative Way of Computing Planetary Longitudes in the Early Western Middle Ages', in Charles Burnett (ed.), *Studies in the History of the Exact Sciences in Honour of David Pingree* (Leiden: Brill, 2004), 181–222. On later attempts to reconcile biblical chronology and astronomy, see C. Philipp E. Nothaft, 'Climate,

176   THE SCIENTIFIC WORKS OF ROBERT GROSSETESTE

As already shown, this omission cannot have been the result of ignorance. Grosseteste's treatise on the liberal arts had already outlined the importance of the planets, and the moon in particular, in relation to alchemy, agriculture, and medicine.[21] Moreover, his handling of this discussion had drawn upon technical matters such as the inter-relationships between the moon and other planets, whose influences the moon both transmitted and modulated. Grosseteste actually stated that the moon's power and effects were themselves affected by its 'aspects' with other planets.[22] For human activity to have a successful outcome it was therefore necessary to know when the moon was in a positive aspect with the fortunate planets. Grosseteste does not name these, presumably because his audience would share this knowledge, but they were usually taken to be Jupiter (the stronger of the two) and Venus. What follows from this is that both Grosseteste and his audience are assumed to have access to the astronomical tables which made it possible to calculate planetary positions and relationships with much greater exactitude than the old techniques made possible. Such tables appeared in Latin translation in the first half of the twelfth century and their appeal was strong.

With reference to Grosseteste's known whereabouts, there is no doubt that astronomical tables were studied and used in Hereford in the late twelfth century, not only by Roger of Hereford but also presumably by those he taught.[23] Whether this was the source of Grosseteste's own knowledge is unclear, but it is worth repeating that Roger possessed expertise which was extremely scarce in England at this time. Roger used, and adapted to a different meridian, the newly-translated Tables of Toledo (first compiled in Arabic, in Toledo, and in the late eleventh century). Some of these tables were ultimately modelled on Ptolemy's own tables, compiled on the basis of his calculations and observations (and therefore dating back to the second century CE and assuming a different geographical reference point). Ptolemy's *Almagest* included

---

Astrology and the Age of the World in Thirteenth-Century Thought: Giles of Lessines and Roger Bacon on the Precession of the Solar Apogee', *Journal of the Warburg and Courtauld Institutes*, 77 (2014), 35–60.

[21] Grosseteste, *De artibus liberalibus* §§11–12; *Knowing and Speaking*, 170–8 and 193–5.

[22] Grosseteste, *De artibus liberalibus* §11.   [23] *Knowing and Speaking*, 29–31.

INTELLECTUAL INHERITANCES   177

detailed instructions for making calculations for different dates and locations, but the intervening millennium with the introduction of different calendrical systems would make these rather laborious to undertake and using the *Almagest* itself was a complex process. It was therefore especially helpful that the translations of the *Almagest* were preceded by translations of planetary tables produced in the Islamicate world, embodying much more up to date observations. These were widely successful, as shown by the numbers of surviving copies and comments.[24]

Through most of the twelfth century the leading set of astronomical tables, at least in England, had been the Latin version of the *Zīj* of al-Khwārizmī (d. *c*.850 CE), as updated for Cordoba in the late tenth century by Maslama al-Majrīṭī (d. *c*.1007 CE).[25] This led to some complications, since the original tables had used the Persian year (of 365 days) and the era of Yazdegird III, but Maslama revised them for the Arabic lunar calendar and the Hijra era. Key tables were also recalculated for the meridian of Cordoba. Since, however, all the mean-motion tables are for Arim the introductory canons for the Latin version continued to define that city as 'the central place of the earth … from which the four ends of the earth have equal distance' and to explain that 'the corrections for the planets and [the reckoning] of time are made with reference' to it.[26] While Arim is now understood to be the central Indian city of Udidjayn, it was essentially an idealized location developed amongst Islamicate authors for the centre of the world.[27] A further complication

---

[24] More than 200 copies of the Latin versions of the Tables of Toledo are cited in F. S. Pedersen, *The Toledan Tables*, Historisk-filosofiske Skrifter 24, 4 vols. (Copenhagen: Det Kongelige Danske Videnskabernes Selskab, 2002); see esp.11–23. No complete listing of manuscripts of al-Khwārizmī's work has been published; but the fact that versions were translated into Latin at least three times in the twelfth century, and that a commentary was itself translated shows strong interest. See for discussion J. Chabás and B. R. Goldstein, *A Survey of European Astronomical Tables in the Late Middle Ages* (Leiden: Brill, 2012); and R. Mercier, 'Astronomical Tables in the Twelfth Century', in C. Burnett (ed.), *Adelard of Bath: An English Scientist and Arabist of the Early Twelfth Century* (London: Warburg Institute, 1987), 87–118.

[25] O. Neugebauer, *The Astronomical Tables of al-Khwarizmi* (Copenhagen: I kommission hos Munksgaard, 1962); For brief details of al-Majrīṭī, see J. Vernet and J. Samsó, 'The Development of Arabic Science in Andalusia', in R. Rashed (ed.), *Encyclopedia of the History of Arabic Science*, Vol. 1 (London: Routledge, 1996), 243–75.

[26] Neugebauer, *The Astronomical Tables*, 10. See also R. Mercier, 'Meridians of Reference in Precopernican Tables', *Vistas in Astronomy*, 28 (1985), 23–7.

[27] Petrus Alfonsi, *Dialogue Against the Jews*, trans. Irven Resnick (Washington, DC: Catholic University of America Press, 2006), 55, n.22.

178 THE SCIENTIFIC WORKS OF ROBERT GROSSETESTE

was introduced by the fact that Islamicate astronomers had found growing problems in Ptolemy's coordinates for fundamentals such as the longitude of the solar apogee and the vernal point (0° celestial longitude, or first point of Aries). For instance, the former was found to have moved from Ptolemy's 5;30° in Gemini to at least 20° in Gemini. A serious problem was posed by the constant advance of precession (the motion in longitude of the fixed stars in relation to the vernal equinox) as proposed by Ptolemy. Grosseteste expressed reservations about the consequences of this in §49 of *On the Sphere*. The solution proposed to such problems was the theory of trepidation, the full modelling of which, with accompanying tables, was attributed, albeit falsely, by most medieval Latin writers, including Grosseteste, to 'Thebit' (Ch. 8).[28] What is perhaps less well known is that the tables of the *Zīj* of al-Khwārizmī/Maslama used sidereal coordinates and therefore allowed for trepidation.[29] It is not at all certain that this subtlety was appreciated by the early translators and users of the tables, many of whom appear to have struggled with the mathematical and calendrical calculations involved.[30] However, it was impossible to miss the fact that the tables for five of the planets included figures for retrogradation. Indeed, chapter thirteen of the introductory 'canons' is entitled 'On the Station and Direct and Retrograde Motion of the Planets' and explains how the figures for 'stations' are found and used, and how the chosen planet's direction of motion at a given date is established. Chapter fourteen moves on to the calculation of the period of time during which the planet will be retrograde.[31]

[28] On Thebit, see F. J. Ragep, 'Thābit's Astronomical Works', *Journal of the History of Astronomy*, 23 (1992), 61–3; and his, 'Al-Battānī, Cosmology and the Early History of Trepidation in Islam', in J. Casulleras and J. Samsó (eds.), *From Baghdad to Barcelona, Studies in the Islamic Exact Sciences in Honour of Prof. Juan Vernet*, Anuari de Filologia, XIX, 2 vols (Barcelona: Instituto 'Millás Vallicrosa' de Historia de la Ciencia Arabe, 1996), i. 267–98. For broader discussion, see J. Dobrzycki, 'The Theory of Precession in Medieval Astronomy', in J. Wlodarczyk and R. L. Kremer (eds.), *Selected Papers on Medieval and Renaissance Astronomy* (Warsaw: Polish Academy of Science, 2010), 15–60.

[29] See Chabás, 'Aspects of Arabic Influence', 27. For broader commentary on the presence of tables relating to trepidation in Andalusī zijes, see D. A. King and J. Samsó, 'Astronomical Handbooks and Tables from the Islamic World (750–1900): An Interim Report', *Suhayl*, 2 (2001), 9–105, at 25.

[30] This is clear from Mercier's findings in 'Astronomical Tables', which show that even a teacher such as Petrus Alfonsi made errors in adapting the tables.

[31] For a translation, see Neugebauer, *The Astronomical Tables*, 32.

INTELLECTUAL INHERITANCES 179

This *zīj* is the one which was translated and brought to England by Adelard of Bath, and was in use at Worcester by 1140, as shown by the manuscript which is now Bodleian Auct. F. 1.9. This is partly in the hand of the monastic chronicler, John of Worcester, and contains a full version of the tables, with annotations probably by John himself. The annotations demonstrate the difficulties experienced in converting dates in the tables into the Christian era and Julian calendar. Further work on the *zīj*, to adapt it for use in England, is demonstrated by the revised version attributed to Robert of Chester.[32] This reorganizes the canons and tables in such a way that related material is grouped together, and has some calendar tables not found in the main version, including a useful one for converting from Arabic to Latin dates. It survives in a manuscript now Madrid Bib. Nac. 10,016, which also provides corrections for the meridian of London. Robert of Chester's expertise was sufficient to enable him to compile tables for London, based on those compiled by Abraham Ibn Ezra for Pisa.[33] Confusingly, these used the canons for the al-Khwārizmī/Maslama Tables even though Ibn Ezra and the London Tables used data apparently from the *zijes* of al-Battānī and al-Ṣūfī.[34] The work of al-Battāni (858–929 CE) was also translated separately by Robert of Ketton. It drew on Ptolemy's *Almagest* and followed Ptolemy on precession. It therefore gave tropical coordinates for the positions of the planets and the vernal point rather than the sidereal ones given by the dominant tables.[35]

[32] Mercier, 'Astronomical Tables', 96–7. A planetary *theorica* datable to the late twelfth century is also linked to the London Tables. See F. S. Pedersen, 'A Twelfth-Century Planetary Theorica in the Manner of the London Tables', *Cahiers de l'Institut du Moyen-Age Grec et Latin*, 60 (1990), 199–318.

[33] Robert's work was known in England in the thirteenth century, although confusion had already arisen. Bodleian Ms Savile 21, which also contains copies of astronomical and astrological materials putatively in the hand of Grosseteste, includes on ff. 91–5v a copy of the short 'canons' for the London Tables. The tables themselves are missing, but the canons are the same as those accompanying the London Tables in B.L. Ms Arundel 377, ff. 7–35. The MS Savile 21 copy is preceded, on ff. 86–91, by a rather confused text which states that the London Tables were composed by Robert of Chester using the *zīj* of al-Battānī, even though this 'accompanying' text refers to tables for Toledo, using the Latin calendar, and starting in 1169. For discussion, see Mercier, 'Astronomical Tables', 96–7.

[34] See Mercier, 'Astronomical Tables', 96–7. For evidence of the complex inter-relationships between 'editions' of such texts, see Pedersen, 'Planetary Theorica', 211.

[35] For al-Battānī's *zīj*, see *Al-Battani sive Albatenii Opus Astronomicum*, ed and trans. C. A. Nallino, 3 vols. (Milan: Ulrico Hoepli, 1899–1907).

180 THE SCIENTIFIC WORKS OF ROBERT GROSSETESTE

Nothing is known of the means by which Roger of Hereford obtained his training, but he clearly knew and understood Raymond of Marseilles' early adaptation of the Tables of Toledo, as well as advanced astrological techniques. He was also able to apply what appears to have been an independently calculated adjustment for the meridian of Hereford.[36] The Toledan Tables themselves were put together by about 1080 and for the meridian of Toledo, as the name suggests. The astronomers working on the project included Ibn al-Zarqāllluh (d. *c.*1100 CE), and usually named as the author of the tables in the Latin versions. They brought together components from both al-Battānī's and al-Khwārizmī's works, as well as using new observations for some tables, especially for the sun. Perhaps surprisingly, they also explicitly adopted a fuller and more complex version of the trepidation theory. Accordingly, the mean motions for the sun and all the other planets are expressed in sidereal longitudes rather than tropical ones.[37] This model is also the one propounded in Grosseteste's *On the Sphere*, which makes it all the more striking that the planetary movements to which it was so closely linked are not mentioned there.

However, Grosseteste's acceptance of this 'modern' theory does not appear to mean that he adopted all the terminology found in the treatise attributed to Thābit ibn Qurra. That treatise uses the technical term 'eighth sphere' to refer to the sphere of the fixed stars. The same term is used in the Toledan Tables. No mention of this term is made in *On the Sphere*. That Grosseteste was acquainted with astronomical tables and their complexities is demonstrated not only by the confidence with which he refers to the calculation of planetary positions in *On the Liberal Arts* but also by his contributions to the compound manuscript, now Bodleian Library MS Savile 21 (Ch. 2, §1). This point is worth emphasizing since some extended remarks in the later *Hexaemeron*

---

[36] The location of Hereford is given in Madrid Bib. Nac. 10,016, the only complete copy of his work (ff. 73–85). Roger's comments on the solar eclipse of September 1178 provide a terminus *post quem* for his work. Roger's version of the tables is not edited by O. Pedersen in *The Toledan Tables*. However, Pedersen deals with over 200 manuscripts of the tables and provides translations of most of the canons edited in volume two.

[37] G. J. Toomer, 'A Survey of the Toledan Tables', *Osiris*, 15 (1968), 5–174, discusses over a hundred manuscripts and the variations in the tables included. A further set of tables made it possible to adjust to tropical longitudes.

INTELLECTUAL INHERITANCES   181

should not be taken as condemning all forms of astrology. The claims of judicial astrologers are rejected, on both theological and practical grounds.[38] On the practical level, however, Grosseteste is clear that their work involves the calculation of planetary positions, aspects, celestial houses, and ascendancies, as well as much more, and that all these must be established for specific places and dates. All this is extremely difficult, if not impossible, to do to the level of precision required. Of course, any idea that the planets dictate human actions is heretical. It is worth noting the greater use of the technical language of astrology in the *Hexaemeron, c.*1235, than in *On the Liberal Arts*, from the turn of the thirteenth century, implying further study over the intervening years. Such work would be very relevant to *On the Sphere*, whose time of writing falls between the two, and to its probable place in Grosseteste's career.

Interestingly, Grosseteste's example of the difficulties experienced by those attempting astrological calculations is drawn not from judicial astrology but from the much more acceptable form of astrology known as 'mundane' or world astrology. The implication may be that Grosseteste was unwilling to make considered judgements on any topic unless he had studied it; study of judicial astrology would clearly be repugnant. The chosen example outlines the complexities of drawing up a chart for the 'revolution of the year' (in other words, the moment of the sun's entry into the sign of Aries, which was closely, if complexly, related to the exact timing of the vernal equinox). As Grosseteste asserts, astronomers could not 'exactly identify this revolution of the year for a given location'. Grosseteste goes further and says that it is also not possible to calculate the exact positions of the planets at any given moment and that 'this is clear to those who have laboured hard with astronomical calculations and tables'.[39] Whether the labour was Grosseteste's own is not actually stated, but it is likely that the comment was based on direct observation if not personal experience.

In this connection items in MS Savile 21 associated with Grosseteste's handwriting and which is datable in part to 1216 by the two horoscopes

---

[38] Grosseteste, *Hexaemeron*, 5.VIII–X, ed. R. Dales and S. Gieben (Oxford: Oxford University Press, 1982).
[39] Grosseteste, *Hexaemeron,* 5.VIII–X.

182   THE SCIENTIFIC WORKS OF ROBERT GROSSETESTE

become important (see Ch. 2, §1). The horoscopes could not have been drawn up without consultation of astronomical tables. Equally, whoever produced them may well have been interested in the influential theories within mundane astrology which concerned the significance of conjunctions involving the outer planets, Saturn, Jupiter, and Mars; and the inter-relations of the planets at the time of the sun's entry into Aries. As noted Ch. 2, §2) the central box of the upper diagram (used to record the subject under investigation) refers to a conjunction of Saturn and Mars and gives the date for the equinox on 22nd March 1216. The lower chart itself shows a conjunction of Saturn and Mars in Libra and accords with the inscription added into the upper chart. The date for this significant conjunction would translate as 4 October 1216, a date on which the sun was also in Libra, further strengthening Saturn which was especially powerful in Libra according to standard textbooks.[40] Such a combination would be rather ominous, unless tempered by more positive factors such as Jupiter and the sun, and/or shown to be narrow in focus by the contextual data provided by the upper chart.

The date for the upper chart itself is clearly intended to be that of the sun's entry into Aries. The information provided in the Julian calendar establishes it as 22nd March 1216 (see Ch. 2, §1.2). Moreover, the chart shows 0° Aries at the top, and is concerned with marking the positions of the planets at the Spring turning point of the year. The level of detail of the information given demonstrates the hard work done with the tables. The conjunction shown in the lower chart demonstrates interest in another central issue in mundane astrology, namely that of conjunctions of the outer planets (and here the two malefics, ominously). Prognostications of major events, such as floods, famines, and periods of fertility or barrenness for agriculture, required the consultation of such pairs of charts, since the revolution of the year, for any given location, provided crucial contextual information against which the impact of significant conjunctions of the powerful outer planets was interpreted.[41]

---

[40] See especially Ptolemy, 'Of Exaltations', *Tetrabiblos*, book 1, cap. 19, ed. and trans. F. E. Robbins (Cambridge, MA: Harvard University Press, 1940), 88–91.

[41] Key works were: Māshā'allāh's *On the Revolutions of the Years of the World*; the treatise *On the Eclipses of Luminaries, Conjunctions of Planets, and Revolutions of Years* which was attributed to Māshā'allāh; Abū Ma'shar's *On the Great Conjunctions/Book of Religions and*

In the case of the horoscopes in MS Savile 21 it is unlikely that their pairing was coincidental, especially given the probable identification of Grosseteste's hand here. The investigation may perhaps have been a means of testing theories in mundane astrology, as well as an exercise in using the Tables. Given political events in England in 1216 (see Ch. 1, §2) it is of course also possible that the related topic of the coming fortunes of the ruler was a matter of interest (and John died on 19 October) although this was precisely the point at which mundane astrology began to share the problems associated with judicial astrology. The work displayed in these horoscopes is painstaking in its charting of the planets, although the upper one required an additional explanation, which is interesting. The dates appear to coincide at least roughly with Grosseteste's work on *On the Sphere* while the problems and misunderstandings show strong echoes of the practical problems emphasized in the later *Hexaemeron*.

Given the standard modern view of astrology as a superstitious and anti-intellectual pseudo-science, it is worth looking briefly at what was involved in drawing up these deceptively simple diagrams. In order to calculate the positions of the planets in relation to the zodiac signs, the astrological houses and their cardinal points, and to one another, several parts of the Toledan Tables would need to be used. First, work with the tables for chronology and calendars would be needed to establish the equivalent dates. Then followed consultation of the tables of mean motions for each planet, to establish its true longitude for the date in question. Full calculation of the houses required use of the tables of ascensions and of houses (although it could also be done by using an astrolabe). If all the data produced was to be fully interpreted, then the tables of the dignities of the planets in each of the signs and their subdivisions (such as decans and terms) were also desirable.[42] Finally, if the full levels of power of the planets involved in the conjunction was to

---

Dynasties; and the *Flores* attributed to the same author. See Abū Maʿshar, *On Historical Astrology (On the Great Conjunctions)*, ed. and trans. K. Yamamoto and C. Burnett, 2 vols. (Leiden: Brill, 2000).

[42] For a survey of astrological terminology, see the very useful compilation by Richard Dunn, 'Glossary', in Richard Dunn, Silke Ackermann, and Giorgio Strano (eds.), *Heaven and Earth United: Instruments in Astrological Contexts* (Leiden: Brill, 2018), 263–76.

184   THE SCIENTIFIC WORKS OF ROBERT GROSSETESTE

be established, then the aspects of each with all the other planets, and their relative placings and interactions, would need to be considered.

The charts in MS Savile 21 are not accompanied by theoretical texts. However, the basic idea that conjunctions of the outer planets had special significance in long term patterns and cycles of history was well established in medieval Europe by the early thirteenth century.[43] The 'great conjunctions' (those of Jupiter and Saturn) occur every twenty years, and form particularly significant patterns every 240 years ('greater conjunctions'). The 'greatest conjunctions' take place only every 960 years, when the zodiac position of the conjunction returns to one close to that of the first conjunction in the series. This pattern is obviously suspiciously regular and was undermined both by the complexities caused by retrograde motions and by inaccuracies in the tables. Nevertheless, it was given credibility by the apparent demonstration that these conjunctions coincided with or presaged known events, including such important ones as Noah's Flood, the birth of Christ, and the coming of Islam. This, together with the testimony of the Bible that God had created the luminaries in part as signs, and that the birth of Christ had been marked by the appearance of a clearly-located star, made it problematic to reject mundane astrology out of hand. Reception of the theory may have been helped by the fact that Seneca had referred to a version of it when he quoted 'Berosus the Babylonian' as saying that a conjunction of all the planets in Cancer or Capricorn would have disastrous consequences for the earth.[44] The Christian astrologer, Ibn Hibinta, who worked in Baghdad shortly after Māshā'allāh's death (c.815 CE) and preserved the fullest version of the latter's work on *Conjunctions, Religions and Peoples*, muddied the waters still further by referring to the work of 'Hermes' on conjunctions and world years.[45] Although Grosseteste's final conclusions appear to be the more negative ones set out in the *Hexaemeron*, the closer arguments for the identification of Grosseteste's handwriting in MS

---

[43] See J. North, 'Astrology and the Fortunes of Churches', *Centaurus*, 24 (1980), 181–211.
[44] D. Pingree, 'Masha'allah's Zoroastrian Historical Astrology', in G. Oestmann, H. D. Rutkin, and K von Stuckrad (eds.), *Horoscopes and Public Spheres: Essays on the History of Astrology* (Berlin: De Gruyter, 2005), 95–100. For Arabic text, translation, and commentary, see E. S. Kennedy and D. Pingree, *The Astrological History of Masha'allah* (Cambridge, MA: Harvard University Press, 1971; reissued 2013).
[45] Pingree, 'Masha'allah's Zoroastrian Historical Astrology'.

INTELLECTUAL INHERITANCES   185

Savile 21 demonstrate that there is indeed evidence of his hard work with calculations and tables. Moreover, this dates from a period which coincides with his work *On the Sphere*. These considerations make the absence of the planets in *On the Sphere* the more intriguing and strengthen the argument that this omission was made by Grosseteste at the service of a specific programme.

3 Tools for Astronomical Observation in Grosseteste's Lifetime

While *On the Sphere* itself is not a guide to carrying out observational astronomy, the observations of others are essential to its structure. In early medieval Europe, observing the heavens was to a large extent an unassisted pursuit, involving few if any artificial tools, but by the early thirteenth century a transformation had occurred with the transmission of new instruments from the Islamicate world. These accompany the expansion in astronomical learning embodied by new astronomical texts and tables. The absence of tools in the earlier period holds true especially if instruments where the observer's gaze was directed towards shadows or sunbeams cast against terrestrial objects rather than the sky itself are left aside. Items in the latter category included various types of sundials, referred to in medieval Latin generically as *horologia*.[46] Grosseteste himself deployed the term when he noted that the *horologium* varies across climes, as do the lengths of the daylight hours (DS §45). This may

---

[46] On early medieval *horologia*, see C. Philipp E. Nothaft, 'Bede's *Horologium*: Observational Astronomy and the Problem of the Equinoxes in Early Medieval Europe (c.700–1100)', *English Historical Review*, 130 (2015), 1079–101; Mario Arnaldi, 'Time Reckoning in the Latin World', in Anthony Turner (ed.), *A General History of Horology* (Oxford: Oxford University Press, 2022), 99–120. On the general subject of medieval astronomical instruments, see Emmanuel Poulle, 'Les instruments astronomiques de l'Occident latin aux XIe et XIIe siècles', *Cahiers de civilisation médiévale*, 15 (1972), 27–40; Emmanuel Poulle, 'L'astronomie du Moyen Âge et ses instruments', *Annali dell'Istituto e Museo di Storia della Scienza di Firenze*, 6 (1981), 3–16; Emmanuel Poulle, *Les sources astronomiques (textes, tables, instruments)* (Turnhout: Brepols, 1981), 22–43; Emmanuel Poulle, *Les instruments astronomiques du Moyen Âge* (Paris: Brieux, 1983); Emmanuel Poulle, 'L'instrumentation astronomique médiévale', in Bernard Ribémont (ed.), *Observer, lire, écrire, le ciel au Moyen Âge. Actes du colloque d'Orléans, 22–23 avril 1989* (Paris: Klincksieck, 1991), 253–81. It is worth noting that Poulle's interpretations occasionally differ from those suggested here.

186 THE SCIENTIFIC WORKS OF ROBERT GROSSETESTE

be no more than a reference to the fact that most sundials and their configurations are latitude-specific (rather than universal) and that their accuracy will deteriorate accordingly if they are used at latitudes for which they were not constructed. At the same time, it is possible that Grosseteste had in mind a more abstract notion of *horologium*, as attested in computistical and astronomical manuscripts of the eighth and later centuries. In these sources, the term *horologium* sometimes serves as a label for graphic representations of the path of the sun and the changing hour-length in the course of the year as well as for schemes for the length of a shadow.[47] Here, too, the key point would have been that the information shown in these diagrams is sensitive to local latitude.

A time-keeping device that involved some degree of actual stargazing was the *horologium nocturnum*, which a local Veronese tradition identifies as an invention made by the ninth-century cathedral canon Pacificus of Verona (d. 844). Its basic features can be inferred from some drawings contained in manuscripts of the tenth to twelfth centuries as well as an associated verse composition, which jointly indicate that the central component of the *horologium nocturnum* was a sighting tube that could be mounted on a support and directed at the celestial pole. The end of the tube was equipped with a disk, the perforations of which aided the observer in tracking the motion of one of the circumpolar stars as an indicator of the hours of the night.[48] An identifiable early medieval astronomer who appears to have had this nocturnal time-measurer in

[47] Examples are discussed in Carla Morini, 'Horologium e *daegmael* nei manoscritti anglosassoni del computo', *Aevum*, 73 (1999), 273–93; Barbara Obrist, 'The Astronomical Sundial in Saint Willibrord's Calendar and Its Early Medieval Context', *Archives d'histoire doctrinale et littéraire du Moyen Âge*, 67 (2000), 71–118; Karlheinz Schaldach, 'Gli "schemi delle ombre" nel Medio Evo latino', *Gnomonica Italiana*, 16 (2008), 9–16.

[48] Joachim Wiesenbach, 'Pacificus von Verona als Erfinder einer Sternenuhr', in Paul Leo Butzer and Dietrich Lohrmann (eds.), *Science in Western and Eastern Civilzation in Carolingian Times* (Basel: Birkhäuser, 1993), 229–50; Francesco Stella, 'Poesie computistiche e meraviglie astronomiche. Sull'*horologium nocturnum* di Pacifico', in Francesco Mosetti Casaretto and Roberta Ciocca (eds.), *Mirabilia. Gli effetti speciali nelle letterature del Medioevo* (Alessandria: Edizioni dell'Orso, 2014), 181–206; Francesco Stella, 'The Sense of Time in Carolingian Poetry: Christianizing the Zodiac and Astronomical Observation in Pacificus of Verona', in Pascale Bourgain and Jean-Yves Tilliette (eds.), *Le sens du temps. Actes du VIIe Congrès du Comité International de Latin Médiéval (Lyon, 10–13.09.2014)* (Geneva: Droz, 2017), 193–219; Laura Cleaver, *Education in Twelfth-Century Art and Architecture: Images of Learning in Europe, c.1100–1220* (Woodbridge: Boydell, 2016), 186–91; Francesco Bertola, 'Tubi astronomici', in Filippomaria Pontani (ed.), *Certissima signa: A Venice Conference on Greek and Latin Astronomical Texts* (Venice: Edizioni Ca' Foscari, 2017), 145–51, at 145–6.

INTELLECTUAL INHERITANCES   187

his arsenal was Gerbert of Aurillac, pope as Silvester II (999–1003), who reportedly built one during a stay with Emperor Otto III in Magdeburg in the 990s.[49]

Cultural contacts between Latin Europe and Muslim al-Andalus from the late tenth century onwards facilitated the introduction of new astronomical tools, the most prominent among which were the planispheric astrolabe and the horary quadrant. Features these instruments shared in common were that they were both portable and could be used to measure the altitudes of celestial objects. In the specific case of the horary quadrant, which appears to have been an Islamicate invention, such altitude measurements were aimed primarily at the sun, whose altitude was indicated by the position of a plumb line on the gradated quarter-circle and subsequently converted into the local time via a diagrammatic grid of hour-lines.[50] Later Latin versions of the quadrant, as first attested in the early thirteenth century, additionally incorporated a shadow square, which made it possible to measure distances, heights, and depths.[51]

The range of potential applications was even greater in the case of the astrolabe, which traces its origins back to the Hellenistic period.[52] Relying on the principles of stereographic projection, the astrolabe was

[49] Thietmar of Merseburg, *Chronicon* 6.100.(61), ed. Robert Holtzmann, MGH SS rer. Germ. N.S. 9 (Berlin: Weidmann, 1985), 392. See Patrick Gautier Dalché, 'Le "tuyau" de Gerbert, ou la légende savante de l'astronomie. Origines, thèmes, échos contemporains (avec un appendice critique)', *Micrologus*, 21 (2013), 243–76.

[50] J. Millàs Vallicrosa, 'La introducción del cuadrante con cursor en Europa', *Isis*, 17 (1932), 218–58; Catherine Jacquemard, Olivier Desbordes, and Alain Hairie, 'Du quadrant *vetustior* à l'*horologium viatorum* d'Hermann de Reichenau. Étude du manuscrit Vaticano BAV Ott. lat. 1631, f. 16–17ᵛ', *Kentron*, 23 (2007), 79–125; C. Philipp E. Nothaft, 'An Overlooked Construction Manual for the *Quadrans Vetustissimus*', *Nuncius*, 34 (2019), 517–34. On the Islamic origins, see King, *In Synchrony with the Heavens*, vol. 2, 111–258.

[51] Nan L. Hahn, *Medieval Mensuration: 'Quadrans Vetus' and 'Geometrie Due Sunt Partes Principales*', Transactions of the American Philosophical Society 72.8 (Philadelphia, PA: American Philosophical Society, 1982); Wilbur R. Knorr, 'Sacrobosco's *Quadrans*: Date and Sources', *Journal for the History of Astronomy*, 28 (1997), 187–222; C. Philipp E. Nothaft, 'The *Liber theoroumacie* (1214) and the Early History of the *Quadrans Vetus*', *Journal for the History of Astronomy*, 51 (2020), 51–74.

[52] On the origins of the astrolabe, see Otto Neugebauer, 'The Early History of the Astrolabe: Studies in Ancient Astronomy IX', *Isis*, 40 (1949), 240–56; Flora Vafea, 'From the Celestial Globe to the Astrolabe: Transferring Celestial Motion onto the Plane of the Astrolabe', *Medieval Encounters*, 23 (2017), 124–48. On its introduction into medieval Europe, see Charles Burnett, 'King Ptolemy and Alchandreus the Philosopher: The Earliest Texts on the Astrolabe and Arabic Astrology at Fleury, Micy and Chartres', *Annals of Science*, 55 (1998), 329–68; Arianna Borrelli, *Aspects of the Astrolabe: 'Architectonica Ratio' in Tenth- and Eleventh-Century Europe*, Sudhoffs Archiv: Beihefte, 57 (Stuttgart: Steiner, 2008).

188 THE SCIENTIFIC WORKS OF ROBERT GROSSETESTE

a two-dimensional representation of the spherical heavens capable of simulating the daily rotation of the firmament. It did so by having its perforated disc, the so-called *rete*, represent the ecliptic and the positions of bright stars, and able to rotate above an inscribed plate representing the local horizon. The combination of *rete* and horizon plate could be applied computationally to a variety of astronomical problems, which included finding the current time as a function of the altitude of the sun or a particular fixed star. Such altitudes could be measured using the *alidade* or sighting-tool mounted on the astrolabe's back. Equipped with two perforated sighting vanes, the alidade allowed taking the solar altitude without having to look directly into the sun, by letting a sunbeam pass through the small holes on both vanes. In addition to the alidade, the backplate of most astrolabes featured other functions, among them a circular calendar scale and the aforementioned shadow square. A creative application of the astrolabe is recorded in a text by Walcher of Malvern, who determined the time of the lunar eclipse of 18 October 1092 by taking the altitude of the moon and using it as input in computing the seasonal and equinoctial hour on his astrolabe. It is the earliest astronomical observation involving this instrument for which evidence survives from medieval Europe.[53]

The popularity of the astrolabe in eleventh- and twelfth-century Latin Europe is indicated by the large number of texts devoted to its construction and use. In addition to various translations from Arabic, new texts were being produced by twelfth-century authors such as Raymond of Marseilles (shortly before 1141), Rudolf of Bruges (1144 or after), and Adelard of Bath (1149/50).[54] It is worth noting that some of these works envisaged observational functions that went beyond the narrow purpose

---

[53] See C. Philipp E. Nothaft (ed.), *Walcher of Malvern*: De lunationibus *and* De Dracone; *Study, Edition, Translation, and Commentary* (Turnhout: Brepols, 2017), 31–6, 114–17, 240–5.

[54] On medieval astrolabe literature, see André Van de Vyver, 'Les premières traductions latines (X^e–XIe s.) de traités arabes sur l'astrolabe', in F. Quicke, P. Bonenfant, Y. Barjon, and L. Jadin (eds.), *1er Congrès international de géographie historique sous le haut patronage de S. M. le Roi de Belges*, Vol. 2 of 2 vols. (Brussels: Secrétariat Général, 1931), 266–90; Emmanuel Poulle, 'L'astrolabe médiéval d'après les manuscrits de la Bibliothèque Nationale', *Bibliothèque de l'École des Chartes*, 112 (1954), 81–103; Emmanuel Poulle, 'Le traité de l'astrolabe d'Adélard de Bath', in Charles Burnett (ed.), *Adelard of Bath: An English Scientist and Arabist of the Early Twelfth Century* (London: Warburg Institute, 1987), 119–32; Paul Kunitzsch, 'Glossar der arabischen Fachausdrücke in der mittelalterlichen europäischen Astrolabliteratur', *Nachrichten der Akademie der Wissenschaften in Göttingen*, phil. hist. Kl., Jg. 1982, Nr. 1 (1982), 459–571; David

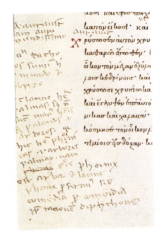

Plate 1. Clockwise from top left: Oxford, Bodleian Library, MS Savile 21, f. 156v, CC-BY-NC 4.0; detail from Cambridge, University Library, MS Ff.1.24, f. 42v, reproduced by kind permission of the Syndics of Cambridge University Library; Cambridge, Corpus Christi College, MS 480, f. 9v, reproduced by kind permission of The Parker Library, Corpus Christi College, Cambridge; detail from Cambridge, University Library, MS Ff.1.24, f. 25r, reproduced by kind permission of the Syndics of Cambridge University Library. Images are not to scale.

Plate 2. Cambridge, Pembroke College, MS 7, f. ii v, reproduced by permission of the Master and Fellows of Pembroke College, Cambridge. Image is not to scale.

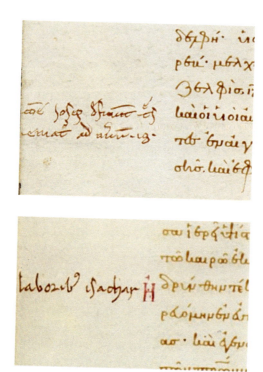

Plate 3. Details from Cambridge, University Library, MS Ff.1.24, f. 107v (above) and Cambridge, University Library, MS Ff.1.24, f. 229v (below). Reproduced by kind permission of the Syndics of Cambridge University Library. Images are not to scale.

Plate 4. Oxford, Bodleian Library, MS Savile 21, f. 157r, CC-BY-NC 4.0. Image is not to scale.

Plate 5. Examples of Grosseteste's Hindu-Arabic numerals in Oxford, Bodleian Library, MS Savile 21, ff. 143r-160v, CC-BY-NC 4.0 (top) and details from Cambridge, University Library, MS Ff.1.24, ff. 203r-261v (bottom), reproduced by kind permission of the Syndics of Cambridge University Library. Images are not to scale.

INTELLECTUAL INHERITANCES 189

of keeping time by measuring altitudes, for instance the use of the stars on the *rete* to investigate the longitudes of planets. According to Raymond of Marseilles, the latter sort of technique was useful for testing the accuracy of astronomical tables, provided the instrument itself was reliable.[55] Indeed, as Raymond himself understood, the accuracy of the method depended strongly on the correct placement of the stars on the astrolabe's rete, which was subject to precession.[56]

Another debilitating factor in the reliability of astrolabes was their typically diminutive size. In order to obtain dependable and precise observational data, astronomers were well advised to turn to some of the larger-scale instruments already employed in antiquity by Ptolemy. These included the mural quadrant, which in contrast to the portable horary quadrant was fixed in the plane of the meridian. Solar altitudes were here measured by having an orthogonal peg cast a shadow against a scale of 90°. One of the first Latin texts to give a detailed account of this technique was Ibn al-Muthannā's commentary on the astronomical tables of al-Khwārizmī, as translated by Hugo Sanctelliensis in the first half of the twelfth century.[57]

For the observation of ecliptic coordinates of stars or planets, Ptolemy's principal tool had been an armillary sphere equipped with sighting implements, which in Latin Europe became known simply as the *instrumentum armillarum*. The ecliptic was here represented by a ring that could be oriented to lie exactly in the path of the sun. Based on the known longitude of the sun, a number of operational steps made it possible to use sighting vanes attached to the innermost ring to

Juste, 'Hermann der Lahme und das Astrolab im Spiegel der neuesten Forschung', in Felix Heinzer and Thomas Zotz (eds.), *Hermann der Lahme. Reichenauer Mönch und Universalgelehrter des 11. Jahrhunderts* (Stuttgart: Kohlhammer, 2016), 273–84.

[55] Raymond of Marseilles, *Tractatus astrolabii*, 16b, ed. with French translation in Marie-Thérèse d'Alverny, Charles Burnett, and Emmanuel Poulle, *Raymond de Marseille: Opera omnia* (Paris: CNRS Éditions, 2009), 92. 'Juxta hunc modum scire poteris de cursuum libris utrum veri an falsi sint, ita dico si verax instrumentum habueris.'

[56] Raymond of Marseilles, *Tractatus astrolabii*, 1g, 7a–b, ed. d'Alverny, Burnett, and Poulle, 54, 70–2.

[57] See the edition in Eduardo Millás Vendrell (ed.), *El comentario de Ibn al-Muṭannā a la Tablas Astronómicas de al-Jwārizmī. Estudio y edición crítica del texto latino, en la versión de Hugo Sanctallensis* (Madrid: CSIC, 1963), 131–2. Another description appears in al-Battānī, *De motu stellarum*, trans. Plato of Tivoli, c.57, in *Continentur in hoc libro: Rudimenta astronomica Alfragani. Item Albategnius astronomus peritissimus de motu stellarum* (Nuremberg: Petreius, 1537), fols. 88r–89r.

190 THE SCIENTIFIC WORKS OF ROBERT GROSSETESTE

determine the position of the moon, fixed stars, and planets.[58] Little is currently known about the role this instrument may have played in Latin astronomy up to Grosseteste's day. An attested case from towards the end of his lifetime is a star table composed by John of London for the year 1246, which shows longitudes and latitudes of bright stars determined, allegedly, not by computation, but *per instrumentum armillarum*.[59]

Besides the armillary astrolabe, Ptolemy mentions and describes a host of other instruments he used to obtain astronomical data.[60] These include the equatorial ring for determining the time of the equinoxes and the parallactic rulers (later known as *triquetrum*) designed for measuring zenith distances. Evidence that these two instruments were put to any practical use in Europe by the time Grosseteste wrote *De sphera* is lacking. It is different for the *camera obscura* or pinhole camera, the principles behind which were first accurately described by Ibn al-Haytham.[61] Its use in observing solar eclipses, without having to look directly into the sun, is recommended at the end of a comprehensive explanation of the theory and practice of astronomical tables (inc. *Diversi astrologi secundum diversos annos...*), which is attributed to Roger of

---

[58] Friedrich Nolte, *Die Armillarsphäre*, Abhandlungen zur Geschichte der Naturwissenschaften und der Medizin 2 (Erlangen: Mencke, 1922); Jarosław Włodarczyk, 'Observing with the Armillary Astrolabe', *Journal for the History of Astronomy*, 18 (1987), 173–95; James Evans, *The History and Practice of Ancient Astronomy* (New York: Oxford University Press, 1998), 125–7, 255–6.

[59] Paul Kunitzsch, *Typen von Sternverzeichnissen in astronomischen Handschriften des zehnten bis vierzehnten Jahrhunderst* (Wiesbaden: Harrassowitz, 1966), 39–46; Kunitzsch, 'John of London and His Unknown Arabic Source', *Journal for the History of Astronomy*, 17 (1986), 52–7; Elly Dekker, 'A Close Look at Two Astrolabes and Their Star Tables', in Menso Folkerts and Richard Lorch (eds.), *Sic Itur ad Astra. Studien zur Geschichte der Mathematik und Naturwissenschaften; Festschrift für den Arabisten Paul Kunitzsch zum 70. Geburtstag* (Wiesbaden: Harrassowitz, 2000), 177–215, at 191–4, 214–15.

[60] See D. R. Dicks, 'Ancient Astronomical Instruments', *Journal of the British Astronomical Association*, 64 (1953–4), 77–85; Derek J. Price, 'Precision Instruments: To 1500', in Charles Singer, E. J. Holmyard, A. R. Hall, and Trevor I. Williams (eds.), *A History of Technology*, Vol. 3, *From the Renaissance to the Industrial Revolution c.1500–c.1700* (Oxford: Oxford University Press, 1957), 582–619, at 586–92; James Evans, 'The Material Culture of Greek Astronomy', *Journal for the History of Astronomy*, 30 (1999), 237–307, at 272–83; Dennis W. Duke, 'Ptolemy's Instruments', in Alan C. Bowen and Francesca Rochberg (eds.), *Hellenistic Astronomy: The Science in Its Contexts* (Leiden: Brill, 2020), 246–58.

[61] David C. Lindberg, 'The Theory of Pinhole Images from Antiquity to the Thirteenth Century', *Archive for History of Exact Sciences*, 5 (1968), 154–76.

INTELLECTUAL INHERITANCES 191

Hereford in several manuscripts, but to a certain Robert of Northampton in the earliest known one, roughly datable to the thirteenth century.[62]

4 Latin Traditions

Alongside observational practice multiple models for discussion of astronomy and the nature of the world within the Latin tradition would have been available to Grosseteste at the time that he composed *On the Sphere*. Most find no place, or only a specific role, in Grosseteste's thinking showing how distinctive was his own programme. This is the case not only for personal observation (Ch. 6, §4.3) but also for allegorical descriptions of the formation of the universe. Although familiar with Macrobius and Martianus Capella, Grosseteste does not engage with their mode of expression: there is no personification of nature or mingling of the classical pantheon and the Christian doctrine of creation from nothing, *ex nihilo*, through the Trinity.[63] Nor does the *Timaeus*, in its Latin reception, form part of his conceptual framework for cosmology; what Grosseteste articulates in *On the Sphere* is far removed from the platonic and neo-platonic notions of what astronomy is and does.[64] To that extent, Grosseteste's exploration of the world eschews the resurgence of treatises in these veins from the early-twelfth century onwards. The commentaries and glosses of William of Conches (d. after 1154) on Macrobius, on Plato, and on Boethius, as well as his *Philosophy of the World* and *Dragmaticon*, are not to be found in Grosseteste's *On the*

---

[62] For the relevant passage, see MSS Oxford, Bodleian Library Savile 21, ff. 42r–60v (s. XIII; attributed to Robert of Northampton), at f. 60v; Oxford, Bodleian Library Digby 168, f. 69vb–83vb (s. XIV, attributed to Roger of Hereford), at f. 83vb; Oxford, Bodleian Library Bodley 300, ff. 1ra–19va (s. XIV/XV; attributed to Roger of Hereford), at f. 19va; Paris, Bibliothèque nationale de France, lat. 15,171, ff. 136r–57v (s. XIV/XV; no attribution), at f. 157v. The passage is quoted in José Luis Mancha, 'Astronomical Use of Pinhole Images in William of Saint-Cloud's *Almanach planetarum* (1292)', *Archive for History of Exact Sciences*, 43 (1992), 275–98, at 275.

[63] On Macrobius and Grosseteste, see Grosseteste, *De artibus liberalibus*, §§3–4, 10; on Martianus and Grosseteste, see Grosseteste, *De artibus liberalibus* §10; *De generatione sonorum*, §§6, 8.

[64] See Calcidius, *On Plato's Timaeus*, ed. and trans. John Magee (Cambridge, MA: Harvard University Press, 2016). For the medieval reception, see William of Conches, *Glossae super Platonem*, ed. E. A. Jeauneau, CCCM 203 (Turnhout: Brepols, 2006), and Édouard Jeauneau, *Rethinking the School of Chartres*, trans. Claude Paul Desmarais from an unpublished original text in French (Toronto: University of Toronto Press, 2009).

192   THE SCIENTIFIC WORKS OF ROBERT GROSSETESTE

Sphere, nor the works of Thierry of Chartres (d. before 1155). Works such as Bernardus Sylvestris's (d. 1178) *Cosmographia*, with its prosimetric invocation of the macro- and microcosmos, and Alan of Lille's (*c*.1128–1202/3) *Plaint of Nature* and *Anti-Claudianus* equally are absent.[65]

4.1 The World Soul and the Elements

An instructive example of Grosseteste's different approach occurs in his use of the phrase world soul [*anima mundi*] in *On the Sphere*, concerning the source of cosmic motion:

Upon these two poles, as we have said, the heavens revolve with all the stars and planets, which are in this same continuously equal and uniform motion once per day and night, and the efficient cause of which movement is the *world soul.* (DS §10)

What might appear, at first glance, like a rather a benign remark might, in the context of the rather turbulent history of the concept of the *anima mundi* in the century previous to Grosseteste's writing, be regarded instead as bold statement. The elision of the *anima mundi* with the Holy Spirit was among the list of heresies attributed to Peter Abelard (*c*.1079–1142) at the Council of Soissons (1121).[66] Abelard's contemporary William of Conches was reprimanded on similar grounds, promoting the World Soul as coterminous with the third person of the Trinity, an idea that he shared with major figures such as Bernardus Silvestris and Thierry of Chartres. Where in William's earliest works such as the *Glosses on Macrobius* his claims for the *anima mundi* were limited to its identification as a vivifying force by the *Glosses on the Consolation of Boethius* he asserts that this force was no less than the Holy Spirit. This text drew the condemnation of no less a figure than St Bernard of

---

[65] Bernardus Silvestris, *Cosmographia*, ed. Peter Dronke (Leiden: Brill, 1978); Alan of Lille, *Anticlaudianus*, ed. Robert Bossuat (Paris: Vrin, 1955); and Alan of Lille, *De planctu Naturae*, ed. Nicholas Häring, *Studi Medievali*, ser. 3, 19 (1978), 797–879.

[66] D. Luscombe, 'The Sense of Innovation in the Writings of Peter Abelard', in H. J. Schmidt (ed.), *Tradition, Innovation, Invention* (Berlin: De Gruyter, 2005), 181–94.

INTELLECTUAL INHERITANCES   193

Clairvaux and William's later work, *Dragmaticon* was free from any reference to the subject.[67]

From the mid-twelfth century onwards discussion of the world soul is uncommon. It was raised, however, in Alexander Neckham's notes on Martianus Capella's *The Marriage of Philology and Mercury* of 1182. Alexander records the heresy of identifying the world soul with the third person of the Trinity while making it plain that he himself does not subscribe to any such an assumption.[68] Neckham spends no great time with the idea, and it goes unmentioned in his later encyclopaedic works *On the Nature of Things* and *In Praise of Divine Wisdom*. Aside from Neckham, it would seem that the only other author audacious enough to rehearse the theory was the anonymous late-twelfth-century author of the Berlin commentary on Martianus, whose intellectual debts seem closer to those of William of Conches.[69] Defenders of the Platonic world soul only emerge once more at the end of the thirteenth century, and even then in a limited way. Robertus Anglicus felt able to resurrect the idea in his 1271 commentary on Sacrobosco's *On the Sphere*, presenting a soul that is the cause of the heaven's uniform motion.[70] Some decades after this Nicholas Trevet's commentary on Boethius' *On the Consolation of Philosophy* makes a similar connection between celestial motion and the world soul, which he describes as the motivating power of the two heavenly orbs that moves all corporeal things. He does however take pains to state clearly that not only was Plato ignorant of the Holy Spirit, but also his admirers wrongly state that the World Soul equates with the Holy Spirit.[71] Since Grosseteste's identification of *anima mundi* as an efficient cause is clearly without pneumatological

---

[67] Bernard McGinn, 'The Role of the Anima Mundi as Mediator between the Divine and Created Realms in the Twelfth Century', in J. J. Collins and M. Fishbane (eds.), *Death, Ecstasy and other Worldly Journeys* (New York: State University of New York Press 1995), 285–316.

[68] Alexander Neckham, *Commentum super Martianum. Commentary on Martianus Capella's 'De Nuptiis Philogiae et Mercurii'*, ed. C. J. McDonough (Firenze: Sismel Edizioni del Gelluzzo, 2006).

[69] Anonymous, *The Berlin Commentary on Martianus Capella's De nuptiis Philologiae et Mercurii*, ed. Haijo Jan Westra (Leiden: Brill, 1994).

[70] L. Thorndike, 'Robertus Anglicus and the Introduction of Demons and Magic into the Commentaries upon the Sphere of Sacrobosco', *Speculum*, 21 (1946), 241–3 at 243.

[71] A. J. Minnis and Lodi Nauta, 'More Platonico Loquitur: What Nicholas Trevet Really Did to William of Conches', in A. J. Minnis (ed.), *Chaucer's Boece and the Medieval Tradition of Boethius* (Suffolk: D. S. Brewer, 1993), 1–33 at 21–2.

194 THE SCIENTIFIC WORKS OF ROBERT GROSSETESTE

association, it fits the tenor of the later-twelfth- and thirteenth-century thinkers better than those of the earlier twelfth century.

Other twelfth-century models for how to describe the world stress elemental composition, as, for example Hermann of Carinthia's *On Essences*, or the alchemical treatise *Clavis sapientiae* attributed to Artephius.[72] This was an area of interest to Grosseteste who covered the elements in some detail in later scientific works, notably *On the Impressions of the Elements*, and in a cosmological sense in *On Light*.[73] *On the Sphere*, however, does not deal with cosmology or geography in this way, and, although traces of the *Clavis* can be detected in Grosseteste's *On the Liberal Arts*, the structure and interest of his later treatise are very different. The same differences apply to the *Book on Inferior and Superior Nature* by Grosseteste's near-contemporary Daniel of Morley (d. 1210).[74] In a similar vein, although Grosseteste deals with some of the same topics, *On the Sphere* is not an encyclopaedic treatise *On the Nature of Things*, and accordingly does not draw on classical, early medieval, or Carolingian works of this title.[75]

4.2 A Tale of Two Spheres

By far the closest analogue to Grosseteste's *On the Sphere* is Sacrobosco's treatise of the same name. Both authors focus their attention on the

---

[72] Hermann of Carinthia, *De essentiis*, ed. and trans. Charles Burnett (Leiden: Brill, 1982); Artefius, *Clavis sapientiae*, in J.-J. Manget, Bibliotheca Chemica Curiosa, 2 vols. (Geneva: Sumpt. Chouet, G. De Tournes, Cramer, Perachon, Ritter, & S. De Tournes, 1702), 2. 2. 2, i. 503–9. See also M. Pereira, *Arcana sapienza* 87–93; and M. Pereira, 'Cosmologie alchemiche', in C. Martello, C, Militello, and A. Vella (eds.), *Cosmogonie e cosmologie nel medioevo. Atti del convegno della Società Italiana per lo Studio del Pensiero Medievale* (S.I.S.P.M.), Catania, 22–4 settembre 2006 (Turnhout: Brepols, 2008), 363–410.

[73] *On the Impressions of the Elements* will be included in Volume III of this series; the current critical edition is Baur, *Die Philosophischen Werke*, 87–9. Robert Grosseteste, *De luce*, ed. Cecilia Panti, 'Robert Grosseteste's De luce: A Critical Edition', in John Flood, James R. Ginther, and Joseph W. Goering (eds.), *Robert Grosseteste and his Intellectual Milieu* (Toronto: Pontifical Institute of Mediaeval Studies, 2013), 193–238. *On Light* will be included in Volume IV of this series.

[74] Daniel of Morley, *Philosophia/Liber de naturis inferiorum et superiorum*, ed. G. Maurach, *Mittellateinisches Jahrbuch*, 14 (1979), 204–55.

[75] Lucretius, *De rerum natura*, ed. Marcus Deufert (Berlin: De Gruyter, 2019); Isidore, *De natura rerum*, ed. and trans. Jacques Fontaine (Paris: Institut d'études augustiniennes, 2002, repr. of Bordeaux 1960 edit.); Bede, *De natura rerum*; Rabanus Maurus, *De universo libri viginti duo*, PL 111, cols. 9–614.

INTELLECTUAL INHERITANCES   195

shape, size, and workings of the celestial spheres and planetary bodies (in the medieval sense), before considering the measurement of the earth. Both treatises engage with the new astronomical learning available by the end of the twelfth century, and both offer programmatic and logical discussion of the subjects under scrutiny. A comparison to a schools text of the same period such as the Pseudo-Bede, *On the Constitution of the Celestial and Terrestrial Worlds*, dating from the twelfth century and copied and annotated up to the thirteenth, shows the greater clarity and precision of the way in which Grosseteste and Sacrobosco set out and discuss the uses of astronomy.[76]

The relationship between the two treatises *On the Sphere* is the subject of scholarly debate, especially relating to their relative chronology (Ch. 1, §1). That Sacrobosco's treatise predated Grosseteste's, as argued in Chapter 1, is taken as a fixed point in this discussion. It should, nevertheless, be pointed out that evidence for Sacrobosco's life is almost non-existent, leaving most of what can be said as suggestive rather than certain.[77] For the present purpose some of broader and more particular differences between the two treatises are reviewed, in order to underline what it is that Grosseteste set himself to explore and under what priorities he set out to teach on the heavenly spheres.

The two texts have different structures. Sacrobosco's is divided into four chapters, focusing on sphericity and the shape of the world; the composition of the celestial and sub-lunary spheres (and that the latter is a microcosm of the former); the ascent and descent of constellations and how different night and day are in different locations; and finally the motion and course of the planets and eclipses.[78] Grosseteste's is, in contrast, a continuous text though possessing a broad division into seven sections on sphericity; celestial and terrestrial coordinates of latitude and longitude and their celestial equivalents; the effects of solar

---

[76] Pseudo-Bede, *De mundi caelestis terrestrisque constitutione liber. La création du monde céleste et terrestre*, ed. Mylène Pradel-Baquerre, Cécile Biasi, and Amand Gévaudan (Paris: Classiques Garnier, 2016); this supersedes *De mundi celestis terrestrisque constitutione: A Treatise on the Universe and the Soul*, ed. and trans. Charles Burnett (London: Warburg Institute, 1985).

[77] O. Pedersen, 'In Quest of Sacrobosco', *Journal for the History of Astronomy*, 16 (1985), 175–220.

[78] Sacrobosco, *De sphera*, 76.

196 THE SCIENTIFIC WORKS OF ROBERT GROSSETESTE

movement on the duration of the day; the ascension of zodiacal constellations; the eccentricity of solar movement on the climate and duration of the day; trepidation of the equinoxes; and the motion of the moon, lunar and solar eclipses (see Ch. 7, §§3–7). Where Sacrobosco's text is studded with classical references, particularly in chapter three on the rising and setting of the zodiacal constellations, and especially from Lucan (*Pharsalia* (also known as the *Civil War*)), Virgil (*Georgics*), and Ovid (*Metamorphoses* and *From Pontus*), Grosseteste's, by contrast, is spartan in this regard. Sacrobosco also records mnemonic verses for those studying, again with a concentration in chapter three. Right and oblique ascensions within the zodiac, with six of each, are to be remembered with these lines:

> They rise aright, oblique descend from Cancer's star
> Till Chiron ends, but the other signs
> Are prone at birth, descend by a straight path.[79]

Grosseteste offers no such level of help for those following his treatise, the structure and flow of the text is, to this extent, more demanding.

What emerges here are differences not so much in terms of fundamental subject matter but of method and individual interest. Grosseteste's concerns within the motions of the spheres are overwhelmingly those which concern life on earth directly, in particular its northern hemisphere, and the appearance of the sky more than the structure of the heavens that gives rise to it. Other ostensibly minor but intriguing differences are Sacrobosco's mention of evidence for the curvature of the sea and the measurement of the circumference of the earth which are omitted by Grosseteste.[80] In a similar divergence Sacrobosco's general distinction between the superlunary and sub-lunary spheres in a discussion of the elements is not mirrored at all in Grosseteste. This sharpens the suggestion that Grosseteste's treatise had a slightly different aim and/or audience. There was, perhaps, no need to include information of this sort given its

---

[79] Sacrobosco, *De sphera*, 103/134 (English Translation): 'Recta meant, obliqua cadunt a sidere Cancri / Donec finitur Chiron; sed cetera signa / Nascuntur prono, descendunt tramite recote'.
[80] Sacrobosco, *De sphera*, 83 and 85.

INTELLECTUAL INHERITANCES   197

presentation in Sacrobosco's treatise if an active relationship from one text to the other is posited. Similarly, Sacrobosco attends to some considerable detail in the three types of 'rising' of constellations and planets, again implying an audience for whom basic, factual, knowledge was at premium.[81] A final difference in method is the deduction from the theory of solar eclipses, by Sacrobosco, that the Passion eclipse must have been miraculous. Grosseteste does not relate his material to issues of Christian doctrine, though the withdrawal of water in the sphere of earth, noted also in Sacrobosco and with the same notion that this was to create a home for terrestrial creatures, echoes the retreat of the waters in Genesis (DS §4).[82]

Another difference between Sacrobosco and Grosseteste is that it is only the former who treats explicitly, albeit briefly, the retrograde motion of the planets, and that along Ptolemaic lines.[83] Ptolemy covers this in books nine to thirteen, and particularly in book twelve of the *Almagest*, where it is explained through the notion of epicycles.[84] Grosseteste was familiar with Ptolemy's work with access in a number of different possible formats (see Ch. 6, §2). In fact, Grosseteste does make reference to the epicyclic hypothesis, but only in addressing the particular example of the epicycle of the moon. Here, however, any connection to retrograde motion is made problematic by the fact that the moon is the only planet with an epicycle that does *not* give rise to retrograde motion (see Ch. 9, §2.4). The absence of discussion of retrograde motion is, as noted above, unlikely to be from ignorance on Grosseteste's part. It was also, as already shown, a phenomenon raised by a number of ancient and medieval sources apart from Ptolemy with which Grosseteste was either familiar, or where familiarity may be suggested. These authors range widely in the extent of their interest and exposition of retrograde motion, and mostly discuss explanations through propulsion by solar rays, and the illusory nature of backwards motion, rather than epicycles.

However, despite this variance, there is a sufficient body of evidence to suggest that knowledge of, and engagement with, various possible mechanisms responsible for retrograde motion was common. Sources range from passing references in Lucan, Seneca, Macrobius, and Pliny, to a

---

[81] Sacrobosco, *De sphera*, 78–9 and 95–8.   [82] Sacrobosco, *De sphera*, 116–17.
[83] Sacrobosco, *De sphera*, 114.   [84] Ptolemy, *Almagest*.

198 THE SCIENTIFIC WORKS OF ROBERT GROSSETESTE

much fuller treatment in Martianus Capella's *Marriage of Philology and Mercury*.[85] Among the distinctive views Martianus advanced was his partially heliocentric ('Capellan') scheme in which Mercury and Venus orbit the sun, rather than the earth.[86] Retrograde motions are correctly assigned to the opposition of the outer planets, with respect to the sun, and attributed causally (through this coincidence) to the force of the sun's rays.[87] In a similar tradition Calcidius, in his translation and commentary on Plato's *Timaeus* referred to matters related to retrogradation a number of times. Most of the discussion leans towards an illusory interpretation of planetary retrograde motion.[88] Calcidius does, however, present epicycles and eccentrics, even referring to detailed geometric diagrams, although his primary example is the sun throughout most of the text, rather than the outer planets. He attributes to Aristotle the opinion that planets would not follow incorporeal paths such as epicycles, but does, eventually, offer full epicyclic explanation of planetary retrograde motion.[89]

Amongst Christian encyclopedists, Cassiodorus' *Institutions* was a significant source for the transmission of ancient teaching about planetary motions.[90] Cassiodorus demonstrated some knowledge of Ptolemy, through reference to the 'Canons' rather than the *Almagest*.[91] Moreover, he mentioned retrograde motion not so much in terms of mechanism,

---

[85] Lucan, *De bello civili libri X. Editio altera.*, ed. David R. Shackleton-Bailey (Teubner: Stuttgart/Leipzig 1997), X.201–20; Seneca, *Naturalium questionum libri*, ed. H. M. Hine (Leipzig: Teubner, 1996), 7.25.1, esp. 7.25; Macrobius, *Commentarii in Somnium Scipionis*, ed. James Willis (Leipzig: Teubner, 1970), XIX.27; Pliny, *Naturalis historia*, 2.59–61, ed. Ludwig von Jan and Karl Friedrich Theodor Mayhoff, 5 vols. (Leipzig: Teubner, 1897–1933; repr. 1967–70). Bruce Eastwood, *Ordering the Heavens: Roman Astronomy and Cosmology in the Carolingian Renaissance* (Leiden: Brill, 2007), 179–87 and 299–303. On Macrobius, Martianus, and Grosseteste, see footnote 63 above.

[86] Martianus Capella, *De nuptiis Philologiae et Mercurii*, ed. James Willis (Leipzig: Teubner, 1983), VIII.857.

[87] Martianus Capella, *De nuptiis Philologiae et Mercurii*, VIII.884.

[88] Calcidius, *On Plato's Timaeus*, 5. 69, 70, 77, 79, 85.

[89] Calcidius, *On Plato's Timaeus*, 5. 85.

[90] Cassiodorus, *Institutiones*, II. praef.2 ed. R. A. B. Mynors (Oxford: Oxford University Press, 1937).

[91] See Cassiodorus, *Institutions of Divine and Secular Learning and On the Soul*, trans. J. Halporn (Liverpool: Liverpool University Press, 2004), 227. The reference is probably to the *Preceptum Canonis Ptolomei*, see David Juste, 'Ptolemy, *Preceptum canonis Ptolomei* (tr. before c.1000)' (update: 18.03.2021), *Ptolemaeus Arabus et Latinus. Works*, http://ptolemaeus.badw.de/work/52 (accessed 13 July 2021).

INTELLECTUAL INHERITANCES   199

but in a list of nomenclature.[92] Isidore was more explicit in his reference to the phenomenon of retrograde motion in his *Etymologies*, and posed the question as to whether retrograde planetary motion was real or illusory.[93] Planetary motion he explored at greater length in *On the Nature of Things*.[94] The motion of the wandering stars, or planets, occurred according to unfixed rules, they were placed at diverse heights, so that those at a farther distance appeared to repeat their courses more slowly, yet nonetheless he asserted that all returned in their own time to complete their proper course.[95] Bede was less equivocal, though drawing heavily on Isidore, and stated in his own *On the Nature of Things* that the seven planets are observed to move with an opposite motion to that by which they are normally borne.[96]

Within the Latin tradition then, there is sufficient evidence to show routes by which Grosseteste would have been familiar with retrogradation in planetary movements, especially with reference to Martianus Capella and Isidore. Furthermore, since it is likely that Grosseteste was in any case familiar with Sacrobosco's treatise he would certainly have known about planetary motion and their retrogradation from this source. The studied absence of the subject implies its deliberate exclusion, which is in turn related to the form and function of the treatise, which will be explored in more detail in Chapter 6. A final Latin inheritance, connected to an earlier twelfth-century thinker Petrus Alfonsi, stands as a reminder of the variety of ways in which astronomical knowledge, ancient and modern, might have been available to Grosseteste and his contemporaries.

4.3 Petrus Alfonsi, Sphericity, Eclipses, and the City of Arim

Early in *On the Sphere* Grosseteste uses the example of the city of Arim as part of the demonstration of the earth's sphericity (DS §§7–8). Since

---

[92] Cassiodorus, *Institutiones*, II.7.2, pp. 154–5.
[93] Isidore of Seville, *Etymologiarum siue Originum libri XX*, III.lxix.10, ed. Wallace M. Lindsay (Oxford: Oxford University Press, 1911).
[94] Isidore *De natura rerum*, XXII.3 and XXIII.   [95] Isidore *De natura rerum*, XXII.1–3.
[96] Bede, *De natura rerum*, 12.

200   THE SCIENTIFIC WORKS OF ROBERT GROSSETESTE

vision proceeds in a straight line, were the earth to be flat, the vision of everyone on the earth would have the same terminal point in the heavens. However,

> it is known by experience that those who are in the land of India in the city of Arim see the northern pole [the Pole Star], and this is the limit of what they can see. And the more human beings depart from that city towards the north, the higher is the pole raised for them, and the limit of what they can see is below the pole.   (DS §7)

That this occurs at all is because the earth is round. Grosseteste goes on to add that the fact that day and night come earlier for eastern parts of the world, and later for western parts, is further evidence for the earth's sphericity.[97] This, he states is known through lunar eclipses. The same eclipse, observable in Arim in the middle of the night, is not visible to those who live further west; an eclipse in the middle of the night for those who dwell in Arim occurs in the evening for westerners, and in the morning for easterners. Arim was a familiar point of reference within astronomical literature (§2), but was also discussed in another source with which Grosseteste may also have been familiar, namely the *Dialogue Against the Jews* by Petrus Alfonsi.[98]

Information about Petrus's early life is scarce and derives principally from remarks within this treatise in the context of his conversion and baptism in 1106.[99] When he was born and when he died are not known. He came, probably, from south-western al-Andalus, and in 1106 lived in

---

[97] In particular it is evidence that the earth is curved east to west as well as north to south; to show a curved surface is rigorously spherical requires demonstration of two independent curvatures, which is precisely what Grosseteste did.

[98] Petrus Alfonsi, *Dialogus*, ed. and trans. (German) Peter Stotz (Florence: SISMEL, 2018). John Tolan, *Petrus Alfonsi and his Medieval Readers* (Gainsville, FL: University of Florida Press, 1993), 103–4, for the association with Grosseteste. The extent to which this work by Petrus Alfonsi had any influence on Grosseteste's attitude towards the Jewish communities of his own day might be explored further. Grosseteste's own views, and actions, towards Jewish communities in England, in particular in Leicester, were harsh, even for the thirteenth century. L. M. Friedman, *Robert Grosseteste and the Jews* (Cambridge, MA: Harvard University Press, 1934) offers an uncompromising condemnation of these views; a more historically nuanced study is to be found in Joseph W. Goering, 'Robert Grosseteste and the Jews of Leicester', in Maura O'Carroll (ed.), *Robert Grosseteste and the Beginnings of a British Theological Tradition* (Rome: Istituto Storico dei Cappuccini, 2003), 181–200.

[99] Alfonsi, *Dialogus*, Prologue, 2–10.

INTELLECTUAL INHERITANCES   201

Huesca, the capital of Aragon.[100] Petrus's early life then would have been dominated by the subjugation and conquest of the Taifa kingdoms of the southern Iberian peninsula which had emerged after the collapse of the Ummayad Caliphate based in Cordoba from the 1030s.[101] The Taifa were threatened from the Christian kingdoms of the north, principally Castile and Aragon, and from the Almoravid movement, originating in the western Maghreb in the Atlas Mountains. Initially asked for aid by various Taifa kingdoms in response to the Castilian capture of Toledo in 1086, the Almoravids by 1102 had conquered al-Andalus, including Valencia, and by 1115 Majorca. These changing circumstances affected the Jewish communities, with greater segregation from civic and court life under the Almoravids, at least initially.[102] While Petrus was not the only Jew to leave Iberia his conversion was more unusual. Petrus's godfather at his baptism was Alfonso I of Aragon, implying, perhaps, service at his court.

The distinguishing feature of Petrus's scholarship was his ability, as John Tolan puts it, 'to bridge several cultures: a Jew from the Arab world of al-Andalus, he converted to Christianity and migrated north into Latin Europe'. Between 1106 and 1116 it is likely, although difficult to prove with certainty, that Petrus moved from Aragon to England, possible moving in a circle of scholars at the court of Henry I, to whom he may have acted as physician. He appears, too, to have taught astronomy in this period and in the southern March and Severn Valley. The prior of Malvern, Walcher, refers to him as 'teacher' in his treatise on eclipses, *On the Dragon (De dracone)*.[103] Another legacy of his expertise in astronomy is his *Letter to the Peripatetics*, directed to unnamed scholars in France, advocating the primary place of astronomy within the liberal arts, and berating their out of date preference for Macrobius rather than the more

---

[100] Tolan, *Petrus Alfonsi*, 3 11.
[101] David J. Wasserstein, *The Caliphate in the West: An Islamic Political Institution in the Iberian Peninsula* (Oxford: Oxford University Press, 1993); Hugh Kennedy, *Muslim Spain and Portugal: A Political History of al-Andalus* (Longman: London, 1996); Brian Catlos, *Kingdoms of Faith: A New History of Islamic Spain* (London: Hurst & Company, 2018).
[102] Tolan, *Petrus Alfonsi*, 6–8.
[103] Walcher of Malvern, *De Dracone*, c. 1, ed. Nothaft, *Walcher of Malvern*, 195. See also Charles Burnett, 'Petrus Alfonsi and Adelard of Bath Revisited', in Carmen Cardelle de Hartmann and Philip Roelli (ed.), *Petrus Alfonsi and his Dialogus: Background, Context, Reception* (Florence: SISMEL, 2014), 77–91, esp. 82–91, for an exploration of Petrus's influence on Adelard.

202 THE SCIENTIFIC WORKS OF ROBERT GROSSETESTE

recent astronomical tables, in this case the *Zīj al-Sindhind* of al-Khwārizmī which Petrus had translated into Latin.[104]

It is, however, the *Dialogue Against the Jews* written in about 1109, soon after Petrus's conversion to Christianity from Judaism, which provides the most intriguing correspondences to three aspects of Grosseteste's *On the Sphere*.[105] The treatise, arranged in twelve books, launched a radical attack on Judaism in its first four books, an attack on Islam in book five, and a defence of Christian belief in the remainder. A central theme to Petrus's text is the assertion of reason in support of authority a mode of argument which he associates with Christianity in opposition to the irrationality of his putative opponents. It is in the first book that Petrus uses astronomical examples in his general criticisms of the Haggadah, the post-biblical rabbinical writings, especially on the issue that the Jews seem to ascribe corporeality to God.[106]

These examples are placed in close argumentative proximity. A discussion of east and west, and their relative location to latitude as part demonstration of the sphericity of the world, is exemplified in a case study of Arim and the difference in the time between observation of a solar eclipse in this city and others to the east and west. The place of Arim at the centre of the world provokes a consideration of the world's habitable zones with Petrus raising, and rejecting the five-clime model, as adopted, for example by Macrobius and Pliny, and taken on in Latin Christendom by Isidore and Bede.[107] Instead the seven-clime model, found in Ptolemy, Islamicate authorities, including al-Farghānī, is adopted.[108] Finally the uninhabitability of the southern hemisphere is addressed, with the solution proposed that of the eccentricity of the sun's orbit.[109]

Grosseteste's discussion of all three of these examples is similar enough to raise a question as to the extent of his familiarity with Petrus's treatise. As described above Grosseteste's description of Arim

---

[104] Tolan, *Petrus Alfonsi*, 55–61 and 66–8; the *Epistola* is edited and translated at 164–81.
[105] Tolan, *Petrus Alfonsi*, 103–4.   [106] Tolan, *Petrus Alfonsi*, 19–27.
[107] Alfonsi, *Dialogus*, I.69–70, pp. 38–9; Macrobius, *Commentarii in Somnium Scipionis*, 2.5–9; Pliny, *Naturalis historia*, 2.71.177; Isidore, *De natura rerum*, 10, p. 209; Bede, *De natura rerum*, 9, p. 199.
[108] Ptolemy, *Almagest*, 2.12; Bahrom Abdukhalimov, 'Aḥmad al-Farghānī and his *Compendium of Astronomy*', *Journal of Islamic Studies*, 10 (1999), 142–58, at 150.
[109] Alfonsi, *Dialogus*, I.36–73, pp. 30–42.

INTELLECTUAL INHERITANCES   203

(DS §§7–8) includes the articulation of sphericity from an analysis of east and west, the use of the experience of eclipses to underline the point, and a certain narrative attention which is given to the inhabitants of the city. The southern hemisphere and the number of climes feature later on in *On the Sphere* (DS §45), where Grosseteste adopts the seven-clime model, although this model, as also noted in the case of Petrus, had not only Ptolemaic authority but more general admission within the Latin tradition as well.[110] Finally, the notion that it is the eccentricity of the sun's orbit which renders the southern hemisphere burnt and lifeless features distinctly in *On the Sphere* immediately before the seven climes (DS §44). There are no linguistic parallels between the *Dialogue Against the Jews* and *On the Sphere* but the overlap in ideas allows for a measure of speculation. Moreover, it would not have been intrinsically unlikely for Grosseteste to have encountered Petrus's treatise. It was a much-copied work, with early, and strong, dissemination in northern France and England.[111] One of the earliest copies was from Gloucester Abbey, now Hereford Cathedral P.2.IV. St Guthlac's, Hereford, was a dependent priory of Gloucester, and one with which Grosseteste is known to have been carried out business on behalf of Bishop William de Vere.[112] Another significant dissemination network for the *Dialogue* was that of the Abbey of St Victor in Paris. Herefordshire, as noted above (Ch. 2, §2.4) boasted a Victorine priory at Shobdon, certainly active in the period from which Grosseteste's residency in, and association with, the diocese can be dated.

Petrus Alfonsi's *Dialogue Against the Jews* highlights the variety of sources for astronomical knowledge available to Grosseteste at the time that *On the Sphere* was composed. From the longer arc of Greco-Arabic discussion, and the parts of that corpus translated into Latin and its reception, to the inheritances of Rome, the early medieval period, and contemporaries, a wide range of models and approaches to the subject existed. The extent to which Grosseteste drew on any of them

---

[110] Ernst Honigmann, *Die sieben Klimata* (Heidelberg: C. Winter, 1929); S. Schröder, 'Die Klimazonenkarte des Petrus Alfonsi', in Ingrid Baumgärtner, Paul-Gerhard Klumbies, and Franziska Sick (eds.), *Raumkonzepte: Disziplinäre Zugänge* (Göttingen: V&R Unipress, 2009), 257–77.
[111] Tolan, *Petrus Alfonsi*, 98.   [112] *Knowing and Speaking*, 20–1.

204 THE SCIENTIFIC WORKS OF ROBERT GROSSETESTE

is intriguing and what informed *On the Sphere* does not, of course, represent the totality of his reading and experience. Nevertheless, how he did use the authorities at his disposal is instructive for any attempt to understand the purpose of *On the Sphere*. Comparing thematic absence and presence opens important insights onto the purpose of the treatise.

6

Astronomy from Liberal Art to Aristotelian Science in Grosseteste's Thought

By the time *On the Sphere* was composed, the available astronomical corpus offered a many-faceted body of material the development of which would take many different directions (Ch. 5). Grosseteste's place within those developments was particular and governed by a highly individual response to the material at his disposal. His purpose in writing the treatise, and the way in which he did so, are complex questions, to which no entirely satisfactory answers have been given. As the Introduction to this volume set out, the treatise has been subject to some level of misunderstanding and misconception, particularly in terms of the expectations placed upon it as an elementary textbook. The purpose and function of the treatise perhaps are better to be understood through an exploration of the idea of astronomy in Grosseteste's works more generally. This provides a new interpretative framework within which the treatise can be analysed in light of its own principles. In this way the horizons of expectation within which *On the Sphere* is interpreted and appraised can be recalibrated.

An influential example of a critical hermeneutic based on assumptions about Grosseteste's purpose in writing is to be found in scholarship comparing Grosseteste's *On the Sphere* and John of Sacrobosco's treatise of the same name. The arguments for regarding the former as dependent on the latter are assumed in the following (see Ch. 1, §1). Sacrobosco's work was a highly successful textbook, providing mnemonic verses, allusions to familiar classical literature, and other pedagogical devices which leave the strong impression of a teaching text. New information was anchored in familiar literary contexts using long-established tools

206 THE SCIENTIFIC WORKS OF ROBERT GROSSETESTE

for memorization. Grosseteste's use of the same title for his treatise and the significant overlap in content might, and have, led to the assumption that his purpose was largely identical to Sacrobosco's.[1] Grosseteste's departures from Sacrobosco's organization and mode of presentation then carry different interpretative valence. Lynn Thorndike, for example, argued, on the basis of relative length and differences of style and organization, that Grosseteste's treatise was a rough and unfinished abbreviation, with some new material added, of 'the highly finished production of Sacrobosco, which from both literary and pedagogical viewpoints is a finely polished little gem.'[2] Grosseteste's *On the Sphere*, in contrast, had a 'less manifest and connected arrangement' than that of his model.[3]

The assumption that Grosseteste intended to produce a comprehensive textbook to teach the rudiments of astronomy in the schools and universities creates, then, consequent expectations of how Grosseteste's text ought to have been organized and what aspects and areas of knowledge the treatise should have covered. It is not, however, self-evident that this was, in fact, Grosseteste's purpose in writing *On the Sphere*. Other comprehensive introductions, including Sacrobosco's *Sphere* and Alfraganus's digest of Ptolemy's *Almagest*, were widely available, as were more specialized treatments of more specific astronomical questions such as Pseudo-Thābit's *On the Motion of the Eighth Sphere*. Moreover, later scribes and scribal directors frequently found it worthwhile to copy Grosseteste's text alongside such comprehensive introductions, which suggests that, for them at least, Grosseteste's treatise may have served a somewhat different end (see Ch. 3, §1 and Ch. 5, §4.2). In order to identify this purpose it is necessary to suspend prior expectations and attend carefully to the particular contributions the treatise makes to the astronomic corpus of its time. To assess these contributions requires a wider consideration of the place of astronomy within Grosseteste's intellectual development.

---

[1] See Thorndike, *The* Sphere *of Sacrobosco*, 10–14; Baur, *Die philosophischen Werke*, 64*; and Panti, *Moti, virtù et motori celesti*, viii, 45, and 69.

[2] Thorndike, *The* Sphere *of Sacrobosco*, 10.

[3] Thorndike, *The* Sphere *of Sacrobosco*, 10.

ASTRONOMY IN GROSSETESTE'S THOUGHT   207

1 Situating *On the Sphere*

Astronomy retained a central role in Grosseteste's scientific writing from his earliest known texts to the rich vein of writing in the 1220s and 1230s. Grosseteste's individual intellectual journey took place within a period of significant change in Latin intellectual culture more generally. The seven liberal arts, the anchor of the intellectual culture of the twelfth century gradually gave way in the thirteenth century to the Aristotelian paradigm of 'a science'.[4] *On the Sphere* sits at the transitional point of this development, still rooted in the mathematical arts of the quadrivium but arguably already embedded in newer traditions of learning. Grosseteste's treatise contains significant parallels to texts both predating and postdating *On the Sphere*, which help place his idea of astronomy in *On the Sphere* in a larger explanatory context. To analyse these parallels is to reveal the features that distinguish Grosseteste's treatise from the rest of the astronomical corpus, and that show his deliberately and carefully constructed contribution to that corpus.

Two works in particular mark the extremities of astronomy seen as a continuum between twelfth-century art and thirteenth-century science in the context of Grosseteste's works. Astronomy as a liberal art features prominently in what was probably the earliest of Grosseteste's shorter scientific works, *On the Liberal Arts*.[5] While the precise dating of this text is difficult to establish, composition around or shortly before the turn of the twelfth century is most plausible, which places its composition more than a decade prior to *On the Sphere*.[6] The concept of a science in the Aristotelian sense was explicitly treated, in great detail, in Grosseteste's *Commentary on Aristotle's Posterior Analytics*. This commentary, the first comprehensive treatment in Latin of Aristotle's notoriously difficult text, is dated to around a decade later than *On the Sphere*.[7] *On the Sphere* itself shows strong continuities with both *On the Liberal Arts* and the *Commentary* (see Ch. 7). The role of astronomy in both of these texts is central to the conceptual framework in which *On the Sphere* emerged.

---

[4] *Knowing and Speaking*, 45–50.   [5] *Knowing and Speaking*, ch. 8.
[6] *Knowing and* Speaking, 15–18.
[7] For the dating of this treatise, see Grosseteste, *Comm. Post. An.*, 18–21.

208 THE SCIENTIFIC WORKS OF ROBERT GROSSETESTE

2 Astronomy as Liberal Art

Astronomy plays a pivotal role in Grosseteste's *On the Liberal Arts*, a
short work justifying the liberal arts curriculum as still valid even within
an intellectual culture turning to the Aristotelian tradition and Latin
translations of Islamicate learning.[8] The liberal arts were defined by
Grosseteste as the seven disciplines that 'purge human works of error
and lead them to perfection'.[9] To Grosseteste, human beings were an
intrinsic part of a universe functioning according to a natural order.
Human nature was unique within the sub-lunar universe in possessing
reason and will, which allowed humans to attain true knowledge and
seek the good that reason perceived.[10] However, human nature, as fallen
and intrinsically tied to corruptible and corrupting matter, could not put
these faculties to their proper use without first purifying them from the
errors that otherwise would thwart the perfecting of the potential inherent in each human being.[11] For Grosseteste the seven liberal arts, the
trivium of grammar, logic, rhetoric and the quadrivium of music, arithmetic, geometry, and astronomy, were the necessary starting point for all
human education, liberating the inquiring agent from intrinsically
human error and equipping them for higher studies.

The liberal arts, as Grosseteste presented them, were fundamentally
practical in purpose. They were not primarily bodies of abstract knowledge to be possessed, but practices ordered to the perfecting of human
works. Human works Grosseteste divided into four basic kinds: the
perception of the mind, the volition of the mind, movement considered
in itself, and movement for some extrinsic purpose. The first two were
perfected through the trivium, the last two through the quadrivial
arts. Arithmetic, geometry, and astronomy served the fourth and last
of the kinds of human works, movement to produce an extrinsic effect.
Astronomy held a key role in governing the application of the more

---

[8] *Knowing and Speaking*, 45–50, 152–95.
[9] Grosseteste, *De artibus liberalibus*, §2: 'In humanis vero operibus erroris purgationes et ad
perfectionem deductiones sunt artes septene, que sole inter partes philosophie ideo censentur
artis nomine quia earum est tantum effectus operationis humanas corrigendo ad perfectionem
ducere.'
[10] *Knowing and Speaking*, 36–45 and 96–111.
[11] Grosseteste, *De artibus liberalibus* §§1–2.

ASTRONOMY IN GROSSETESTE'S THOUGHT   209

abstract mathematical arts to the material world, 'teaching us to discover the position of the world and intervals of time by the movements of the stars.'[12] Two principal tasks fell, then, to astronomy: providing a framework within which order and position could become intelligible and definable, and providing the basis for accurate time measurement.

The liberal arts were designed to allow human beings to flourish *qua* human beings, and also *qua* natural beings working together with the rest of nature. The arts of arithmetic, geometry, and astronomy were not primarily aimed at correcting and perfecting exclusively human acts, but rather at equipping human beings in their co-operation with nature as a whole.[13] This aspect received further emphasis as Grosseteste moves from presenting astronomy in itself to explaining how this discipline is an essential tool for natural philosophy:

> Natural philosophy needs the service of astronomy more than [it needs] any of the other [arts]; for there are no works belonging to nature and to us—as for instance the planting of plants, the transmutation of minerals, the healing of illness—that may be excluded from the service of astronomy. For lower nature does not act except when celestial power moves it and draws it out from potency into act.[14]

Grosseteste, following Abū Ma'shar's *Introduction to Astronomy*, goes on to point out that the moon is the mediator of the power by which celestial

[12] Grosseteste, *De artibus liberalibus* §8: 'Cum autem per motus nostros preter ipsos motus aliquid intendimus, aut coniuncta dividimus, aut divisa coniungimus, aut ordinem aut situm damus, aut figuras extrahimus. In hiis configurandis arithmeticam et geometriam constat esse rectificantes. Quibusdam tamen rebus non damus <ordinem et situm> absque errore nisi precognito mundi situ, et quedam opera nostra non exstant usquequaque ordinata nisi certis temporum spatiis fuerint mensurata. Propter hoc predictis tribus accessit astronomia, mundi situm et spatia temporum nostrorum motibus docens dignoscere.'

[13] Grosseteste, *De artibus liberalibus* §9: 'Cum igitur sint opera quedam nature tantum, quedam nostra tantum, quedam vero nostra et nature, et he sole philosophie partes que dicte sunt opera tantum nostra et etiam opera nature et nostra inquantum nostra sunt rectificent et perficiant, et artis sit definitio seu dispositio quod sit regula nostre operationis: merito he sole artis vocabulo nuncupantur. He septem naturalis et moralis <philosophie> sunt ministre.'

[14] Grosseteste, *De artibus liberalibus* §11: 'Astronomie ministerio plus ceteris eget philosophia naturalis. Nulla enim est operatio que nature sit et nostra—utpote vegetabilium plantatio, mineralium transmutatio, egritudinum curatio—que possit ab astronomie officio excusari. Non enim agit natura inferior nisi cum eam movet et de potentia in actum educit virtus celestis.'

210   THE SCIENTIFIC WORKS OF ROBERT GROSSETESTE

bodies cause natural phenomena in the sub-lunar world.[15] The relative position of the moon to the other celestial bodies has crucial causal significance. In consequence any human activity that relies on harnessing natural processes for human aims, for example, agriculture, metallurgy, and medicine, therefore requires careful timing to ensure that the most favourable conditions possible are in place when the activity is carried out.[16] Astronomy is the discipline that tracks the relative movements of the celestial bodies measured in time, and consequently the art that allows human beings to pursue a deeper study of natural processes, which is to say, to study natural philosophy.

It is important in this context to note the conceptual distinction between the arts of the quadrivium and what Grosseteste calls natural philosophy. For a modern readership, it may fall intuitively to regard the mathematical arts as being already part of natural philosophy; for Grosseteste this was not necessarily so. In stressing how essential the liberal arts are for doing natural philosophy, Grosseteste at the same time introduced, implicitly, a distinction between the two. Aristotelian natural philosophy, of which Grosseteste was an early proponent, understood the study of nature primarily as an inquiry into the causes for natural change, that is, every movement and change resulting from the nature of things rather than from random chance or rational deliberation.[17] The quadrivial arts as presented by Grosseteste provided methods for measuring change, and were therefore essential *tools for* natural philosophy; but they did not in and of themselves reveal the causes of change, and so they were not strictly speaking *parts of* natural philosophy as such.

Astronomy, however, occupied an ambiguous position with regards to the quadrivial arts and natural philosophy. Since Aristotle held that all change can be reduced to movement, natural philosophy first of all studied the movement of natural bodies.[18] Boethius, in his book *On the Institutions of Arithmetic*, a source for Grosseteste's *On the Liberal Arts*,

---

[15] *Knowing and Speaking*, 169–72.   [16] Grosseteste, *De artibus liberalibus* §11.
[17] See for instance Helen S. Lang, *The Order of Nature in Aristotle's Physics* (Cambridge: Cambridge University Press, 1998); cf. James A. Weisheipl, 'The Concept of Nature', in William E. Carroll (ed.), *Nature and Motion in the Middle Ages* (Washington, DC: Catholic University of America Press, 1985), 1–24.
[18] See for instance Aristotle *De caelo* I.1, 268a1–6.

ASTRONOMY IN GROSSETESTE'S THOUGHT   211

had defined arithmetic as dealing with quantity per se, music as treating quantity in relation to some other quantity, geometry as measuring immobile magnitudes, and astronomy as studying mobile magnitudes.[19] In other words, astronomy emerges as the only quadrivial discipline in which change or movement plays a part. It is therefore continuous with natural philosophy in a unique way, demonstrating how to apply the abstract teachings of music, arithmetic, and geometry to the concrete and observable world.[20] Astronomy is therefore at once the easiest art to apply to the observable world, and the most difficult art to delimit and define. It ranges from the art of describing the movement of the heavens by applying the other quadrivial arts, to supporting speculations about the causes and effects of these movements.

While astronomy in this sense was the highest of the quadrivial arts it could also be defined as the lowest. The various disciplines could be ordered not only according to usefulness but also according to logical priority, that is, according to how fundamental the principles of one discipline was to the others. While astronomy's particular openness towards the study of nature rendered it preeminent among the quadrivial arts in terms of its practical usefulness for natural philosophy, the dependence of its basic principles on the conclusions of other quadrivial arts placed it, hierarchically, below them. Since arithmetic deals with quantity simply speaking while the other three deal with quantity in some specific way, arithmetic is necessary for the others and consequently prior to them. In his explicit discussions of a hierarchical ordering of sciences and disciplines in *On the Liberal Arts*, Grosseteste

---

[19] Boethius, *De institutione arithmetica* I.1, ed. H. Oosthout and J. Schilling, CCSL, 94A (Turnhout: Brepols, 1999), 10: 'Horum ergo illam multitudinem, quae per se est, arithmetica speculatur integritas, illam uero, quae ad aliquid, musici modulaminis temperamenta pernoscunt, immobilis uero magnitudinis geometria notitiam pollicetur, mobilis uero scientiam astronomicae disciplinae peritia uindicauit [Of these, then, arithmetic integrity investigates quantity in and of itself; the measures of music, moreover, comes to know fully [quantity] in relation to some other; geometry promises knowledge of immobile magnitude; while expertise in the astronomical disciplines has laid claim to knowledge of [magnitude in as much as it is] mobile]'. See below for Grosseteste's development of Boethius's line of thinking in this respect.
[20] Grosseteste seems to have developed doubts about the possibility of astronomy as natural philosophy later in life, when the conflicting schools of thought regarding planetary motion appeared to undermine one another leaving agnosticism the only honest option; see Robert Grosseteste, *Hexaemeron* III.VI.1, ed. Richard C. Dales and Servus Gieben (Oxford: Oxford University Press, 1982), 106.

212   THE SCIENTIFIC WORKS OF ROBERT GROSSETESTE

followed the pattern set down by Boethius who had shown how music and geometry depend on number in an absolute sense, and how arithmetic therefore is prior to both.[21] Astronomy, dealing with celestial movements governed by harmonies and ratios discovered by music, which movements describe shapes measured by geometry, is in this sense dependent all the other three and prior to none of them.[22] The quadrivial arts, on Grosseteste's Boethian account, constituted an ordered unity.

At the time Grosseteste was writing *On the Liberal Arts*, however, the unity of the arts was already under threat.[23] As each discipline expanded with the growth of higher education in the twelfth century, it became increasingly difficult to justify the effort required to acquire each of the liberal arts.[24] Many found it tempting to study the most useful disciplines in isolation from the whole of which they formed parts, and to take short-cuts through the morass of accumulated learning pertaining to each art.[25] As late as 1220, Grosseteste's older contemporary and acquaintance Gerald of Wales, for instance, would warn against the danger posed to the unity and completeness of the arts by the new learning becoming available, pre-eminently the corpus of natural philosophy, the *libri naturales*, centred around the recent Latin translations of Aristotle's natural works.[26] Already in *On the Liberal Arts* Grosseteste was combining the old learning with the new, providing (intentionally or not) an example of how the destruction of the old by the new might be avoided.[27]

The Latin translations of the Islamicate traditions of astronomy posed particular challenges for the discipline of astronomy. In the same way that the translations of the Aristotelian *Libri naturales* became

[21] *Knowing and Speaking*; Boethius, *De institutione arithmetica* I.1, ed. Oosthout and Schilling, 12–13.
[22] Boethius, *De institutione arithmetica* I.1, ed. Oosthout and Schilling, 13.
[23] For the quadrivial framework, see in particular Guy Beaujouan, 'The Transformation of the Quadrivium', in Robert L. Benson and Giles Constable (eds.), *Renaissance and Renewal in the Twelfth Century* (Cambridge, MA: Harvard University Press, 1982), 467–83.
[24] Stephen C. Ferruolo, *The Origins of the University: The Schools of Paris and their Critics, 1100–1215* (Stanford: Stanford University Press, 1985), 140–55.
[25] See for instance Cédric Giraud and Constant Mews, 'John of Salisbury and the Schools of the 12th Century', in Christophe Grellard and Frédérique Lachaud (eds.), *A Companion to John of Salisbury* (Leiden: Brill, 2014), 31–62.
[26] Ferruolo, *The Origins of the University*, 182–3.   [27] *Knowing and Speaking*, ch. 13.

ASTRONOMY IN GROSSETESTE'S THOUGHT   213

synonymous with natural philosophy in Latin intellectual culture, so astronomy centred on the astronomical doctrines of Claudius Ptolemy (Ch. 5, §§1–2; Ch. 9, §1). Ptolemy had attempted to provide a mathematical model that could explain the observable movements of the celestial bodies within the framework of Aristotelian physics, but in order to provide mathematical explanations for observed phenomena he had to introduce important changes such as epicycles and eccentrics into Aristotle's model. The potential tension between these two models became undeniable after Michael Scot's Latin translation of al-Biṭrūjī's *On the Movements of the Heavens* in 1217, since this work explicitly offered a metaphysical alternative to Ptolemy's mathematical model, but *On the Sphere* appears to sit comfortably within what is still a broadly Ptolemaic framework.[28]

Even if the new astronomical learning was anchored in the Ptolemaic tradition, however, it still presented a large mass of information to be processed, and contained internal tensions that needed to be addressed. The Latin translations of the late twelfth century made the Ptolemaic tradition available through a number of works.[29] Gerard of Cremona had not only translated Ptolemy's *Almagest* into Latin from Arabic by *c.*1175 but also a number of Islamicate treatises summarizing and revising Ptolemy's doctrines. These include al-Farghānī's (Latinized as Alfraganus) *Liber de aggregationibus scientie stellarum* (which had already been translated by John of Seville somewhat earlier); Abū Muḥammad Jābir ibn Ḥayyān's (Geber) *De astronomia*; the Toledan astronomical tables with explanations by the mathematician al-Zarqālī (Arzachel) in Latin; the anonymous treatise *Theorica planetarum* sometimes attributed to Gerard as his original work.[30] Although not known to be a translation

---

[28] Edgar Laird, 'Robert Grosseteste, Ptolemy, and Christian Knowledge,' in *Robert Grosseteste and His Intellectual Milieu*, ed. John Flood, James R. Ginther, and Joseph W. Goering (Toronto: Pontifical Institute of Mediaeval Studies, 2013), 131–52.

[29] For a comprehensive overview of the trajectories of translation, see Charles Burnett, 'Translation and Transmission of Greek and Islamic Science to Latin Christendom', in David Lindberg and Michael Shank (eds.), *Cambridge History of Science*: Vol. 2, *Medieval Science* (Cambridge: Cambridge University Press, 2013), 341–64; cf. Charles Burnett, *The Introduction of Arabic Learning into England* (London: The British Library, 1997).

[30] Paul Kunitzsch, *Der Almagest: Die 'Syntaxis mathematica' des Claudius Ptolemäus in arabisch-lateinisch Überlieferung* (Wiesbaden: Otto Harrassowitz, 1974); cf. Paul Kunitzsch, 'Gerard's Translations of Astronomical Texts, Especially the Almagest', in Pierluigi Pizzaiglio (ed.), *Gerardo da Cremona*, Annali della Biblioteca statale e libreria civica di Cremona 41

214   THE SCIENTIFIC WORKS OF ROBERT GROSSETESTE

Pseudo-Thābit's (Thebit) important adjustment of Ptolemy's account of the movement of the stars and planets in *On the Motion of the Eighth Sphere* was also important in this wider Ptolemaic inheritance.[31] In addition, the Latin work of uncertain attribution now referred to as the *Almagesti minor* provided what has been dubbed 'a Euclidization' of Ptolemy's basic doctrines.[32] In this work, the close affinity between Euclidian geometry and Ptolemaic astronomy became particularly evident, and succinct geometrical passages from the *Almagesti minor* came to be used as glosses for the somewhat more discursive exposition of Ptolemy in the *Almagest*.[33]

It is quite possible that Grosseteste knew all these texts by the time he wrote *On the Sphere*. He refers directly to Ptolemy's *Almagest* (§49) and Pseudo-Thābit's *On the Motion of the Eighth Sphere* (§50) in *On the Sphere*. The doctrines of Ptolemy in question here might also have reached him via Alfraganus or Geber; but Grosseteste was demonstrably familiar with Ptolemy's own text by the time he wrote the *Compotus*, and there is little direct reason to doubt him at his word here. Pseudo-Thābit's work is in the part of MS Savile 21 most likely to contain Grosseteste's own handwriting, so there is no need to doubt this reference either. Furthermore, Grosseteste uses vocabulary that links his account to one or both of Alfraganus and the anonymous *Theorica planetarum* (for instance §§40 and 54). Much of the Ptolemaic tradition would also have reached Grosseteste through the text of Sacrobosco. However, since Grosseteste uses Islamicate sources not mentioned by Sacrobosco, and, as will be shown below, imposes his own individual

---

(Cremona: Libraria del convengno editrice, 1992), 71–84. John of Seville's translation was printed as Alfraganus, *Compilatio Astronomica* (Ferrara: Andrea Gallus, 1493) and also in al-Farghani, *Differentie*, trans. John of Seville, ed. Francis J. Carmody (Berkeley: University of California Press, 1943); Gerard of Cremona's translation is found in Alfraganus, *Il 'libro dell'aggregazione delle stelle'*, ed. Romeo Campani (Città di Castello: S. Lapi, 1910); Geber filius Affla Hispalensis, *De astronomia* (Nüremberg: Iohannes Petreius, 1534); Fritz Saaby Pedersen (ed.), *The Toledan Tables: A Review of the Manuscripts and the Textual Versions with an Edition*, 4 vols. (Copenhagen: Reitzel, 2002); *Theorica Planetarum*, ed. Francis J. Carmody, *Theorica planetarum Gerardi* (Berkeley: University of California Press, 1942).

[31] Pseudo-Thābit, *De motu octave sphere*, ed. Francis J. Carmody, *The Astronomical Works of Thabit b. Qurra* (Berkeley/Los Angeles: University of California Press, 1960), 102–7; cf. José María Millás Vallicrosa, 'El "Liber de motu octave sphere" de Ṭābit ibn Qurra', *Al-Andalus*, 10 (1945), 89–108.

[32] *Almagesti minor*, ed. Henry Zepeda (Turnhout, Brepols, 2018), 21–7.

[33] *Almagesti minor*, 81–4.

ASTRONOMY IN GROSSETESTE'S THOUGHT   215

argumentative structure on the material, it is unlikely that his knowledge was purely derivative. There is solid evidence for Grosseteste's use of the *Almagesti minor* in the *Compotus*, and some suggestions that the former work also influenced *On the Sphere*.

While the new learning presented challenges in how to absorb and organize its considerable body of material it also offered routes by which this could be achieved. The *Almagesti minor* provides a good example. As the title indicates, the *Almagesti minor* or 'Lesser Almagest' selected, compressed, and re-ordered Ptolemy's treatise, clarifying the expression of his geometrical account of the universe to make clear the internal order and dependency of the individual postulates. However, there were also other resources, foremost amongst which was Aristotle's *Posterior Analytics*. In this work Aristotle discussed the acquisition, processing, and organization of knowledge in its most profound sense. As noted above (§1) Grosseteste's commentary, the first complete Latin commentary on this text, dates to the mid-to-late 1220. Although this is about decade or so after the completion of *On the Sphere*, the date of the commentary has, strictly speaking, no bearing on when Grosseteste first became familiar with the *Posterior Analytics*. Grosseteste's eventual command of Aristotle's text might plausibly indicate a longer acquaintance. The Latin translation of Aristotle's Greek original had been in circulation in northern Europe since the middle of the twelfth century. This circumstance, coupled with the demonstrable knowledge of significant parts of the Aristotelian corpus from the earliest stages of his writing career, notably *On the Soul*, and the *Physics*, make it not unreasonable to posit that the *Posterior Analytics* was known to Grosseteste when he wrote *On the Sphere*.[34] A great number of Aristotle's examples in the *Posterior Analytics* were taken from the discipline of astronomy, and Grosseteste in turn relied on these astronomical examples in his exegesis of the Aristotelian concept of a science in the *Commentary*. There are therefore strong grounds for presenting this Aristotelian concept of a science alongside Grosseteste's early works in order to bring out more clearly the identifying features of Grosseteste's treatment of astronomy in *On the Sphere*.

---

[34] *Knowing and Speaking*, chs. 7 and 13.

216 THE SCIENTIFIC WORKS OF ROBERT GROSSETESTE

3 Astronomy as an Aristotelian Science: The *Posterior Analytics*

The notion of 'an Aristotelian science', in the sense of a structured body of knowledge providing a comprehensive explanatory account within a given domain, was laid out in the first of the two books of Aristotle's *Posterior Analytics*.[35] The complexity of the argument laid out by Aristotle was compounded for scholars in Latin Christendom by the near impenetrability of James of Venice's very literal Latin translation, which itself had to address the challenge of translating Aristotle's flexible terminology.[36] Translation into modern languages has brought additional challenges, exemplified in the still ongoing scholarly debates on how to interpret and translate the core concept of the work, expressed in the term ἐπιστήμη (*epistêmê*).[37] In the *Posterior Analytics*, the usage of ἐπιστήμη spans a semantic arc from the denotation of a certain cognitive state of knowledge and understanding in an individual soul to description of a structured body of knowledge about a specific topic, which poses particular problems for translating the term into English.[38] While the Latin equivalent of the Greek term *scientia* admits of a semantic range similar to the Greek term, English has no applicable noun that can

---

[35] For a comprehensive reading of the *Posterior Analytics* as a whole, see in particular David Bronstein, *Aristotle on Knowledge and Learning* (Oxford: Oxford University Press, 2016).

[36] James of Venice's translation is printed in *Aristoteles Latinus* IV.1–4, ed. L. Minio-Paluello and B. G. Dod (Turnhout: Brepols, 1968), 5–107. For an overview of the medieval reception of this text, see Sten Ebbesen, 'The Posterior Analytics 1100–1400 in East and West', in Joël Biard (ed.), *Raison et démonstration. Les commentaires médiévaux sur les Seconds Analytiques* (Turnhout: Brepols, 2015), 11–30. For glosses and commentaries accompanying this translation, see for instance David Bloch, 'James of Venice and the Posterior Analytics', *Cahiers de l'Institut du Moyen-Âge Grec Et Latin*, 78 (2008), 37–50; and David Bloch, 'Monstrosities and Twitterings: A Note on the Early Reception of the *Posterior Analytics*', *Cahiers de l'Institut du Moyen-Âge Grec Et Latin*, 79 (2010), 1–6.

[37] For discussions on how to translate the term, see for instance Jonathan Barnes (trans. and comm.), *Aristotle: Posterior Analytics*, 2nd ed. (Oxford: Oxford University Press, 1993), 82; and Myles Burnyeat, 'Aristotle on Understanding Knowledge', in E. Berti (ed.), *Aristotle on Science: 'The Posterior Analytics'*, Proceedings of the Eighth Symposium Aristotelicum (Padua: Editrice Antenore, 1981), 97–139, reprinted in Myles Burnyeat, *Explorations in Ancient and Modern Philosophy*, Vol. 2 (Cambridge: Cambridge University Press, 2012), 115–44. For discussions of this term in a medieval context, see José Antonio Valdivia Fuenzalida, 'La contingence et la science. À propos de la réception des Seconds Analytiques au XIIIe siècle', *Scripta Mediaevalia* 11 (2018), 43–79; and Richard A. Lee Jr., *Science, the Singular, and the Question of Theology* (New York: Palgrave, 2002), 7–15.

[38] Burnyeat, 'Aristotle on Understanding Knowledge', 115.

ASTRONOMY IN GROSSETESTE'S THOUGHT 217

be both countable, as in 'a science', and uncountable, as in 'scientific understanding', both of which examples have been used to translate ἐπιστήμη at various points. The term anchors both a theory of human knowledge and what might be referred to as a theory of science and scientific disciplines, and unites these two strands into a single coherent account.

Grosseteste's organization of his argument in *On the Sphere* departs at various points from his models within the astronomical corpus. These departures can be linked to his use of the Aristotelian account of ἐπιστήμη, expressed in most detail in the *Commentary on Posterior Analytics*. The argument implicit in what follows is that Grosseteste shows clear signs in *On the Sphere* of being familiar with the argument of Aristotle's text, which he later commented on. Grosseteste's main sources for the *Commentary* were available in Latin well before the composition of *On the Sphere* began. It is consequently necessary to keep a discussion of Aristotle's doctrines in the form demonstrably available to Grosseteste in the 1210s distinct from a discussion of the *Commentary*. This section therefore presents a summary of Aristotle's argument, while §4 below analyses Grosseteste's reception of this body of thought to demonstrate the depth and sophistication of Grosseteste's grasp of the Aristotelian framework. The interpretive framework ensuing from these two sections will then be applied to *On the Sphere* in Chapter 7.

3.1 ἐπιστήμη as Understanding

Whatever else it may be, ἐπιστήμη was to Aristotle the paradigmatic and most complete kind of human knowledge. Throughout his preserved works, Aristotle insisted that to know something fully is to understand, and to be ably explain, *why* this something is so.[39] It is possible to have knowledge in a more general sense, γνῶσις (*gnôsis*), of a matter of fact,

---

[39] Two particularly clear examples can be found in Aristotle, *Posterior Analytics*, trans. James of Venice, in Lorenzo Minio-Paluello and Bernard G. Dod (eds.), *Aristoteles Latinus* IV.1–4, (Turnhout: Brepols, 1968), I.2 and AL *Physica*, I. 1.

218   THE SCIENTIFIC WORKS OF ROBERT GROSSETESTE

for example that a lunar eclipse is the darkening of the moon; but understanding *why* the moon is darkened, that is, that the earth screens the moon from the light of the sun is to know in a much more profound way what an eclipse *is*. Such knowledge by way of understanding the reason why is what Aristotle calls ἐπιστήμη.[40]

This means that ἐπιστήμη necessarily depends on, and proceeds from, knowledge (γνῶσις) already possessed: this is stated in the very first sentence of the *Posterior Analytics*.[41] In this work, Aristotle shows how ἐπιστήμη can be reached from a starting point of knowledge already possessed in a partial and qualified way through a certain mode of reasoning, that is, demonstration, ἀπόδειξις (*apodeixis*). Aristotle defines demonstration as 'a syllogism that produces understanding' (ἐπιστήμη): 'συλλογισμός ἐπιστημονικὸς' (*sullogismos epistêmonikos*) in Aristotle's Greek and 'sillogismus faciens scire' in the Latin translation by James of Venice. The structural requirements of a demonstration are taken from formal logic: a demonstration consists of an ordered set of two premises containing three terms with strictly defined roles, that is, the minor, major, and middle terms. The middle term provides a link between the minor and major terms, the so-called extremities of the argument. In the minor premise, the middle term is predicated of the minor term, while in the major premise the major term is predicated of the middle term. This allows a validly inferred conclusion in which the major term is predicated of the minor. This formal syllogistical structure was known in Latin through the older logical corpus translated by Boethius; what the *Posterior Analytics* added was a procedure for separating syllogisms or demonstrations that produce understanding from other inferences that fulfil the formal criteria for syllogistic logic, but fail to produce ἐπιστήμη.[42]

---

[40] See Burnyeat, 'Aristotle on Understanding Knowledge', 120–3.

[41] Aristotle, *Posterior Analytics* I.1, 71a1–5; cf. discussions in Bronstein, *Aristotle on Knowledge and Learning*, 11–27, and Gail Fine, *The Possibility of Inquiry: Meno's Paradox from Socrates to Sexus* (Oxford: Oxford University Press, 2014), 179–225.

[42] For an overview of Boethian logic and its development in the period leading up to the composition of *On the Sphere*, see for instance Christopher J. Martin, 'The Development of Logic in the Twelfth Century', in Robert Pasnau and Christina Van Dyke (eds.), *The Cambridge History of Medieval Philosophy*, Vol. 1 of 2. (Cambridge: Cambridge University Press, 2010), 129–45.

ASTRONOMY IN GROSSETESTE'S THOUGHT 219

Aristotle frequently used astronomical examples to illustrate his abstract and general account of demonstration. For instance, in *Posterior Analytics* I.13, Aristotle provides a proof demonstrating that the moon waxes because it is spherical; that is to say, the moon is capable of waxing and waning by virtue of being spherical, since only what is spherical will wax and wane depending on the angle between a light source and an observer with the spherical object at the vertex.[43] If 'the moon' is posited as a minor term (A), 'waxes' as a major term (C), and 'is spherical' as a middle term (B), this produces the following demonstration (Example 1):

Minor premise: The moon (A) is spherical (B)
Major premise: What is spherical (B) waxes (C)
Conclusion: The moon (A) waxes (C)

However, a demonstration in the sense in which Aristotle uses the term in the *Posterior Analytics* cannot simply be reduced to a formal logical figure. In order to produce understanding or ἐπιστήμη, the middle term has to provide more than a logically valid link between the minor and major terms; the middle term has to be the proximate or direct cause for why the major term could be predicated of the minor term. ἐπιστήμη is knowledge by way of understanding the reason why, and in the example demonstration above the sphericity of the moon is the reason or cause why the moon waxes. Contrast the following example, where 'the moon' is the minor term (A), 'spherical' is the major term (C), and 'waxes' is the middle term (B) (Example 2):

Minor premise: The moon (A) waxes (B)
Major premise: what waxes (B) is spherical (C)
Conclusion: The moon (A) is spherical (C)

This is a valid inference following the same logical form as the previous example, yet it does not produce ἐπιστήμη. The middle term 'waxes' is not the cause or the reason why the moon is spherical; the moon waxes because it is spherical, but its waxing does not cause its sphericity. This

---

[43] Aristotle, *Posterior Analytics* I.13, 78b4–11.

220 THE SCIENTIFIC WORKS OF ROBERT GROSSETESTE

argument then, establishes the fact that (το ὅτι) the moon is spherical, whereas the previous argument provides the reason why (το διότι) the moon waxes. The first example, then, is a demonstration properly speaking, while the second is a demonstration only in a qualified and weaker sense. In James of Venice's Latin translation a demonstration of a fact, το ὅτι, was translated as a demonstration 'ipsum quia', while a demonstration properly speaking, a demonstration giving the reason why, το διότι, was translated as a demonstration 'propter quid'.[44] This is a distinction central to Grosseteste's reasoning both in *On the Sphere* and elsewhere, as will be discussed below.

A demonstration το διότι or *propter quid*, then, required much more than a specific logical form, particularly when it came to what could play the role of a middle term. Aristotle distinguished between causes properly speaking, which provided the relevant sort of explanation, and accidental and contingent conditions that account for individual things and events. Knowledge that was transient or accidental was intrinsically inferior to ἐπιστήμη, which was limited to an understanding of the stable and enduring intelligible structures of the world as it is. In the *Physics*, Aristotle listed four different but complementary kinds of causes that yield understanding of why something is as it is: the 'that-from-which' or material cause; the formal cause by which a thing is what it is; the original source of motion and rest or efficient cause; and that for the sake of which something exists, the final cause.[45] While all four causes could be a middle term of a demonstration, the formal cause plays a central role for Aristotle's theory of knowledge.[46] The form actualizes the potency of matter, determines what sort of efficient and final causality is applicable to each kind of thing, and is the property that makes each thing intelligible to the human mind.[47] A form, however, is a universal property

---

[44] See for instance Aristotle, *Posterior Analytics* I.13, 78a22: 'Τὸ δ' ὅτι διαφέρει καὶ τὸ διότι ἐπίστασθαι', translated by James of Venice as 'Sed quia differt et propter quid scire'; ed. L. Minio-Paluello and B. Dod, *Aristoteles Latinus* IV.1, 29.

[45] Aristotle, *Physica* II.3, 194b16–195a2.

[46] For a discussion of the applicability of all four causes to the role of a midle term, see Aristotle, *Posterior Analytics* II.11.

[47] See Aristotle, *Posterior Analytics* I.3 and 2.2, and Bronstein, *Aristotle on Knowledge and Learning*, 55–6; cf. *Metaphysics* Z 16–17, 1040b5–1041b33 and T. H. Irwin, *Aristotle's First Principles* (Oxford: Clarendon Press, 1988), 245–7 and 252–5.

ASTRONOMY IN GROSSETESTE'S THOUGHT   221

shared by all individual members of a certain species. The Aristotelian term for both 'form' and 'species' is εἶδος, for the form is that by virtue of which a thing belongs to a species; the form is both a classificatory principle and a principle of intelligibility. If ἐπιστήμη is reached by way of an explanation providing a cause, and all causes can be reduced, at least in a qualified sense, to the formal cause, this means that ἐπιστήμη cannot be had of individuals, but only of kinds.

The principles from which a demonstration proceeds, then, are determined by the kind of thing with which the demonstration is concerned, and must be constitutive of that kind of thing and explanatory of its properties. A demonstration proper shows both the reason why something is the case, and that this holds universally and cannot be otherwise for the kind in question. Demonstration, therefore, must proceed from 'what is primitive in the kind with which the proof is concerned—and not every truth is appropriate', otherwise the demonstration would lack explanatory power.[48] Moreover, the terms and premises of the demonstration must belong to the kind in question by virtue of itself, that is, they must belong to the kind intrinsically and universally for that kind, and not incidentally or partially: 'Since in each kind whatever holds of something in itself and as such holds it from necessity, it is clear that scientific demonstrations [αἱ ἐπιστημονικαὶ ἀποδείξεις] are concerned with what holds of things in themselves and that they proceed from such items.'[49] In this way, then, the basic concept of knowing-through-understanding constituting the core of Aristotle's idea of ἐπιστήμη can also serve as a foundation for the construction of a systematic understanding of a given kind of thing. In other words, understanding

[48] Aristotle, *Posterior Analytics* I.6, 74b25–6: 'τὸ πρῶτον τοῦ γένους περὶ ὃ δείκνυται· καὶ τἀληθὲς οὐ πᾶν οἰκεῖον'. English translation Barnes, *Aristotle: Posterior Analytics*, 10. James of Venice's translation: 'primum genere circa quod demonstratur; et verum non omne proprium', *AL* IV.1, 17.

[49] Aristotle, *Posterior Analytics* I.6, 75a28–31: 'Ἐπεὶ δ᾽ ἐξ ἀνάγκης ὑπάρχει περὶ ἕκαστον γένος ὅσα καθ᾽ αὑτὰ ὑπάρχει καὶ ᾗ ἕκαστον, φανερὸν ὅτι περὶ τῶν καθ᾽ αὑτὰ ὑπαρχόντων αἱ ἐπιστημονικαὶ ἀποδείξεις καὶ ἐκ τῶν τοιούτων εἰσίν'. English translation Barnes, *Aristotle: Posterior Analytics*, 12. James of Venice's translation: 'Quoniam autem ex necessitate sunt circa unumquodque genus quecumque per se sunt, et secundum quod unumquodque est, manifestum est quoniam de his que per se sunt demonstrative scientie et ex talibus sunt', *AL* IV.1, 19.

222 THE SCIENTIFIC WORKS OF ROBERT GROSSETESTE

can be methodically expanded to constitute a comprehensive system of understanding, that is, a science.

3.2 ἐπιστήμη as a Science

ἐπιστήμη can be defined as an understanding of the attributes of a kind issuing from demonstrations proceeding from the basic principles and causes of that kind, ordered in an explanatory framework where the cause providing the reason why for each attribute is revealed. An exhaustive explanatory framework for some kind of thing, then, is what is meant by the term 'a science' or ἐπιστήμη in the Aristotelian sense. The same term can be used both for the understanding the soul attains by way of a demonstration and for the abstract explanatory framework so produced (that is 'a science') because the structure is the same in both cases. Just as a demonstration productive of understanding must assume three terms, so a system of understanding of a given kind of thing must assume three elements:

> Every demonstrative science is concerned with three things: [1] what it posits to exist (these items constitute the kind of which it studies the attributes which hold of it in itself); [2] the so-called common axioms, i.e. the primitives from which demonstrations proceed; and thirdly [3], the attributes, where it assumes what each of them means.[50]

Returning to the sample demonstration above of the moon's waxing on account of its sphericity, the moon would belong to the kind of thing, in this case celestial bodies, constituting the domain of the science in question (astronomy) (1); its sphericity would be a causal property

---

[50] Aristotle, *Posterior Analytics* I.10, 76b11–16: 'πᾶσα γὰρ ἀποδεικτικὴ ἐπιστήμη περὶ τρία ἐστίν, ὅσα τε εἶναι τίθεται (ταῦτα δ' ἐστὶ τὸ γένος, οὗ τῶν καθ' αὑτὰ παθημάτων ἐστὶ θεωρητική), καὶ τὰ κοινὰ λεγόμενα ἀξιώματα, ἐξ ὧν πρώτων ἀποδείκνυσι, καὶ τρίτον τὰ πάθη, ὧν τί σημαίνει ἕκαστον λαμβάνει.' English translation Barnes, *Aristotle: Posterior Analytics*, 15. James of Venice's translation: 'Omnis enim demonstrativa scientia circa tria est, et quecumque esse ponuntur (hec autem sunt genus, cuius per se passionum speculativa est), et que communes dicuntur dignitates, ex quibus primis demonstrant, et tertium passiones, quarum quid significet unaqueque accipit', *AL* IV.1, 24.

ASTRONOMY IN GROSSETESTE'S THOUGHT 223

derived from the primitives or first principles intrinsic to celestial bodies (2); and waxing would be an attribute (3) that can be demonstrated as holding of the kind intrinsically. Each demonstration within the science of astronomy would hold to this same pattern.

An important point of considerable significance for the interpretation of Grosseteste's mode of reasoning in *On the Sphere* follows from this definition of a science. Science in the Aristotelian sense is not primarily an activity centred on gathering new knowledge through observation. Within this framework, observation is central to acquiring the knowledge from which scientific understanding or ἐπιστήμη issues, but scientific activity properly speaking begins only when the observational data have been gathered. In the second and final book of the *Posterior Analytics* Aristotle moved from discussing demonstration to offering an account of how human beings come to grasp the universal causes that can function as middle terms in a demonstration properly speaking; but this is a pre-requisite for science and not intrinsic to science of itself.[51] Just as the purpose and end of inquiry is no mere notice of facts (γνῶσις) but understanding of causes (ἐπιστήμη), so scientific activity is not simply getting to know (γιγνώσκειν), but understanding (ἐπίστασθαι). Turning the liberal art of astronomy into an Aristotelian science would not entail, primarily, acquiring new information, but rather the imposition of a specific explanatory structure on information already acquired. *On the Sphere* maps well onto this process (see Ch. 7).

3.3 The Hierarchical Ordering of Sciences and the Concept of Subalternation

The Aristotelian account of a science does not only allow for a hierarchical ordering of principles, causes, and effects within a single science. It also allows, under specific conditions, a hierarchical ordering of the individual sciences to one another. Aristotle took great pains to specify the conditions under which different sciences could have overlapping

---

[51] For a refutation of the idea that the *Posterior Analytics* provide anything like a scientific method in the modern sense, see Burnyeat, 'Aristotle on Understanding Knowledge', 127–8.

224 THE SCIENTIFIC WORKS OF ROBERT GROSSETESTE

principles, since his account of a science as an exhaustive explanatory structure dealing with a single kind of thing otherwise could appear to create discrete sciences without any intrinsic relation to one another. Since demonstrations must proceed from principles intrinsic to each kind, it is not possible to cross between kinds in a demonstration without violating the criteria for how ἐπιστήμη is reached.[52] That is, all three terms in a demonstration must come from the same kind, otherwise they would not be the sort of truths that are appropriate for demonstrations. A geometrical demonstration by way of principles belonging intrinsically to arithmetic but not to geometry would not prove anything about geometry. When this nonetheless does not create monadic and discrete sciences without any reciprocal order, this is because some principles are so fundamental that they form part of the axioms and first principles of many sciences. Aristotle distinguished between the principles that are proper to a single science and those that are common to many, with the qualification that common principles will behave differently as they are adapted to the specific kind studied by a given science. Since different kinds of things are related as genera, that is, broader family groups, and species, that is, more narrowly defined kinds of things, so are also sciences related by the generic and specific principles that are the primitives of each science.

Since each science must *assume* its first principles as a basis for demonstrations concerning its attributes, it cannot *demonstrate* the validity of its own principles.[53] Some sciences, however, take some or most of their principles from other, higher sciences. In this way, optics falls under geometry and music falls under arithmetic, since optics and music derive their principles from geometry and arithmetic respectively.[54] Such a relationship between two sciences came to be called 'subalternatio' in Latin ('subalternation' in English). In such cases, the higher science demonstrated the reason why for the lower, while the lower demonstrated the specific matters of fact pertaining to the kind of thing it studied.

[52] Aristotle, *Posterior Analytics* I.7, 75a38–b20.
[53] Aristotle, *Posterior Analytics* I.9, 76a16ff.
[54] Aristotle, *Posterior Analytics* I.13, 78b35–9.

ASTRONOMY IN GROSSETESTE'S THOUGHT 225

Crucially for present concerns, Aristotle described astronomy, ἀστρολογική ἐπιστήμη (*astrologikê epistêmê*, the science of astrology), as a science that demonstrates the reason why, while a subalternate science studies the matters of fact pertaining to the observable movements and patterns, τὰ φαινόμενα (*ta phainomena*, Latin *apparentia*) of celestial bodies.[55] This latter science appears to be what Aristotle immediately goes on to call 'nautical astronomy', ἀστρολογία ναυτική (*astrologia nautikê*, Latin *astrologia navalis*). Grosseteste certainly took the study of celestial observations or *apparentia* and nautical astronomy to be identical in his commentary.[56] While the subalternate science studies the heavens in order to navigate more accurately and securely, the subalternating science of astronomy proper, or mathematical astronomy (ἀστρολογία μαθηματική, *astrologia mathematikê*) demonstrates the reason why celestial bodies possess the attributes that can be observed.

The notions of subalternation and of the difference between demonstration proper are intertwined in Aristotle's presentation, and come to their most explicit expression in what is now counted as the thirteenth chapter of *Posterior Analytics* Book I. While Aristotle here establishes a general point, most of his examples are astronomical, and astronomy as a science is specifically mentioned among the sciences that demonstrate the reason why. In this passage, then, the discipline of astronomy emerges as a paradigmatic example of a science that provides the causal explanations of what can be observed concerning celestial phenomena, within a system of sciences that admits of interrelationships ranging from a partial overlap in first principles to a full subalternation of one science to another. In other words, Aristotle's scheme provided exactly what Grosseteste would need in order to structure the wealth of

---

[55] Aristotle, *Posterior Analytics* I.13, 78b35–79a2: 'τοιαῦτα δ' ἐστὶν ὅσα οὕτως ἔχει πρὸς ἄλληλα ὥστ' εἶναι θάτερον ὑπὸ θάτερον, οἷον τὰ ὀπτικὰ πρὸς γεωμετρίαν καὶ τὰ μηχανικὰ πρὸς στερεομετρίαν καὶ τὰ ἁρμονικὰ πρὸς ἀριθμητικὴν καὶ τὰ φαινόμενα πρὸς ἀστρολογικήν. σχεδὸν δὲ συνώνυμοί εἰσιν ἔνιαι τούτων τῶν ἐπιστημῶν, οἷον ἀστρολογία ἥ τε μαθηματικὴ καὶ ἡ ναυτική, καὶ ἁρμονικὴ ἥ τε μαθηματικὴ καὶ ἡ κατὰ τὴν ἀκοήν.' James of Venice's translation: 'Huiusmodi autem sunt quecumque sic se habent ad invicem, et quod alterum sub altero est, ut speculativa ad geometriam et machinativa ad stereometriam et armonica ad arithmeticam et apparentia ad astrologicam. Fere autem univoce sunt harum quedam scientiarum, ut astrologia mathematica que et navalis est, et armonica mathematica que est et secundum auditum.', *AL* IV.1, 31. For the overlapping meanings of astrology and astronomy in this body of texts, see *Knowing and Speaking*, 166–7.

[56] Grosseteste, *Comm. Post. An.* I.12, ed. Rossi, 194–5.

226 THE SCIENTIFIC WORKS OF ROBERT GROSSETESTE

information about celestial phenomena into a framework leading to understanding and not mere accumulation of facts.

4 Grosseteste on Aristotelian Scientific Understanding

There can be no doubt that Grosseteste eventually integrated the ideas of scientific demonstration from the *Posterior Analytics* into his own thought.[57] His treatise *On the Rainbow* explicitly sets out to provide an account of the rainbow demonstrating the reason why, *propter quid*, which takes as its starting point comments Aristotle made in the passage from *Posterior Analytics* I.13 quoted above.[58] The mode of reasoning put into practice in *On the Rainbow* had been laid out in detail in the *Commentary on the Posterior Analytics*. As the first comprehensive commentary on this text in Latin, seemingly composed before Averroes's commentary was available, Grosseteste's *Commentary* relied heavily on James of Venice's Latin translation of Aristotle's Greek text. This translation appears to have been accompanied by glosses translated from Greek commentaries, most probably composed by the sixth-century philosopher and theologian John Philoponus; the influence of these glosses can be discerned on Grosseteste's work.[59] In addition, Grosseteste also used an epitome of Aristotle's work by Themistius (*c.*317–388 CE), translated into Latin from Arabic by Gerard of Cremona in the mid-twelfth century.[60] In order to show how the Aristotelian notion of a science can be

---

[57] For studies of Grosseteste's *Commentary*, see for example, Jeremiah Hackett, 'Robert Grosseteste and Roger Bacon on the *Posterior Analytics*', in Pia Antolic-Piper, Alexander Fidora, and Matthias Lutz-Bachmann (eds.), *Erkenntnis Und Wissenschaft / Knowledge and Science. Probleme der Epistemologie in der Philosophie des Mittelalters / Problems of Epistemology in Medieval Philosophy* (Berlin: De Gruyter, 2004), 161–212; W. R. Laird, 'Robert Grosseteste on the Subalternate Sciences', *Traditio*, 43 (1987), 147–69; David Bloch, 'Robert Grosseteste's Conclusiones and the commentary on the Posterior Analytics', *Vivarium*, 47 (2009), 1–23; Christina Van Dyke, 'The Truth, the Whole Truth, and Nothing but the Truth: Robert Grosseteste on Universals (and the *Posterior Analytics*)', *Journal of the History of Philosophy*, 48 (2010), 153–70.

[58] A new edition of Grosseteste's treatise *On the Rainbow* will form part of Volume V of the present series.

[59] For Grosseteste's sources for the Commentary, see Hackett, 'Grosseteste and Bacon on *Posterior Analytics*', 165–6.

[60] This translation is edited in J. R. O'Donnell, 'Themestius's Paraphrasis of the *Posterior Analytics* in Gerard of Cremona's Translation', *Mediaeval Studies*, 20 (1958), 242–315.

ASTRONOMY IN GROSSETESTE'S THOUGHT   227

picked out also in *On the Sphere* it is necessary first to outline Grosseteste's use of it in the *Commentary*.

4.1 Subalternation and the Science of Astronomy

It should be noted at first that Grosseteste in his *Commentary* used the structures of the liberal arts tradition to frame and clarify his explanation of the Aristotelian notion of subalternation of sciences:

> The subject of arithmetic is number simply speaking, in as much [number] is able to receive dispositions that are absolute and not said in reference to something else; however, when dispositions said in reference to something else are conjoined to number and one composite comes to be from these, then the subject of music is established. For the subject of music is not number to which relation accrues, but the composite made from number and relation; and number is not predicated of this composite, because a part is not predicated of its whole.[61]

The hierarchy of priority that Boethius had set up (§3.1) is here placed more explicitly within the Aristotelian system of subalternation, by specifying what Boethius had left indeterminate. Music cannot simply be reduced to the science of numbers, since the subject of music is not number per se, but a composite of which number is a part. The same, Grosseteste claims, is true of geometry. It is possible to give numerical value to all geometrical magnitudes, 'and yet it is not truly said that lines and surfaces are numbers, but when number crosses over [from its own genus] into them [sc. to lines and shapes], lines and shapes somehow

---

[61] Grosseteste, *Comm. Post. An.* I.12, ed. Rossi, 195: 'Subiectum enim arithmetice est numerus simpliciter secundum quod est receptibilis dispositionum absolutarum et non ad aliquid dictarum; cum autem cum numero coniungantur dispositiones ad aliquid dicte et fit ex eis unum compositum, iam constituitur subiectum musice. Non enim est subiectum musice numerus cui accidit relatio, sed compositum ex numero et relatione, et de hoc composito non predicator numerus, quia pars non predicator de suo toto.'

228   THE SCIENTIFIC WORKS OF ROBERT GROSSETESTE

come to be in the nature of number (...).'[62] In other words, it is possible to rank and organize the quadrivial arts based on the logical primacy of their principles, but they cannot simply be collapsed into subdisciplines of arithmetic.

The Aristotelian conceptual scheme elaborated in the *Posterior Analytics*, then, did not in Grosseteste's *Commentary* replace or break the liberal arts and their reciprocal structure underlying *On the Liberal Arts*. The new learning, however, allowed for a much more nuanced and profound account of how individual sciences might be related. Grosseteste returned several times to the system of the quadrivial arts to explain the complex details of the notion of subalternation. As shown above (§3.2), Aristotle stated that each science posited a set of first principles and primitive causes from which all demonstrations within that science ultimately derived. These principles could be primitives within the kind with which the science was concerned or derived from one or more related sciences. If a science received its first principles wholesale from a higher science, the subalternation was so complete that the two sciences could share a single name; such was the case with the science of musical sound, which was fully subalternated to mathematical music.[63] That is to say, the composite of number and relation that defined music as a science could be predicated of both purely mathematical study of relational numbers and of the study of sounding relational numbers. While music in turn is subalternated to arithmetic, this subalternation is only partial, as shown in the quotation above; number cannot be predicated of relational number in an unqualified sense. In this way, Aristotle's account of a science allowed Grosseteste to develop a complex and profound grasp of how various sciences are related.

This notion of the hierarchical relations pertaining between different sciences also allowed Grosseteste to formulate three essential requirements for a unified science:

---

[62] Grosseteste, *Comm. Post. An.* I.12, ed. Rossi, 195–6: 'nec tamen vere dicitur quod linee et superficies sint numeri, sed cum descendit numerus in hec aliquo modo fiunt hec in natura numeri sicut supra dictum est.' The phrase 'supra dictum' refers to I.7, 136–7.
[63] Grosseteste, *Comm. Post. An.* I.12, ed. Rossi, 195: 'armonica secundum auditum et armonica mathematica utraque dicitur armonica.'

ASTRONOMY IN GROSSETESTE'S THOUGHT 229

a unity of the subject upon which demonstration is constructed; axiomatic principles, unified in this subject, from which demonstration is made; and that the subject has species or parts or intrinsic attributes from which the demonstrated conclusion is composed.[64]

The Aristotelian account of a science, then, provided Grosseteste with resources both for making explicit the unity of a single science and for emphasizing the ways in which a given science is related to other sciences. Grosseteste put this flexible and precise conceptual scheme to use to clarify the complex and somewhat ambiguous position astronomy occupied in relation to other arts and sciences. On the one hand, astronomy was clearly subalternated to mathematics in general and to the other quadrivial disciplines more specifically, within the liberal arts scheme as well as the new Aristotelian natural philosophy (see §§2 and 3.3 above). On the other hand, Grosseteste recognized that the subject of astronomy, mobile magnitude, also brought it into contact with the same new learning.[65] Grosseteste accepted and emphasized Aristotle's insistence that each science has its own set of questions proper and unique to it; but this raised a potential objection in the case of astronomy:

Now if someone were to object that both the natural philosopher and the astronomer investigate and conclude, albeit through different middle terms, that the earth and the moon are spherical, and therefore overlap in terms of questions, it should be responded to them that the natural philosopher deals with these [sc. the earth and the moon] in as much as they are natural corporeal mobile substances, while the astronomer deals with them only in as much as they are mobile magnitudes; for the mobile magnitude is the subject of the astronomer.[66]

---

[64] Grosseteste, *Comm. Post. An.* I. 18, ed. Rossi, 260: 'unitas subiecti super quod erigitur demonstratio, principia inmediata unificata in subiecto illo ex quibus fit demonstratio, et quod subiectum habeat aut species aut partes aut per se accidentia ex quibus complectatur conclusion demonstrativa.'

[65] This was already indicated in Grosseteste, *De artibus liberalibus*, §11; *Knowing and Speaking*, 88–9.

[66] Grosseteste, *Comm. Post. An.* I.11, ed. Rossi, 174–5: 'Si autem obiciat quis quod tam naturalis quam astronomus interrogat et concludit, licet per diversa media, quoniam terra et luna sunt spherice et ita communicant in questionibus, respondetur ei quod naturalis accipit hec in quantum sunt substantie corporee naturales mobiles, astronomus vero accipit hec in quantum sunt solum magnitudines mobiles; magnitudo enim mobilis est subiectum astronomi.'

230 THE SCIENTIFIC WORKS OF ROBERT GROSSETESTE

In other words, even if astronomy and natural philosophy concerned the same subject and sometimes reached identical solutions, they were distinct sciences with their own distinct questions and modes of demonstration. While this passage indicates an intrinsic relationship between natural philosophy and astronomy, however, it is not immediately clear from the passage just quoted that one of these sciences is subalternated to the other. What is left indeterminate here is made clear elsewhere in Grosseteste's oeuvre. The order that on Grosseteste's view pertained between the science of nature and the science of stars and planets only emerges in the light of his explication of Aristotle's distinction between demonstrations *propter quid* and demonstrations *quia*.

4.2 Demonstration *Propter Quid* and Demonstration *Quia*

The concept of subalternation is intrinsically linked to the doctrine of demonstration. It is far from coincidental that Grosseteste's most elaborate account of subalternation occurs in the context of his commentary on *Posterior Analytics* I.13, which also contains Aristotle's most explicit treatment of the difference between demonstration properly speaking, *propter quid* or 'demonstration of the reason why', and demonstration in a less proper sense, *demonstratio ipsum quia* or 'demonstration of the fact that'. Grosseteste used the occasion to sum up the requirements Aristotle gave for demonstration properly speaking, and to deepen the understanding of what this implies by a thorough analysis contrasting the proper and the looser kinds of demonstration:

> From the beginning of the book up to this point, Aristotle has demonstrated that demonstration is from [principles] that are primitive, true, immediate, prior, better known, causal, necessary, intrinsic in and of themselves, universal, eternal, and incorruptible, and from appropriate [sc. to each science] principles, questions, and conclusions; and all these conditions are not brought together at one and the same time except in what is above all and most properly called demonstration,

ASTRONOMY IN GROSSETESTE'S THOUGHT 231

which produces knowledge most properly called, according to how knowing is defined at the start [sc. of the book].[67]

This passage shows that Grosseteste was well aware of how Aristotle's account of demonstration followed from his definition of the highest form of knowing as understanding causal explanations. A demonstration properly speaking not only produced knowledge of this kind, it also provided a structure through which one could analyse one's own reasoning and in this way make a reasoned judgement that one's knowledge was of this kind.[68] Grosseteste knew full well that such analysis relied on more than the logical form of demonstrative reasoning:

knowledge acquired through demonstration is either acquired through the proximate cause of the thing known or it is not acquired through the proximate cause of the thing known. The one that is acquired through the proximate cause, then, is called 'knowledge of the reason why' [scientia propter quid], and this is knowledge most fully and properly speaking, and the demonstration by which this knowledge is acquired is demonstration in its fullest sense. The [knowledge] that is not from the proximate cause is called 'knowledge of the fact that' [scientia quia], and this is called knowledge in a secondary way, and the demonstration by which this is acquired is called demonstration in a secondary way.[69]

The terms of a demonstration, then, had to be connected through real relations relevant to the kind of thing under investigation. As for

---

[67] Grosseteste, Comm. Post. An. I.12, ed. Rossi, 188: 'A principio libri usque ad locum istum demonstravit Aristoteles quod demonstration est ex primis et veris et immediatis et prioribus et notioribus et causis et necessariis et per se inherentibus et universalibus et perpetuis et incorruptibilibus et ex propriis tam principiis quam interrogationibus et conclusionibus; et he omnes conditiones non aggregantur simul nisi in demonstration maxime et propriissime dicta, que acquirit scientiam propriissime dictam, secundum quod diffinitum est scire in principio.'

[68] See Hackett, 'Grosseteste and Bacon on Posterior Analytics', 168.

[69] Grosseteste, Comm. Post. An. I.12, ed. Rossi, 189: 'Scientia itaque acquisita per demonstrationem aut est per proximam causam rei scite aut aut est acuisita non per proximam causam rei scite. Que autem est acquisita per proximam causam vocatur scientia propter quid, et hec est scientia maxime et propriissime dicta, et demonstratio qua hec scientia acquiritur est maxime demonstratio. Illa autem que non est er proximam causam dicitur scientia quia, et hec est scientia dicta per posterius, et demonstratio qua hec acquiritur est demonstration dicta per posterius.'

232 THE SCIENTIFIC WORKS OF ROBERT GROSSETESTE

Aristotle, this allowed Grosseteste to show the continuity between the understanding the mind reached as a product of demonstrative reasoning, and the ordered systems of such understanding of a given kind of thing that constituted a science.

A concrete example of Grosseteste's grasp not only of Aristotle's ideal of demonstration but also how astronomy as a science combined principles particular to astronomy with principles derived from other sciences can be found in his treatment of Aristotle's sample demonstrations of attributes of the moon quoted above (§3.1, Examples 1 and 2). Grosseteste summarized and clarified the point of these lunar examples and other astronomical examples used by Aristotle which is to illustrate the difference between demonstration *propter quid* and *ipsum quia*. If the cause of the attribute to be demonstrated is known to someone,

> then for them the effect is demonstrated through the cause, and there will be knowledge and demonstration of the reason why, as it is for the one who knows through astronomical demonstration that the planets are near and through natural demonstration that the moon is round; for, through this they will demonstrate and know the reason why the planets do not twinkle and why the moon admits of waxing and waning. They, however, who do not know the aforesaid through demonstration, but who nevertheless know from sense perception that the planets do not twinkle and that things that do not twinkle are near, and that the moon admits of waxing and waning and that things that admit of waxing and waning are round, will demonstrate and know from the effect that the planets are near and that the moon is round.[70]

Demonstrations *propter quid*, then, draw on causal principles such as the relative proximity of the planets to earth or the sphericity of the moon to

---

[70] Grosseteste, *Comm. Post. An.* I.12, ed. Rossi,190: 'tunc apud ipsum demonstrabitur effectus per causam, et erit scientia et demonstratio propter quid, sicu test apud ipsum qui novit per demonstrationem astronomicam quod planete sunt prope et per demonstrationem naturalem quod luna sit circularis; per hoc enim demonstrabit et sciet propter quid planete non scintillant et luna recipit incrementa. Qui autem non novit predicta per demonstrationem, novit tamen per sensum quod planete non scintillant et quod non scintillant sunt prope et quod luna recipit incrementa et quod recipientia incrementa sunt circularia, demonstrabit et sciet per effectum quod planete sunt prope et quod luna est circularis.'

ASTRONOMY IN GROSSETESTE'S THOUGHT   233

demonstrate the reason why for effects such as the planets not twinkling or the moon's waxing, while middle terms derived from sense perceptions only establish matters of fact and not the reason why.

It is important to note that Grosseteste goes beyond Aristotle in spelling out explicitly what the argument of this part of the *Posterior Analytics* assumes. He provides thorough proofs for the premises 'what is near does not twinkle' and 'what is spherical admits of waxing and waning', in both cases drawing heavily on geometry and geometrical optics to find principles providing causal explanations of the astronomical phenomena in question. As Jeremiah Hackett has pointed out, Grosseteste relied heavily on Euclid to formulate his own understanding of Aristotle's main arguments in the *Posterior Analytics*.[71] Euclidian geometry provided Grosseteste with a paradigmatic case of Aristotelian demonstrative reasoning leading to understanding and, since so many of Aristotle's examples were astronomical, justifying the basic premises of Aristotle's examples also allowed Grosseteste to bring out the ways in which astronomy was subalternated to geometry. Geometry was also of fundamental importance to Ptolemy's astronomical method as well as later developments and critiques of the *Almagest*. What characterizes Grosseteste's use of geometry in his *Commentary* is the detailed and deliberate way in which he uses geometrical demonstrations to construct demonstrative reasoning as described by Aristotle. That is to say, far from reducing Aristotle's account of demonstrative reasoning to geometry, Grosseteste enlists geometry in the service of illustrating the formal and abstract structure of demonstrative reasoning as well as in proving fundamental principles and premises underlying in particular the astronomical demonstrations Aristotle invoked as examples. As will be shown in the next chapter, this was fundamental to Grosseteste's reasoning also in *On the Sphere* (see Ch. 7, §2).

Geometry, moreover, is not the only science to which astronomy is subalternated in the last quotation from the *Commentary* above. Grosseteste's examples of *propter quid* demonstrations here are taken from Aristotle's account (see §3.1, Example 1), but in specifying how the middle terms of the sample demonstrations could become known,

---

[71] Hackett, 'Grosseteste and Bacon on *Posterior Analytics*', 166.

234 THE SCIENTIFIC WORKS OF ROBERT GROSSETESTE

Grosseteste again specified what Aristotle left indeterminate.[72] The fact that the planets are near, which on Aristotle's account explains why they do not twinkle, is known through astronomical demonstration, while the fact that the moon is round, which explains why it admits of waxing, is known through natural reasoning. This distinction between astronomical inquiry and natural philosophy is central to early paragraphs of *On the Sphere* (DS §§5–9), as will be discussed in Chapter 7 below. However, neither these passages in *On the Sphere* or the passage from the *Commentary* quoted above make clear what 'natural demonstration' means, and how precisely it differs from astronomical demonstration. It is therefore necessary to look more widely at Grosseteste's works to find an explanation of this important distinction.

The most explicit discussion in Grosseteste's works of what defines and distinguishes natural philosophy and natural reasoning from other related sciences can be found in his *Notes on Aristotle's Physics*.[73] These notes or glosses are dated to the mid- to late 1220s, but it is clear that Grosseteste was familiar with the text of the *Physics* well before that time (§2). In the *Physics*, Aristotle had delineated the proper purview of the natural philosopher, the student of φύσις (*phusis*) or nature, from other disciplines, particularly mathematics.[74] For Aristotle, there was no universal nature but only the specific natures of natural kinds of thing.[75] The study of nature, then, meant the study of the principles of change and completion of change specific to each natural kind. The branches of mathematics that are nearest to natural philosophy, such as music, optics, and astronomy, were the most challenging to fit into such a distinction of sciences.[76] Since these branches of mathematics study mathematical

---

[72] Cf. Aristotle, *Posterior Analytics* I.13, 78a30–b13.

[73] Published under the title *Commentarius in VIII libros physicorum Aristotelis* by Richard Dales (Grosseteste, *Comm. Pys.*). However, since these glosses or comments on the *Physics* were never elaborated into a finished commentary form like the *Commentary on Posterior Anaytics*, Neil Lewis has suggested that '*Notes on the Physics*' is a more appropriate title; see Neil Lewis, 'Robert Grosseteste's Notes on the Physics', in Evelyn A. Mackie and Joseph Goering (eds.), *Editing Robert Grosseteste* (Toronto: University of Toronto Press, 2003), 103–34.

[74] Aristotle, *Physica* II.2, 193b22–194b15.

[75] For Aristotle's concept of nature, see for instance Lang, *The Order of Nature*; cf. *Knowing and Speaking*, 279–81.

[76] Aristotle, *Physica* II.2, 194a7–12.

ASTRONOMY IN GROSSETESTE'S THOUGHT   235

entities not in and of themselves but as they occur in nature, they seem to
elide a clear separation of mathematics from natural philosophy.

In his commentary on this passage, Grosseteste attempted to systematize and clarify Aristotle's account. He set up his clarification of
the distinction between mathematical and natural sciences by separating
out three different subjects of study: natural bodies, quantifiable magnitudes intrinsic to natural bodies, and the properties that belong to
magnitudes considered in isolation from natural bodies. Pure mathematics, Grosseteste explained, concerns the last of these three, that is,
magnitude fully abstracted from matter and movement, and what can be
demonstrated to pertain to abstract magnitude *qua* abstract magnitude.[77]
Natural philosophy, in contrast, primarily concerns the first subject
listed, that is, corporeal and mobile body and what can be demonstrated
to pertain to a given corporeal and mobile body by virtue of its nature.[78]
The second subject, magnitude as occurring in natural bodies, falls under
the purview both of natural philosophy and of specialized mathematical
sciences such as astronomy and optics. Astronomy, alongside optics and
harmonics, therefore, occupies a liminal position between mathematics and
natural philosophy:

> Now, the astronomer demonstrates shaped magnitudes concerning
> natural bodies, but not in as much as [these magnitudes] belong to
> [natural bodies] by virtue of those bodies being natural. For the
> astronomer does not show how being spherical is a property of
> the moon by virtue of the moon being a natural body; it suffices for
> the astronomer to show that the moon is spherical through an effect
> or through the cause of sphericity—the cause of sphericity simply
> speaking transcends [individual] nature. The astronomer therefore
> shares with the natural philosopher both the subject and the predicate

---

[77] Grosseteste, *Comm. Phys.*, 36: 'Dico itaque quod tria sunt, corpus scilicet physicum,
magnitudines que accidunt corporibus physicis, et accidencia magnitudinum pure.
Mathematici magnitudines abstrahunt a motu et a materia et subiciunt magnitudines abstractas
et de his demonstrant accidentia per se magnitudinibus.'

[78] Grosseteste, *Comm. Phys.*, 36–7: 'Physicus vero non demonstrat per se accidentia magnitudinibus de magnitudinibus inquantum accidunt simpliciter magnitudinibus, sed de corporibus physicis demonstrat magnitudines figuratas secundum quod accidunt corporibus physicis
ex parte ea qua physica sunt.'

236   THE SCIENTIFIC WORKS OF ROBERT GROSSETESTE

of the conclusion being demonstrated; but the natural philosopher demonstrates that the predicate belongs to the subject by virtue of its nature, while the astronomer pays no heed to whether it belongs by nature or not.[79]

It is important to note that only the reasoning of the natural philosopher here fulfils the criteria for demonstration properly speaking set out in the *Commentary on Posterior Analytics*. The natural philosopher argues from the immediate, or proximate, cause of the sphericity of the moon and therefore demonstrates the reason why the moon is spherical; the astronomer, arguing either from the effect or from a more remote cause, only demonstrates the matter of fact and not the reason why, and this reasoning is only called demonstration in a less proper sense.

This passage from the *Notes on the Physics* therefore explains both what is meant by the term 'natural demonstration' in the *Commentary on Posterior Analytics*, and why Grosseteste there is stating that only natural demonstration gives the *propter quid* of the sphericity of the moon while astronomical arguments here only establish the matter of fact. This implies, as Grosseteste makes more explicit in the *Notes on the Physics*, that astronomy in some respects is subalternated to natural philosophy as well as to mathematics.[80] Grosseteste mentions in passing in the *Commentary on Posterior Analytics* that a natural demonstration of the sphericity of the moon can use the perfect homogeneity of celestial bodies as a middle term, but he stops short of setting out such a

---

[79] Grosseteste, *Comm. Phys.*, 37: 'Astrologus uero de corporibus physicis demonstrat magnitudines figuratas, sed non inquantum accidunt eis ex ea parte qua corpora sunt physica. Non enim ostendit spericum accidere lune ex ea parte qua luna est corpus naturale, sed sufficit ei ostendere lunam esse spericam aut per effectum aut per causam spericitatis. Causa <autem> spericitatis simpliciter naturam transcendit. Astrologus igitur et in subiecto et in predicato conclusionis demonstrare communicat cum physico, sed physicus demonstrat predicatum accidere subiecto per naturam; astrologus vero non curat an accidat a natura an non.'

[80] Grosseteste, *Comm. Phys.*, 37: 'Propterea subiectis pure mathematicis superadduntur accidencia naturalia et fit subiectum compositum ex mathematico et naturali, et demonstratur accidens mathematicum de tale subiecto composito secundum quod accidit ei propter accidens naturale quod est in subiecto. Utpote ex linea et radiositate componitur linea radiosa et demonstrantur de ea accidencia et figurationes lineae quae accidunt ei *ex parte* radiositatis, et propter hoc magis physicum quam mathematicum est hoc. Et forte astrologia in quibusdam conclusionibus suis est huic simile.'

ASTRONOMY IN GROSSETESTE'S THOUGHT 237

demonstration in full in that work.[81] As will be shown in the next chapter, however, such a demonstration can be found in *On the Sphere* (Ch. 7, §7.2).

Within this framework, then, Grosseteste had the resources to formulate a notion of astronomy as a science in the Aristotelian sense of ἐπιστήμη that remained in strong continuity with the treatment of astronomy in *On the Liberal Arts*. This can be seen both from Grosseteste's treatment of subalternation, where astronomy is shown to be partially subalternated to the other quadrivial arts as well as to natural philosophy while in turn subalternating the lesser science of nautical astronomy to itself, and from the way Grosseteste engaged with and explicated the predominantly astronomical examples Aristotle had used to illustrate his account. In other words, the Aristotelian commentaries show Grosseteste in possession of a confident grasp of astronomy as a science both in an abstract and a formal way, and, more concretely, a grasp of how astronomical demonstrations could be made and recognized as such. The continuity between this mature notion of astronomy as science and his earlier exposition of astronomy as a liberal art, moreover, suggests that this intellectual development holds importance also for *On the Sphere*. How this can be discerned in the text itself is laid out in Chapter 7 below.

4.3 The Role of Observation

A final note to be made concerning Grosseteste's use of the Aristotelian idea of science is on the role of observation within the conceptual scheme Grosseteste sets out. As mentioned above (§3.2), Aristotelian scientific reasoning is distinct from the process of acquiring a grasp of the universal principles from which true demonstration must proceed. Grosseteste maintained this distinction in his *Commentary*, and again his use of examples is illuminating for how he conceived astronomy. In the second

[81] Grosseteste, *Comm. Post. An.* I.12, ed. Rossi, 190: 'Et forte eiusdem conclusionis est scientia propter quid et scientia quia in eadem scientia, licet Aristoteles de hoc non ponat exemplum. In scientia enim naturali potest demonstrari quod luna sit circularis tam per hoc quod ipsa est corpus omogeneum quam per hoc quod recepit incrementa.'

238   THE SCIENTIFIC WORKS OF ROBERT GROSSETESTE

book of the *Commentary*, Grosseteste follows Aristotle's account of how human beings come to grasp the causal principles that function as middle terms in scientific demonstrations. Since the formal cause, the account of what a thing is, held paramount importance for demonstrations leading to ἐπιστήμη, the relevant kind of cause would be identical to a definition.[82] The example Aristotle used to illustrate the identity of the relevant cause with a definition was again astronomical, this time the eclipse of the moon. Grosseteste explains:

> Since the cause of lunar eclipse is the interposition of the earth along a straight line between the sun and the moon, seeking to know whether there is an eclipse of the moon or whether the moon is eclipsed in this way is to seek, as a middle term, whether there is an opposition of the earth along a straight line between the sun and the moon; and this is to seek the definition of an eclipse. For to be eclipsed is to have a shadowy body placed as an obstruction that prevents the reception of illumination, and when we know by way of such a middle term, we at the same time know both the fact that something is so and the reason why something is so.[83]

In contrast, if the cause of an eclipse is not known, it is still possible to know the fact that an eclipse is taking place:

> For instance, when we know that the moon is eclipsed from the fact that it casts no shadow at the time of the full moon, even when there is nothing placed in the middle between us and the moon by which its illumination may be kept from reaching the earth, we only know that it is eclipsed and not yet the reason why; and similarly we know that there is an eclipse but do not yet know what an eclipse is.[84]

---

[82] This is elaborated in Aristotle, *Posterior Analytics* II, 8–10.

[83] Grosseteste, *Comm. Post. An.* II.2, ed. Rossi, 333: 'Cum causa defectus lune sit interpositio terre secundum lineam rectam inter solem et lunam, querere an defectus lune sit vel an luna sic deficiat, est querere sicut medium an oppositio terre sit secundum lineam rectam inter solem et lunam; et hoc est querere diffinitionem defectus. Deficere namque est habere umbrosum interpositum quod prohibeat receptionem illuminationis, et cum scimus per tale medium, tunc simul cogniscimus quia est et propter quid est'.

[84] Grosseteste, *Comm. Post. An.* II.2, ed. Rossi, 333–4: 'sicut cum scimus quod luna deficit propter hoc quod in plenilunio non proicit umbram; cum tamen nichil sit medium quod

ASTRONOMY IN GROSSETESTE'S THOUGHT 239

Grasping the fact of the eclipse is not yet the perfect knowledge that is acquired through an understanding of the cause, but it may serve as a spur for seeking a cause and a definition of the phenomenon in question:

we may proceed to seek to know the reason why there is an eclipse and what an eclipse is, and this is to seek a middle term that is the definition of an eclipse, such as the placing of the earth between the sun and the moon or the opposition of the sun along the diameter of the world. The aforementioned middle term by which we know the fact that the moon is eclipsed does not provide the cause, but only a sign; for the fact that the moon at the time of the full moon does not create shadow of standing things on the surface of the earth, even if the air between us and the moon is pure and without a cloud placed between, is a sign of the eclipse and not its cause—indeed, *because* it is eclipsed, *therefore* it cannot create shadow.[85]

This distinction between cause and sign, then, is crucial for understanding Grosseteste's Aristotelian ideal of scientific reasoning. Signs reveal the existence of a phenomenon in need of explanation, but in order to explain this phenomenon one needs to seek its cause.

This cause, however, cannot simply be the product of observation and sense perception. Grosseteste makes this point more emphatically and explicitly than Aristotle had done. Grosseteste notes that Aristotle may appear to be positing that universals can be the objects of sense perception, which was a position held by many.[86] However, Grosseteste

interpositum est inter nos et lunam per quod prohibeatur eius illuminatio a terra, scimus solum quod deficit et nondum scimus propter quid. Et similiter scimus quia defectus est et nondum scimus quid est defectus'.

[85] Grosseteste, *Comm. Post. An.* II.2, ed. Rossi, 334: 'possumus adhuc querere propter quid deficit et quid est defectus, et querere hoc est querere medium quod est diffinitio defectus, sicut obiectionem terre inter solem et lunam aut oppositionem soli secondum diametrum mundi. Hoc enim forte vocat Aristoteles conversionem lune, scilicet oppositionem qua luna opponitur soli per diametrum mundi vel reditionem lune ad oppositum solis per diametrum mundi. Medium autem supra dictum quo scimus quia luna deficit non dicit causam, sed signum solum; quod enim luna in hora plenilunii non facit umbram rerum erectarum super superficiem terre, cum tamen aer sit purus inter nos et lunam et non sit nubes interposita, signum est defectus et non causa immo quia deficit ideo non potest umbram facere.'

[86] Grosseteste, *Comm. Post. An.* I.18, ed. Rossi, 267, commenting on *Posterior Analytics* I.31, 87b33ff.

240 THE SCIENTIFIC WORKS OF ROBERT GROSSETESTE

goes on to state, it follows from careful attention to Aristotle's argument that this is impossible. Even if human beings could perceive the cause explaining the phenomenon they seek to understand, they would still not perceive it *qua* cause, since sense perception only perceives the particular while knowing a cause is a grasp of the universal. Again, the concrete example illustrating this point is the lunar eclipse:

> For instance if we were to be elevated to the moon at the time of a lunar eclipse, and we were to see clearly the eclipse of the moon and its cause, that is, the earth placed between the moon and the sun, yet we would not have attained *scientia*, because sensation would not shows us anything except *this* eclipse and *this* positioning of the earth at *this* time, and it would not show us the cause of the eclipse in a universal sense.[87]

Causes or universals, then, are grasped from repeated sense impressions, but this grasp is an act of intellect that is distinct in kind from the act of sense perception. Grosseteste's point here is not that observation has no role in human inquiry, but rather that the grasp of universals cannot be reduced to sense perception, and that scientific reasoning requires a prior grasp of universals and therefore is distinct in kind from observational inquiry.

5 Grosseteste's Account of the Discipline of Astronomy as a Framework for *On the Sphere*

The expectations and assumptions within which Grosseteste's *On the Sphere* is to be read require a measure of recalibration. Above all there is little in Grosseteste's general intellectual development and his particular treatments of astronomy that supports the expectation that *On the Sphere* was intended as an elementary textbook for university teaching. The profound relationship astronomy bore to other sciences and disciplines in *On the Liberal Arts* was deepened and developed by a more

---

[87] Grosseteste, *Comm. Post. An.* I.18, ed. Rossi, 267–8: 'si essemus elevati usque ad lunam in hora eclipsis lune, et videremus manifeste eclipsim lune et causam eius, scilicet, terram interpositam inter lunam et solem, nec tamen esset nobis acquisita scientia, quia sensus non ostenderet nobis nisi hanc eclipsim et hanc interpositionem terre in hac hora, et non ostenderet nobis universaliter causam eclipsis'.

ASTRONOMY IN GROSSETESTE'S THOUGHT 241

explicit and profound engagement with Aristotelian thinking, but the *Commentary* can, and perhaps should, be seen more as an example of organic development of earlier ideas than as a sharp rupture with the past and a rejection of old learning for new. This places *On the Sphere* in the middle of a process of structuring learning already possessed, and not in a context of teaching the rudiments of astronomy. Grosseteste's early discussions of astronomy were within a liberal arts context that was already showing signs of Aristotelian influence, while the mature discussion of astronomy as a science in the Aristotelian sense in the *Commentary on Posterior Analytics* drew still on the Boethian liberal arts framework within which Grosseteste had worked decades earlier. The period in which *On the Sphere* was written, then, is characterized by a progressive working out of a stable set of astronomical principles within an Aristotelian framework.

As already shown (Ch. 1, §3.2; Ch. 2, §2.3), there is little direct evidence that Grosseteste was engaged in university teaching during this period. It can be strongly suggested that he was serving Hugh Foliot, often in the diocese of Hereford. As part of this work Grosseteste can be shown to have imposed a clear and intelligible order on the rapidly and chaotically growing body of penitential literature dominating pastoral care around the time of the Fourth Lateran Council. The picture emerging from a historical study of Grosseteste's activity in the period between *c.*1214 and 1221 is also unsupportive of the idea that Grosseteste would focus on writing elementary university textbooks in this period. To the contrary, the historical context supports the trajectory of Grosseteste's intellectual development and the place of astronomy in it outlined above rather better. Astronomy, like pastoral care, had seen an exponential growth in available texts, which raised the need for organization and systematization that could provide understanding of this ever more complex body of learning. Penitential manuals like *The Temple of God* shows that Grosseteste used a diagrammatical and highly structured approach to pastoral care, in which the detailed instructions to priests were seen to follow from a limited set of basic principles. The *Commentary* shows that Grosseteste was, by that point, in possession of resources for organizing astronomical learning in a similar way. Chapter 7 will explore the extent to which such expectations may be of use in understanding the special characteristics of Grosseteste's main astronomical work.

7

Structure, Scope, and Sources of
On the Sphere

The complex clarity of Grosseteste's *On the Sphere* arguably emerges most fully if the treatise is read within a matrix formed by, on the one hand, the ongoing development of astronomical learning in all its various facets, and on the other Grosseteste's own intellectual development towards the Aristotelian notion of special sciences and their interrelations. The task of this chapter, therefore, is to situate *On the Sphere* in relation to these two developments, through an analysis of the treatise with a particular focus on what material Grosseteste selected for inclusion and on how this material was given argumentative form within the overall structure of the text. The patterns that emerge from this analysis provide a foundation for a more profound understanding and appreciation of Grosseteste's treatise.

Some preliminary points should be made to clarify what follows. The first of these concerns the first aspect mentioned above, the sources of Grosseteste's astronomical knowledge as it appears in *On the Sphere*. Comparison with other astronomical texts will be an important analytical tool in what follows, and the framework and purpose of these comparisons should first be described. While the breadth of astronomical texts available to Grosseteste at the time *On the Sphere* was written has been laid out above (Ch. 6, §2), as the present focus shifts from context to text it should now be noted that these sources operate on several different levels in Grosseteste's treatise. On the surface, the most popular textbooks that defined the scope of the discipline of astronomy are important influences throughout *On the Sphere*. Foremost among these is John of Sacrobosco's *On the Sphere*. As mentioned above (Ch. 1, §1; Ch. 5, §4.2; Ch. 6, §1), it is overwhelmingly likely that Grosseteste used Sacrobosco's work in composing his own; it is undeniable

STRUCTURE, SCOPE, AND SOURCES OF *ON THE SPHERE*   243

that there is a considerable overlap throughout the two works. However, while Sacrobosco's influence on Grosseteste is pervasive, it is rarely deep. That is to say, while the two texts cover much of the same material and share similar characteristics on a generic level, a detailed analysis of Grosseteste's *On the Sphere* in comparison with Sacrobosco's treatise will show significant differences of organization, selection, and mode of argument. Sacrobosco's treatise, it will be argued, was important mainly as a point of departure, highlighting a pool of concerns and questions from which Grosseteste selected topics for inclusion in his own work. In what follows, the substantial differences underlying more superficial similarities between the two treatises will help to bring out the particular characteristics of Grosseteste's *On the Sphere*.

Another level of influence on Grosseteste's astronomical thought can be found in the Islamicate texts available in the early thirteenth century. Grosseteste was clearly familiar with Islamicate astronomy not only through Sacrobosco but also through extensive study of his own.[1] He shows independent use of Alfraganus's widely copied compendium of Ptolemaic astronomy, to which Sacrobosco was also deeply indebted.[2] Underpinning Alfraganus's overview is the more diffuse presence of Ptolemy's *Almagest*, the well-spring of the main current of the astronomical corpus, to which Grosseteste nevertheless explicitly referred (see DS §49; also Ch. 6, §2). In addition, moreover, Grosseteste could draw on several more specialized treatises focusing on specific aspects of astronomy, such as Pseudo-Thābit's *On the Motion of the Eighth Sphere* and the anonymous *Almagesti minor* and *Theorica planetarum*, as well as criticisms of Ptolemy's calculations such as Geber's *De astronomia* (Ch. 6, §1) Such texts may have a less pervasive influence than Sacrobosco, but frequently also a more authoritative one. Attention to how Grosseteste incorporates such information into his argument adds depth and precision to a grasp of his own individual purposes in writing. While Grosseteste appears to have placed great trust in the factual accuracy of Islamicate account, he very rarely adopted wholesale the

---

[1] This claim will be substantiated in what follows but is strengthened by the identification of Grosseteste's hand in Bodleian MS Savile 21 (Ch. 2, §3).

[2] For Grosseteste's use of Alfraganus, see for instance Panti, *Moti, virtú e motori celesti*, 73–4; see also, Thorndike, *The* Sphere *of Sacrobosco*, 15–21.

244 THE SCIENTIFIC WORKS OF ROBERT GROSSETESTE

argumentative structures of his sources. On the contrary, he consistently re-ordered and adjusted the astronomical knowledge he found elsewhere. The simple fact that Grosseteste chose not to adopt the mode and sequence of presentation he found in his sources opens the possibility that he wrote with a particular purpose in mind, different from and contributing to the already extant astronomical corpus. The analysis presented here will have a qualitative more than a quantitative focus, concerned more with exploring what Grosseteste did with the sources at his disposal than with an exhaustive list of source texts used.[3] A pattern does indeed emerge from such an analysis, one that is too consistent and too specific not to be deliberate.

The pattern of argument that can be discerned in the structure of *On the Sphere* points towards a third, and in some senses fundamental, level of textual and intellectual influence on the treatise. This follows the second key aspect of this chapter, Grosseteste's own intellectual development. Underlying the treatise is a structure of argument that aligns with two foundational influences on Grosseteste's work as a whole: Euclid and Aristotle. Euclidian geometry and Aristotelian cosmology and physics frame the understanding of the universe informing *On the Sphere*. The argumentative structure of the treatise can be shown to follow Aristotelian principles of demonstration, of which Euclid's geometry was to become a paradigmatic example in Grosseteste's *Commentary on Posterior Analytics* (Ch. 6, §4.2). While Grosseteste's *Commentary* cannot (and will not) be taken as direct evidence for his thinking as he wrote *On the Sphere*, the strong structural similarity between the argument of the latter and the principles of argument as adumbrated in the former can be taken as evidence that Grosseteste was familiar with Aristotle's text by the time *On the Sphere* was being composed. While some instances of more direct overlap between the *Commentary* and *On the Sphere* will be highlighted in what follows, the pattern of similarities between Grosseteste's argument and Aristotle's ideals as explained in the *Posterior Analytics* is arguably the most illuminating factor in bringing out the individual characteristics of

---

[3] A full source apparatus for Grosseteste's *On the Sphere* is found in Cecilia Panti's critical edition in *Moti, virtú e motori celesti*, 289–319, and will not be repeated here. The present chapter is indebted to Panti's study.

Grosseteste's contribution to the astronomical corpus. It should be noted, moreover, that the *Commentary* is important in and of itself not only for demonstrating Grosseteste's reception of Aristotle but also for the crucial role awarded to Euclidian geometry both as a paradigmatic example of scientific reasoning and as one of the sciences to which astronomy is subalternated. In this way, the *Commentary* shows a continuity with Grosseteste's earlier work *On the Liberal Arts* that makes it possible to triangulate the position of *On the Sphere* relative to Grosseteste's intellectual development.

Nevertheless, the argument presented here does not presuppose a direct textual link between *On the Sphere* and the *Commentary*, the former written *c.*1219 the latter *c.*1225. Rather, the implicit argument here is that the simplest and most economical explanation of the patterns of argument intrinsic to *On the Sphere* arose from Grosseteste's reading of Aristotle as he continued to develop the framework already evident in *On the Liberal Arts*. The *Commentary* text is used here as evidence of how Grosseteste came to interpret Aristotle's work and as a point of comparison against which the features of *On the Sphere* may be more clearly discerned. Comparisons with both *On the Liberal Arts* and Grosseteste's more complete thoughts on *Posterior Analytics* therefore allow the vision of astronomy in *On the Sphere* to be identified more clearly, in light of its own principles and in tension, positive and negative, with its sources. The intrinsic structure of *On the Sphere* remains the gravitational centre of the present approach, however, while other works are introduced to the analysis only to the extent that they contribute to bring out this structure. The natural place to start is with the opening of *On the Sphere*, which not only sets out the main scope of the treatise but also situates it in relation to geometry in a way that resonates with other works by Grosseteste.

1 *On the Sphere* and the Definition of Sphericity: Astronomy as Subalternated to Geometry

On the Liberal Arts places astronomy among the mathematical arts that make the natural world intelligible by determining order and position (*ordo et situs*) and drawing out geometrical shapes (*figurae*), and in

246 THE SCIENTIFIC WORKS OF ROBERT GROSSETESTE

which the specific purpose of astronomy was to determine the position of the world and the measurement of time by observing the movement (*motus*) of celestial bodies.[4] *On the Sphere* invokes this same schema in its opening statement:

> Our purpose in this treatise is to describe the shape [*figura*] of the world machine and the [relative] position |*situs*| and shapes of its constituent bodies, and the movements [*motus*] of higher bodies and the shapes of their orbits. Therefore, since the machine of this world is spherical, we should state at the beginning what a sphere is.   (DS §1)

Grosseteste, then, identified four constituent parts in what astronomy studies: geometrical shape or *figura*; relative and ordered position or *situs*; local movement or *motus* within the spatial matrix marked out by the *situs* of the universe; and time defined against the backdrop of local movement. These four aspects of astronomy served to connect abstract and mathematical principles of geometry to the temporal, observable universe. Moreover, the aspects picked out as programmatic for *On the Sphere* embrace the main aspects of astronomy as a liberal art. From this unity of definition, however, the two treatises follow different paths of inquiry. *On the Liberal Arts* does not discuss astronomy for its own sake beyond the definition provided, and moves on to focus on the important ways in which astronomy supports central areas of natural philosophy: alchemy, planting, and medical practice.[5] *On the Sphere*, by contrast, is concerned with the intrinsic principles and structures of the discipline of astronomy. Although the overlap between the treatises is limited it should be noted that astronomy in *On the Sphere* remains within the remit marked out for it in *On the Liberal Arts* and retains the strong bonds to the other mathematical arts as well as to natural philosophy.

The sequence in which the four aspects of astronomy recur in the introductory paragraph of *On the Sphere* is important, however, for understanding the design of the work as a whole. The opening paragraph moves from the shape of the world to the positions and shapes of its

---

[4] Grosseteste, *De artibus liberalibus* §8; *Knowing and Speaking*, 84–5.
[5] See *Knowing and Speaking*, 166–95.

STRUCTURE, SCOPE, AND SOURCES OF *ON THE SPHERE* 247

constituent bodies relative to each other and to the whole, and finally to the movements of the bodies and the shapes these movements describe. This sequence recalls the way astronomy is subalternated to the other mathematical disciplines within the quadrivial system as described by Boethius and elaborated by Grosseteste both in *On the Liberal Arts* and, later, in the *Commentary on Posterior Analytics*. Relative position, *situs*, is only intelligible within an already constituted spatial framework; it is therefore dependent on a grasp of the *figura* of the universe, that is, a geometrical exposition of that framework. Local movement or *motus*, in turn, depends for its intelligibility on the changing local positions of bodies measured against a stable framework of unmoving reference points as provided by *situs*. This causal and conceptual dependence is borne out by the structure of *On the Sphere* as a whole as well as the individual sections of which the work as a whole consists. Finally, as Grosseteste makes explicit in *On the Liberal Arts*, the passing of time is then measured with reference to the regular movements of the unchanging celestial bodies. While this measuring of time is not mentioned in the introduction to *On the Sphere*, it is very much a feature of the text as a whole, as will be demonstrated below.

As the programmatic statement opening the treatise suggests, the initial paragraphs of *On the Sphere* are deeply geometrical (Ch. 4). The abstract geometrical principles of a system of concentric spheres are laid out first, and then predicated of specific spheres as dictated by cosmology (DS §§2–4). That is to say, the reader is first invited, in a language redolent of Euclidian geometry, to imagine geometrical shapes and outlines of increasing complexity, before the shapes and relative positions so described are assigned to individual bodies. Grosseteste first posits Euclid's definition of a sphere from *Elements* XI, before expanding on this definition with a description that also recalls the definition of a sphere provided by Theodosius in his supplement to Euclid's *Elements* known in Latin as *Spherica*.[6] Only when the geometrical principles have been fully worked out does Grosseteste proceed to justify the claim that the universe as a whole and its main constituent bodies are all spherical, as analysed below (DS §§5–9).

---

[6] Cf. Panti, *Moti, virtú e motori celesti*, 89.

248 THE SCIENTIFIC WORKS OF ROBERT GROSSETESTE

This anchoring of the science of astronomy in principles established by geometry follows naturally from the idea of a hierarchical subalternation of sciences. It also sets up the demonstrations of sphericity that follow, since these demonstrations proceed from the definition of a sphere as one of their primitive principles. It is important to note that this geometric anchoring was not in any way in opposition to the Aristotelian method of scientific demonstration. On the contrary, Grosseteste frequently used examples from Euclid to explain and illustrate Aristotle's account of demonstration in his *Commentary on Posterior Analytics*.[7] Grosseteste presented Euclidian geometry as a paradigmatic example of the sort of demonstration that produced understanding. The geometrical emphasis at the start of *On the Sphere* links Grosseteste's account in this treatise both to the presentation of the quadrivium in *On the Liberal Arts* and to the account of Aristotelian science he found in *Posterior Analytics* and which he elaborated in his *Commentary*.

Grosseteste could draw on several models for this use of geometry, and of Euclid in particular, as a way to bring clarity and organization to astronomical discussions. Ptolemy himself had emphasized the mathematical approach of his model, but his calculations and his discursive and meandering mode of exposition drew sharp criticism from some Islamicate astronomers. Of particular importance was the twelfth-century al-Andalusī astronomer Jābir ibn Aflaḥ's *Iṣlāḥ al-Majisṭī* (*Correction of the Almagest*), which was known in Latin as *De astronomia* under the Latinized name of Geber.[8] The *Almagesti minor* had shown how astronomy could be presented within a more formalized geometrical framework, proceeding from postulates and principles to clearly demonstrated proofs.[9] It has been demonstrated that Grosseteste knew this work well by the time he wrote the *Compotus*, and there are indications that he also used it for

---

[7] Jeremiah Hackett, 'Robert Grosseteste and Roger Bacon on the Posterior Analytics', in Pia Antolic-Piper, Alexander Fidora, and Matthias Lutz-Bachmann (eds.), *Erkenntnis und Wissenschaft. Probleme der Epistemologie in der Philosophie des Mittelalters / Knowledge and Science: Problems of Epistemology in Medieval Philosophy* (Berlin: De Gruyter, 2004), 166, 172–3.

[8] Richard P. Lorch, 'The Astronomy of Jābir ibn Aflaḥ', *Centaurus*, 19 (1975), 85–107.

[9] *Almagesti minor*, (ed.) Henry Zepeda, (Turnhout, Brepols, 2018), 21–7.

STRUCTURE, SCOPE, AND SOURCES OF *ON THE SPHERE* 249

the composition of *On the Sphere* (Chs. 5 and 6).[10] While the organization of *On the Sphere* differs sharply from the *Almagesti minor*, the clarity and geometric precision of the latter as a way to consolidate the information contained in the *Almagest* may have been one influence on Grosseteste's method in laying out his own treatise. Islamicate astronomers similarly revised and improved Ptolemy's mathematical and geometrical basis, displaying a mathematical sophistication that Grosseteste openly acknowledged in the *Compotus*.[11]

In *On the Sphere*, however, Grosseteste consistently avoids laying out the complex calculations supporting the claims to which he refers, only mentioning specific numerical values when they are crucial for understanding his argument (although the presence of such calculations, in the background, is implied at several points (see Ch. 4)). In opening his treatise with a geometrical definition of a sphere, Grosseteste appears to have been following the example of Sacrobosco, who presents the same basic definition with explicit references to Euclid and Theodosius.[12] These generic similarities, however, should not be allowed to conceal a more significant difference underneath. Sacrobosco moves from terse geometrical definitions to speak of the divisions of the sphere into nine concentric spheres and into right and oblique spheres, without explicitly connecting these divisions to the definitions that preceded them. Grosseteste, by contrast, uses the geometric definition to outline a diagrammatic image of the celestial and elemental spheres. In this way the reader is encouraged to form a mental picture of the basic *figura* of the universe that will be mapped out in ever greater detail as the treatise progresses. By steering a middle course between the complex geometrical demonstrations of texts like the *Almagesti minor* or Geber's *De astronomia* and the terse definitions of Sacrobosco, therefore, Grosseteste constructs a pedagogical framework for helping the reader understand new elements of astronomy as they are introduced.

---

[10] See Grosseteste, *Compotus*, 18.

[11] See for instance Grosseteste, *Compotus* c. I, 56: 'Et super modum istum motuum solis et stellarum sustentati sunt fundatores kalendarii nostri. Verumtamen in quantitate anni secuti sunt doctrinam Abrachis, que remotior est a veritate quam doctrina Ptolomei vel Albategni.'

[12] Thorndike, *The* Sphere *of Sacrobosco*, 76.

250 THE SCIENTIFIC WORKS OF ROBERT GROSSETESTE

2 *On the Sphere* and the Demonstration of Sphericity: Astronomy Subalternated to Natural Philosophy

Having stated as fact that celestial and elementary bodies are spherical, and provided a geometrical account of what this entails, Grosseteste proceeds to show on what grounds the geometric definition of sphericity can be predicated of the universe and the celestial and elementary bodies it contains.[13] Grosseteste's account is distinctive in structure, and to a certain extent also in content, and quite different to that provided by the sources of astronomical learning available to him. A close reading of *On the Sphere* §§5–9 shows that the distinctive features separating Grosseteste's argument from such parallel accounts all point in the same direction, that is, towards a deliberately constructed series of demonstrations of the reasons why the heavens and the earth are spherical.

To see how this is so, it is important to turn first of all to the way Grosseteste framed his arguments for the sphericity of the heavens and the earth: 'Both natural reasoning and astronomical observations show that the aforementioned bodies are spherical' (DS §5). Grosseteste did not explain within the context of *On the Sphere* what he meant by the term 'natural reasoning', *rationes naturales*. However, as mentioned above (Ch. 6, §4.1), Grosseteste addressed the difference between the arguments of the astronomer and the arguments of the natural philosopher in his glosses on Aristotle's *Physics* as well as in his *Commentary on Posterior Analytics*. In the Aristotelian example discussed by Grosseteste, the natural philosopher demonstrates the sphericity of the moon from the causal principles inherent in the moon's nature, while the astronomer demonstrates either from the effect of the moon's sphericity, or from the cause of sphericity in a more abstract sense, which transcends individual natures.[14] A natural demonstration based on the moon's nature could establish the reason why the moon is spherical using the homogeneity of the lunar body as a middle term, while the effect of admitting of waxing

---

[13] Such an arrangement makes particular sense in light of the Aristotelian account of demonstration; the terms of a demonstration must be understood individually before one can conclude that one can be predicated of the other; see for instance Aristotle, *Posterior Analytics* II.1–2.

[14] Grosseteste, *Comm. Phys.*, 37.

STRUCTURE, SCOPE, AND SOURCES OF *ON THE SPHERE* 251

and waning could be the middle term in an astronomical demonstration *quia* of the matter of fact alone.[15] It is unlikely to be coincidental, therefore, that the examples of 'natural reasoning and astronomical observations' in *On the Sphere* consistently follow the pattern set out in the Aristotelian commentaries.

Grosseteste proposed three separate *naturales rationes* in *On the Sphere* (DS §§5–6); the first is an argument from the nature of celestial bodies, while the other two argue from the natural movements proper to elementary and celestial bodies respectively. Natural reasoning by definition lies outside of the purview of astronomy, and Grosseteste's presentation is consequently somewhat compressed. The first argument reads in full as follows:

> For since the form [of a thing] comes from the nature of that thing, and each of the abovementioned natural bodies is of a single nature, any part of which shares the same name and definition as the whole, it was necessary that each [body] should have a uniform shape, any part of which should be like the whole. There is no such [shape] except the spherical.   (DS §5)

Even in this clipped form, however, Grosseteste arguably provided enough information to make it possible to expand his argument into a fully worked-up demonstration *propter quid*. Spelled out in detail, the argument requires two steps. The major term of the first step is (C) having a single, simple form, that is, a form in which all parts are like the whole. 'Form' here means or includes 'shape', and 'simple' means 'not composite'. The minor term is 'the aforementioned bodies', that is, the celestial and elemental bodies the geometrical layout of which had been described in §§2–4. The middle term is the perfectly simple or non-composite nature of each of these bodies. Step one, then, can be presented in the form of a demonstration as follows:

---

[15] Grosseteste, *Comm. Post. An.* I.12, 190: 'Et forte eiusdem conclusionis est scientia propter quid et scientia quia in eadem scientia, licet Aristoteles de hoc non ponat exemplum. In scientia enim naturali potest demonstrari quod luna sit circularis tam per hoc quod ipsa est corpus omogeneum quam per hoc quod recepit incrementa.'

252 THE SCIENTIFIC WORKS OF ROBERT GROSSETESTE

Minor premise: A celestial or elemental body (A) has a single, simple nature (B)

Major premise: A single, simple nature (B) has a single, simple form (C)

Conclusion: A celestial or elemental body (A) has a single, simple form(C)

The proof of the major premise is the axiom that the form of a thing comes from that thing's nature. This first step, then, establishes that each of the bodies described in *On the Sphere* (DS §§2–4) have a single, simple, homogenous form. Step two uses the conclusion of step one as a premise, and the major term of step one as its middle term:

Minor premise: A celestial or elemental body (A) is a body of a single, simple form (*figura uniformis*) (D)

Major premise: A single, simple bodily form (D) is spherical (E)

Conclusion: A celestial or elemental body (A) is spherical (E)

The proof of the major premise of step two is taken from Aristotle's *On the Heavens*, where it is demonstrated that only the spherical shape is a perfectly simple and homogenous three-dimensional shape.[16] In this way, this line of argument as a whole remains fully in line with the demarcation of the scope of natural philosophy laid out in Grosseteste's *Commentary on the Physics* (Ch. 6, §4.1), demonstrating sphericity of celestial and elemental bodies from the properties they possess by virtue of their nature. The middle terms around which the demonstrations pivot in both examples identify natural properties that explain how the major terms can be predicated of the minor terms. Although compressed, therefore, the argument here conforms to the criteria Grosseteste sets up for natural reasoning elsewhere.

Moreover, both steps in this argument are demonstrations *propter quid*, in which the middle terms that prove the conclusions provide the reason why the conclusion is true by pointing to immediate causal factors belonging to the subject by virtue of its kind (Ch. 6, §§3.1 and 4.2). At several points in the *Commentary on Posterior Analytics*

---

[16] Aristotle, *De caelo* II.4, 286b10–32.

STRUCTURE, SCOPE, AND SOURCES OF *ON THE SPHERE* 253

(again Ch. 6, §4.2), Grosseteste emphasized that, for certain conclusions, natural philosophy could demonstrate the reason why, whereas astronomy could only demonstrate the matter of fact. As will become clear in the following, the reasoning presented in *On the Sphere* §§5–9 gives a very good example of how this principle could work in practice. Grosseteste's argument from nature for the sphericity of celestial and elemental bodies, therefore, is wholly consonant with Aristotelian notions of demonstration, subalternation, and the specific position of astronomy in relation to natural philosophy and geometry.

It is important to note, however, that Grosseteste's mode of arguing here not only conforms to Aristotelian ideals, but also explicitly anticipates Grosseteste's developments and clarifications of Aristotelian doctrine later put down in the Aristotelian commentaries. While the argument from nature analysed here is founded on ideas Aristotle elaborated in the *Physics* and the *Posterior Analytics*, Grosseteste went beyond Aristotle's own words in significant ways. Aristotle himself consistently used the sphericity of the moon alone as an example, and Grosseteste followed suit in his commentaries. Moreover, Aristotle never suggested what a natural demonstration of the sphericity of the moon, the earth, or the heavens might look like, but only that such a demonstration was possible in principle. Grosseteste, however, would suggest in the *Commentary on Posterior Analytics* that the sphericity of the moon might be demonstrated from the natural homogeneity of the lunar body, as noted above.[17]

In *On the Sphere*, the point is universalized to apply to all celestial and elementary bodies. This is not an unwarranted expansion of Aristotle's example, however, and the argument is substantially the same across Grosseteste's works as well as in Aristotle's own treatises. In all cases, sphericity is predicated of the subject of the demonstration by way of the simple and uniform nature of that subject. In other words, the same middle term is used to demonstrate the sphericity of all celestial and elemental bodies in *On the Sphere* as in the demonstration of the sphericity of the moon in Aristotle's works and Grosseteste's commentaries on them. The terminology is somewhat different, 'uniform' in *On the Sphere* and 'homogenous' in the *Commentary on Posterior Analytics*,

---

[17] See above, footnote 15.

254   THE SCIENTIFIC WORKS OF ROBERT GROSSETESTE

but this purely verbal difference rests on a deeper conceptual unity where the single, simple form of the body in question follows from its single, simple nature. What holds of the moon when it comes to shape holds of all celestial and elemental bodies equally, and the principles at play and the form of demonstration are identical in both cases. The 'natural reasoning', then, that Grosseteste invokes to demonstrate the sphericity of celestial bodies in *On the Sphere* is specifically adduced as an example of a demonstration *propter quid* in the *Commentary* and of 'natural demonstration' both in the *Commentary* and in the *Notes on the Physics*. This suggests that the personal reception and interpretation of Aristotle's thought that came into full expression in the commentaries was already well advanced by the time *On the Sphere* was being written.

The two subsequent arguments proposed under the heading of 'rationes naturales' follow the same pattern. The first of these demonstrates the sphericity of elementary bodies from the movements natural to each element. Grosseteste's text reads

> Moreover, because every heavy object tends towards the centre, and the deeper place is the place around the centre, it is necessary that the two heavy bodies [earth and water] have a spherical shape; and similarly the two light bodies [fire and air], because the higher place is the one the furthest from the centre, and everything light tends towards what is higher.   (DS §5)

Again, the argument hinges on natural attributes, the main framework this time being taken from Aristotle's account of the elemental spheres in *On the Heavens*. Grosseteste's argument here is somewhat elliptical and leaves some premises implicit, but he does provide the causal explanation of the sphericity of elemental bodies that can be used as a middle term of a fully worked-up demonstration. This causal explanation is anchored in the idea of the natural movement intrinsic to the four elements. Earth and water are heavy by nature and fall down, while air and fire are light by nature and rise up.[18] It is tacitly assumed that the elements will

---

[18] See Aristotle, *De caelo* III.2. For a detailed account of Aristotle's theory of elemental movement, see Helen S. Lang, *The Order of Nature in Aristotle's Physics* (Cambridge: Cambridge University Press, 1998), 166–262.

STRUCTURE, SCOPE, AND SOURCES OF *ON THE SPHERE*   255

distribute themselves evenly as close as possible to the terminus of their natural movement, which assumption is valid within the definition of elementary bodies at play in the Aristotelian tradition.[19] Again, then, the premises form part of the natural properties of bodies that astronomy assumes but cannot prove, and which are demonstrated by natural philosophy. A perfectly evenly distributed body the centre of which is at the centre of the universe must by necessity be spherical, and a perfectly evenly distributed body contained within a perfect sphere must also by necessity itself be spherical. The implicit demonstration for the sphericity of the heavy elementary bodies can be schematized as follows:

Heavy elementary bodies (A) distribute themselves evenly around the centre (B)
What distributes itself evenly around a centre (B) is spherical (C)
Heavy elementary bodies (A) are spherical (C).

The middle term (B) is alluded to, but not formally defined, in Grosseteste's argument. It may seem like Grosseteste did not regard it as within his current scope to spell out in detail the natural demonstrations explaining astronomical matters of fact.

Finally, Grosseteste provides a demonstration for why the quintessence, in which all celestial bodies are contained, by necessity must be spherical. Once more, the argument proceeds by way of natural movement as a middle term:

Concerning the fifth essence the Philosopher shows that it is spherical because it is necessary that rectilinear movements, which belong to heavy and light elements, are brought back to circular movement, which by necessity belongs to the fifth essence. However, if it is moved in a circle, it is by necessity spherical, because if it were angular there would be by necessity an empty place.   (DS §6)

Again, the basic principles from which the conclusion is demonstrated are assumed for the purposes of astronomy, while being proved by

---

[19] See for instance Aristotle, *De caelo* II.14, 297b10–14.

256 THE SCIENTIFIC WORKS OF ROBERT GROSSETESTE

natural philosophy. Here, the Aristotelian framework elaborated in *On the Heavens* is explicitly invoked.[20] This third natural demonstration assumes the impossibility of a vacuum outside the extremity of the universe, which can be demonstrated from the basic principles of Aristotelian physics and cosmology.[21] It follows necessarily from the simplicity of the quintessence that it must move with a perpetual and perfect circular movement; and the only shape that can be moved with perfect circularity without creating a vacuum is the sphere. The underlying structure of this argument is therefore once again a demonstration *propter quid*, with the necessary perfect circularity of the movement of the quintessence as the explanatory middle term, and with the impossibility of a vacuum as the proof of the major term.

Only once natural demonstrations *propter quid* have been given for the sphericity of the whole universe and its constituent parts does Grosseteste move on to arguments from observation. Grosseteste offers three arguments; one for the roundness of the earth on the north-south axis (DS §7), one for the roundness of the earth on the east-west axis (DS §8), and one for the sphericity of the heavens (DS §9).[22] For the first of these, the proof hinges on the varying observed positions of the Polar Star at different latitudes; for the second on how the same observed celestial event occurs at different times of day on the east-west axis; and for the third on how the observed size of the stars remains constant from their rising through their highest point to their setting. None of these observations could work as middle terms of demonstrations *propter quid*, since none of them offer a causal explanation of the phenomenon under scrutiny; the roundness of the earth and the heavens are not caused by these observable phenomena. The entire passage of *On the Sphere* discussing the sphericity of celestial bodies, therefore, follows the distinction between natural philosophy and astronomy alluded to in Aristotle's *Physics* and explained in Grosseteste's *Notes* on that text as well as his *Commentary*. Natural reasoning provides *propter quid* demonstrations while astronomical arguments from observable effects only

---

[20] Aristotle, *De caelo* II.4.
[21] Aristotle, *De caelo* II.4, 287a11–22. The impossibility of a void is discussed in Aristotle's *Physica* IV.8, while the claim that circular movement is primary is established in VIII.9.
[22] See also Chapter 4 above.

STRUCTURE, SCOPE, AND SOURCES OF *ON THE SPHERE*   257

demonstrate the matter of fact. Seen from this perspective, Grosseteste's account of the sphericity of the world emerges as a tightly structured and deliberately ordered line of reasoning, organized according to Aristotelian notions of scientific inquiry directed towards understanding causes, and an Aristotelian hierarchy of individual sciences linked through subalternation.

The distinctive structure and focus of Grosseteste's discussion of the sphericity of the world stands out against the backdrop of parallel accounts elsewhere in the astronomical corpus. There is no obvious parallel or model for Grosseteste's decision to put natural demonstrations front and centre on the question of the sphericity of the world. To the extent that the astronomical texts known to Grosseteste address proofs of the sphericity of the world at all, they mainly follow the arguments advanced in Ptolemy's *Almagest*. Ptolemy grounded his own account of the sphericity of the heavens in the observation that the stars could be seen to move in perfectly regular circles around the fixed centre of the Pole Star, similar to the observations described in *On the Sphere* §9.[23] Ptolemy similarly established the sphericity of the earth from observations of the same basic kind as those summarized by Grosseteste in §§8–9.[24] The sequence, structure, and kind of arguments presented by Ptolemy are also found in those Islamicate compendia of, and critical responses to, the *Almagest* that address the sphericity of celestial and elemental bodies. Alfraganus repeats and clarifies the Ptolemaic observations for the sphericity of the heavens.[25] A detailed and highly critical exposition of Ptolemy's arguments is provided by Geber; whose main concern is to reveal his predecessor's geometrical shortcomings and offer a better geometrical basis for astronomy. He offers no independent arguments for the sphericity of the heavens and the earth.[26] In this sense, while it is difficult to determine which specific text or texts Grosseteste drew on for his proofs of sphericity from astronomical observations, the basic phenomena he mentioned are

---

[23] Ptolemy, *Almagest* I.3.   [24] Ptolemy, *Almagest* I.4.

[25] Alfraganus, *Il 'libro dell'aggregazione delle stelle'*, trans. Gerard of Cremona, (ed.) Romeo Campani (Città di Castello: S. Lapi, 1910), 64–8.

[26] [Geber filius Affla Hispalensis] Jābir ibn Ḥayyān', *De astronomia* (Nüremberg: Iohannes Petreius, 1534), II, 20–2.

258 THE SCIENTIFIC WORKS OF ROBERT GROSSETESTE

common features of the Ptolemaic tradition. Moreover, being based on observation these arguments are also necessarily demonstrations *quia*, not *propter quid*. Admittedly, Ptolemy and his critic Geber do both end their respective discussions of the sphericity of the heavens by mentioning that ether is composed of parts that are similar to one another, which is only possible in a spherical body; but here this is not placed in the context of formal causality which Grosseteste used to provide a middle term for a *propter quid* demonstration in DS §5.

Among the source texts closer to Grosseteste's time and milieu, John of Sacrobosco once again offers useful generic parallels beyond which significant contrasts can be discerned. Sacrobosco closely and explicitly follows Alfrganus's account, but, similarly to Grosseteste, Sacrobosco prefaces this observational account with a three-fold *ratio* providing reasons for the sphericity of the heavens.[27] The three reasons Sacrobosco adduced are labelled 'likeness, convenience, necessity'.[28] The 'likeness' argument is based on the notion that the world is created in the likeness of an archetype, which has no beginning and no end; therefore the perceptible world has a shape lacking beginning and end. The 'convenience' argument states that since the world has to contain all things, and the sphere is the most capacious of all solid bodies, it was most convenient that the world should be a sphere. The 'necessity' argument states that since an angular world would entail both vacuum and a body without a place, both of which are impossible, the world is necessarily spherical.

Importantly, none of these reasons can be cast in the form of *propter quid* demonstrations, and none of them argue from the natures of the bodies described. The first two rely on implicit premises that fail to meet the criteria Aristotle set up for scientific demonstrations, in that they invoke principles that do not follow necessarily from the assumptions constituting the kind of thing astronomy studies (see, Ch. 6, §3.2). The final reason is based on an argument from Aristotle's *On the Heavens* and resembles Grosseteste's demonstration of the sphericity of the quintessence. However, this argument requires as a premise that celestial bodies are moved in a circle, which Grosseteste demonstrates with reference to Aristotelian notions of elemental and celestial movement,

---

[27] Sacrobosco, *De sphera*, 80–1.
[28] Sacrobosco, *De sphera*, 80: 'similitudo, commoditas, necessitas'; English translation at 120.

STRUCTURE, SCOPE, AND SOURCES OF *ON THE SPHERE* 259

but Sacrobosco merely assumes. While Grosseteste, then, constructed *propter quid* demonstrations from natural properties in order to show the reason why the world is spherical, Sacrobosco's reasoning offers no model for grounding astronomical fact in proofs from natural philosophy in this way.

In fact, the most intriguing parallel to Grosseteste's reasoning on the sphericity of the heavens is found in Grosseteste's near-contemporary Daniel of Morley's (*c.*1140–1210) *Philosophia*. Daniel argued from the unity and simplicity of celestial bodies to the conclusion that the sphere was the only bodily shape that met these criteria, paralleling the first of Grosseteste's arguments from nature.[29] Daniel explicitly positioned his arguments within the framework set out in Aristotle's *On the Heavens* and *Physics*.[30] After setting out the Aristotelian arguments Daniel added that corroborating observations could be found in Alfraganus's *Aggregatio*. Again, it is difficult to determine whether Grosseteste derived some of his premises from Daniel's account or whether they both derived them from a common source, for which the obvious candidate would be Aristotle's *On the Heavens*. Grosseteste's account went significantly beyond that of Daniel, adding the framework of a clear distinction between natural philosophical demonstrations and astronomical arguments from observation as well as individual premises and arguments. It remains probable, therefore, that Grosseteste's strict adherence to Aristotelian demonstration is his own structure imposed on his material.

A close reading of Grosseteste's account of the sphericity of the heavens and the earth, then, shows that, far from being at the mercy of his sources, Grosseteste was able to select important items of knowledge from extant learning and impose a strict argumentative structure closely modelled on the Aristotelian account of demonstration and subalternation. He added concrete comparisons between the astronomically significant city of Arim to the basic Ptolemaic arguments from observation, comparisons that he quite probably drew from Petrus Alfonsi (see Ch. 5, §4.3), and he subordinated the arguments from observation to *propter quid* demonstrations relying on principles drawn from higher sciences,

---

[29] Daniel of Morley, *Philosophia/Liber de naturis inferiorum et superiorum*, ed. G. Maurach, *Mittellateinisches Jahrbuch*, 14 (1979), 204–55, at 233.
[30] Daniel of Morley, *Philosophia* 119, ed. Maurach, 232–3.

260 THE SCIENTIFIC WORKS OF ROBERT GROSSETESTE

primarily geometry and natural philosophy. This arguably reveals central concerns shaping the mode of argument characterizing Grosseteste's *On the Sphere*. It is of course crucial that everything presented as true should be consonant with actual and potential observations, but Grosseteste's contribution to astronomical learning is not based on new or more accurate observations. What Grosseteste offers is a structured argument issuing not simply in knowledge of matters of fact but in understanding of why these matters of fact are so. The passages on sphericity in *On the Sphere* (DS §§5–9) appear as a deliberate attempt at constructing an Aristotelian science of astronomy based on shared facts established by observation and disseminated through the Ptolemaic tradition and beyond.

3 *Situs*: The Basic Reference Points of the Spherical Universe

The conclusion on which Grosseteste ends *On the Sphere* §9 allows him to assign a definite orientation to the spherical universe as a whole, with the poles as the fixed points defining the axis around which the firmament rotates in an unchanging circular movement (DS §10). Grosseteste largely follows Sacrobosco's terminology and his etymological explanations of these terms when speaking of the polar points of the axis of the universe. There are some additions, most notably the puzzling claim that a fixed axis is called *magal'* in Hebrew. This claim is erroneous and may be a scribal corruption of the Arabic term *mihwar*, usually transliterated as *meguar* in Latin, which is found in some manuscript copies of *On the Sphere*.[31]

Later, in the work now known as *On the Six Differences*, Grosseteste would discuss in greater detail whether definite orientations exist in the sphere of the universe, a discussion that belonged to natural philosophy more than to astronomy.[32] Explaining the causes of the movement of the heavens lies beyond the scope of astronomy, and in *On the Sphere*,

---

[31] See Panti, *Moti, virtù e motori celesti*, 95–6.
[32] A new edition and translation of *De sex differentiase—On the Six Differences* will be published in Volume III of this series.

STRUCTURE, SCOPE, AND SOURCES OF *ON THE SPHERE* 261

Grosseteste is content to close his introductory outline of the spherical universe with the initially puzzling comment that the efficient cause of the regular and diurnal movement of the heavens is the world soul. While the term *anima mundi* appears to introduce theologically suspect Neoplatonic ideas into the discussion, it is worth noting that the notion of efficient causality is incompatible with the Neoplatonic doctrine of the self-moving world soul (see Ch. 5, §4.1).[33] Efficient causality, as defined in Aristotle's *Physics*, was the causing of movement in some other thing, and the brief mention of the world soul as efficient cause is compatible with Grosseteste's later account of the causes of movement in works such as *On Corporeal Movement and On Light* and *On Supercelestial Movement*.[34] In the context of *On the Sphere*, this comment can therefore be regarded as an oblique nod to natural philosophical considerations lying outside the compass of the science of astronomy.

The role of movement in *On the Sphere*, then, can be reduced to the aforementioned distinction between geometry as the science of immobile magnitudes and astronomy as the science of mobile magnitudes. Astronomy studies the patterns, but not the causes, of celestial movements, and Grosseteste was careful to introduce the notion of movement gradually into his discussion. His account moves from abstract geometry (DS §§2–4), through demonstrating how these geometrical shapes can be predicated of celestial bodies (DS §§5–9), to a basic outline of the underlying principles of the observable movement of the firmament around a fixed axis anchored in the polar points (DS §10).

Grosseteste used the fixed axis and its polar extremities as a foundation for further geometrical specification of points and lines of reference within the universe as a whole. Both terminology and the explanations of terms are still recognizably grounded in the Ptolemaic and Latin traditions transmitted through Sacrobosco, but the sequence by which crucial celestial points and circles are introduced is Grosseteste's own.[35] While

---

[33] See for instance Helen S. Lang, *Aristotle's* Physics *and Its Medieval Varieties* (Albany: SUNY Press, 1992), 36–44; James A. Weisheipl, *Nature and Motion in the Middle Ages* (Washington, D.C.: Catholic University of America Press, 1985), 75–97.

[34] See for instance Panti, *Moti, virtù e motori celesti*, 329–45. *De motu corporali et de luce* will form part of Volume IV of the present series.

[35] The circles and points described by Grosseteste in these paragraphs are found spread out over pp. 77–80 in Thorndike's edition of Sacrobosco.

262 THE SCIENTIFIC WORKS OF ROBERT GROSSETESTE

Sacrobosco provided a discursive account of the great circles of the world by their order of significance, Grosseteste once again chose to anchor the new phenomena he introduced in the foundational model he has established up to this point. In this way, the new elements are immediately given fixed and identifiable points of reference. Once again, then, Grosseteste invites the reader to imagine the world as a geometric figure, on which circles may be drawn through both poles at 90 degrees to one another (DS §11), and a third circle, the equinoctial circle or equator, being drawn at an equal distance to both poles at 90 degrees to both the first two (DS §12). The equinoctial is so called because when the sun describes this circle by the rotation of the firmament, the day is equal to night in all places; a first nod, in passing, to the measuring of time as defined by movement that would come more to the foreground as the treatise progresses.

Based on this geometrical foundation, then, Grosseteste described accurately and concisely the *situs* of the ecliptic circle and the zodiacal belt of which the ecliptic is the central line, and the tropical circles described by the movement of the northernmost and southernmost points of the zodiac against the firmament (DS §§12–15). Geometrical visualization is then called upon for the account of the Arctic and Antarctic circles described by the poles of the zodiac against the firmament (DS §16). This allows Grosseteste to map out with great precision the orbit of the sun's own movement under the ecliptic, which together with the movement of the firmament produces a composite, oscillating spiral movement around the earth between the tropical circles determined by points on the zodiac (DS §§17–18). In accordance with the pattern of the treatise so far, geometry sets up the framework within which relative positions are established, before movement is added to complete a causal and geometrical explanation of what can be seen. Here too, then, it can be observed that Grosseteste follows the precise definition of astronomy given in the liberal arts corpus and repeated in his *Commentary*: astronomy uses the principles by which geometry measures immobile and abstract magnitudes to study mobile magnitudes. It is worth noting that Grosseteste's description of the zodiac is keyed to his account of the two movements of the sun, in such a way that the geometry of the ecliptic and the zodiac marks out the parameters of

STRUCTURE, SCOPE, AND SOURCES OF *ON THE SPHERE* 263

solar movement through the diurnal cycle of the movement of the firmament and the annual cycle of the proper movement of the sun. This allows Grosseteste to demonstrate with great clarity the consequences that follow.

The structural and argumentative strengths arising from how Grosseteste brought out the geometry of the relationship between the zodiac and the movement of the sun emerges more discernibly through a comparison with alternative accounts of this topic. The information Grosseteste conveys is found in Sacrobosco and the latter's model Alfraganus, who corrected Ptolemy's calculations of the points of intersection between the ecliptic and the *colures*.[36] However, Grosseteste once again rearranges the information to make explicit how his account relies on geometrical principles and points of reference already established. He compresses the account of the zodiac, including only the information needed to determine and trace the movement of the sun. As his description of the universe develops and takes on a more definite form, each new feature is introduced in a way that not only communicates knowledge but also allows the reader to understand the geometrical connections more deeply.

Grosseteste then proceeded to add an observer-relative dimension to the universal account followed so far, based on the concept of 'horizon'. After a brief explanation of the meaning of the term, a more substantial explanation is given from a geometrical perspective (DS §19). No matter where one stands on the surface of the earth, straight lines in all directions from the observer, tangent to the earth's surface, will divide the heavens into two halves, one visible and the other hidden by the mass of the earth. Since the size of the earth is as a dimensionless point compared to the heavens, both halves are exactly equal in size. The horizon can be precisely defined as the line that separates the visible half from the one hidden by the mass of the earth, the 'limiter of vision'. Since each actual horizon is relative to an observer, 'as many horizons are possible as there are places on the earth and the circumference'. Nevertheless, the general definition Grosseteste provides is geometrical and universal.

---

[36] See Panti, *Moti, virtú e motori celesti*, 97.

264 THE SCIENTIFIC WORKS OF ROBERT GROSSETESTE

4 Intervals of Time

At this point, Grosseteste had laid the foundations required for an intelligible discussion of the perceived effect of celestial, and particularly solar, movements on the tracking of intervals of time at different locations of the earth.[37] As he moves towards an account of the perceptible difference in the length of day and night at various latitudes and at various times, he explicitly links this increased emphasis on movement and time to the spatial grid of determinate positions set out in the treatise so far: 'From the position of the horizons and parallels mentioned above it is easy to see what happens in every position on earth in terms of equality and inequality of days and nights' (DS §20). The extended passage spanning *On the Sphere* §§20–47 forms an explicit unity, as highlighted by Grosseteste's emphasis on how the various aspects he discusses contribute to a comprehensive explanation of the perceptible length of the day as observable in the different regions of the earth. While the measurement of time becomes increasingly important to the treatise, *On the Sphere* does not contain computus or pursue computistical aims. The astronomical phenomena described in these passages although relevant for computus, are adduced here by way of explaining their causal foundations, not their calendrical implications. Grosseteste shows how phenomena that are relative to the observer and change over time, such as the length of day and night and the rising of the signs of the zodiac, can be explained by universal and unchanging causes.

The account remains dependent on Sacrobosco and Alfraganus, and probably also the *Almagesti minor*, for his information, but Grosseteste's arrangement of this information into a structured argument is entirely his own.[38] This emerges in particular in how Grosseteste incorporates the rising of the zodiac and the eccentricity of the sun into his discussion of length of days, unifying in a single argumentative structure aspects that are treated separately in the astronomical corpus. The account below follows Grosseteste's argumentative structure based on a two-fold causality of the inequality of days and nights and draws on the explanation provided above (Ch. 4).

---

[37] Cf. *De artibus liberalibus* §8.   [38] Cf. Panti, *Moti, virtù e motori celesti*, 100.

4.1 The Sun and the Zodiac

The first of Grosseteste's two causal factors for the inequality of days and nights is closely tied to the zodiac, since the sun's own orbit follows the ecliptic which is also the centre of the zodiacal belt. The time during which the sun is visible above the horizon at different places therefore follows the rising and setting of the zodiacal signs, which follows a consistent and universal pattern. Since daylight depends on the length of time that the sun is visible above the horizon, and the horizon is observer dependent, this combines both universal and relative causal factors into a unified whole.

Grosseteste emphasized the fundamental difference between the experience of those who live under the equinoctial circle and the perceptions of those who live at any other latitude. Under the equinoctial circle the sun and the other stars always rise perpendicularly to the horizon and follow a straight course across the sky, and consequently every day and every night will be of the same length. North or south of the equinoctial circle, however, the sun's course under the ecliptic causes variations that become more pronounced the further from the equinoctial one moves. It should be emphasized that Grosseteste's argument at all points refers back to the parallel latitudes and the definition of horizon with which this part of the treatise began (DS §19). The complexities of the inequalities of days and nights at different latitudes can therefore be demonstrated to arise from an ordered set of determinate geometric definitions. Again, the objective *situs* of the universe as a whole causes observable effects relative to the position of the observer.

Since the rising and setting of the sun at different latitudes is linked to the rising and setting of the signs of the zodiac, Grosseteste used his description of the spherical universe (DS §§2–20) to demonstrate the causes for the variations of the zodiac at different places on the earth (DS §§31–7). As before, geometry provides the basis, a geometric description of the *situs* of the universe the matrix, and movement of celestial bodies becomes the efficient cause of variation over time and space on earth. In moving from secure first principles of geometry to the particulars of the observable universe, Grosseteste was providing the most complete causal explanation possible within a broadly Aristotelian framework of inquiry.

266 THE SCIENTIFIC WORKS OF ROBERT GROSSETESTE

In this way, the account of the zodiac is incorporated into the main argumentative structure in a strong way that emphasizes causal connections and explains variations in observation patterns. Grosseteste stressed this point explicitly and deliberately as he moved from the first to the second of the two causes he had identified for the inequality of days and nights:

> Since we have spoken of the rising and setting of signs, whose rightness and obliquity is one cause of why natural days are unequal to one another, it remains to add [something] concerning another cause of the inequality that comes from the fact that the sun is eccentric, so that from the combination of these two causes of the inequality the whole reason for why natural days are unequal to one another may become evident.   (DS §38)

There can be no doubt, therefore, that the argumentative structure focused on causal connections was deliberately chosen by Grosseteste, or that he saw the account of the rising and setting of the zodiacal signs as part of this causal structure.

4.2 The Sun's Eccentricity

While the eccentricity of the solar orbit was universally recognized, Grosseteste's emphasis of this feature of solar movement as a causal factor for the inequality of days and nights appears to lack clear textual models. Grosseteste returns to the terminology of central astronomical texts to speak of the centre of the eccentric of the sun as the 'circle of the displaced cusp', *circulus egresse cuspidis*, an expression found in John of Seville's translation of Alfraganus as well as in the *Theorica Planetarum* tentatively attributed to Gerard of Cremona.[39] However, unlike his sources, Grosseteste only introduces these terms and the referents once he has set up a geometrical matrix within which these terms become determinate and intelligible within a wider causal structure.

---

[39] Cf. Panti, *Moti, virtú e motori celesti*, 110–11.

STRUCTURE, SCOPE, AND SOURCES OF *ON THE SPHERE* 267

As in the preceding cases, Grosseteste's causal explanation is grounded in geometry determined by what he has established of the *situs* of the universe so far. Once more the reader must mentally picture the spherical universe and draw up a line between the 18th degree of Gemini, via the centre of the universe, to the corresponding degree of Sagittarius opposite (DS §39; see the corresponding paragraph in Ch. 4). On this line, at a specified distance from the centre towards Gemini, lies the centre of the orbit of the sun. The consequences of this geometrical computation for the relative positions of the earth and the sun are then drawn out (§40), before the phases of movement and their consequences for the length of days and nights at different latitudes are laid out (DS §§41–5; see Ch. 4).

The varying lengths of days and nights are not the only effects of the eccentricity of the sun, however. Since the sun is closer to the earth during its passage through the southern signs of the zodiac and further away when it is in the northern signs, Grosseteste supposed that the southern hemisphere is significantly hotter than the northern and considered to be uninhabitable. The inhabitable part of the world is divided into 7 climes (DS §§46–7; see Ch. 4). It is worth noting that while these passages introduce what we would recognize as climatic differences, the definition of a 'clima' used by Grosseteste is based on perceptible differences in time measurement at different latitudes. In other words, even on this topic Grosseteste stays within his larger subtopic of how universal and constant processes can lead to different perceptible results at different locations. At this point of the treatise, the geometrical principles that dominate the opening sections recede somewhat into the background, always assumed but not often explicitly mentioned, to give space to the ever more complex conclusions about movement and variation over time that constitutes the aim of astronomy. In this way, the treatise as a whole resembles its constituent parts in providing *propter quid* demonstrations from first principles to causal conclusions about particulars.

5 The Fixed Stars

As he moved on to discuss the fixed stars, Grosseteste pointed out that the fixed stars are so called because the always retain the same positions

268   THE SCIENTIFIC WORKS OF ROBERT GROSSETESTE

relative to one another, but that as a group they possess their own natural motion in addition to that of the firmament (DS §48). At this point Grosseteste discussed the different models of the movement of the fixed stars offered in Ptolemy's *Almagest* and Pseudo-Thābit's *On the Motion of the Eighth Sphere* (DS §§49–50; see Ch. 8 below). These models existed in parallel within the astronomical corpus, and Grosseteste appears to have found Pseudo-Thābit's argument important enough to copy *On the Motion of the Eighth Sphere* in his own hand (Ch. 2, §4). Grosseteste would also highlight significant challenges to Ptolemaic doctrines elsewhere, for instance in his *On Supercelestial Movement*.[40] For his juxtaposition of the two models in *On the Sphere*, Grosseteste once more called on geometrical imaging to picture two zodiacal circles, one in the firmament, and another, the visible zodiac, in the sphere of fixed stars below the firmament, offset from the first zodiac and moving with its own natural movement. The *situs* and proper movement of this second zodiac then explains the variations observable in the heavens over time (DS §§51–3). Grosseteste did not go into great detail here; he appears content to have provided enough discussion to allow the reader to see the basic coherence between the observable variations of the visible zodiac and what has been said in the foregoing. The complexities in the Ptolemaic foundation and the criticism of Pseudo-Thābit are assumed but not addressed explicitly.

6 The Moon

The final part of the treatise specifies the relative position and movements of the lowest of the celestial bodies, that is, the moon, and demonstrates the explanatory force of this account for the variations in the moon's movement over time (DS §§54–63). As on all other topics, Grosseteste starts by using geometry to situate the relative position of the moon and its orbit in relation to what has already been established. The movement of the moon is complex, but clearly explained in relation to the zodiac and the sun (DS §54; see Ch. 4). Grosseteste then draws out

---

[40] Panti, *Moti, virtú e motori celesti*, 345.

STRUCTURE, SCOPE, AND SOURCES OF *ON THE SPHERE* 269

the consequences of the general statement of the movement of the moon, with its ever-changing position relative to the sun, the zodiac, and the earth (DS §55–6). The terminology Grosseteste employs here suggests that he was using one or both of the *Theorica planetarum* or John of Seville's translation of Alfraganus at this point.[41] In contrast to his sources, however, Grosseteste's account is concentrated on the geometric layout of the movements of the moon, leaving out most of the calculations producing the numerical values of the patterns described in the text. The result is a clear exposition of the terms and concepts necessary to understand the movements of the moon both absolutely and in relation to movement patterns already established.

Building on this account, the final part of the treatise explains the effects these movements have on what can be observed from the earth over time, in particular lunar and solar eclipses (DS §§57–63). The explanation of lunar eclipses here steers a middle course between two distinct and opposed tendencies in the astronomical corpus available at the time *On the Sphere* was composed. On one end of the spectrum, Alfraganus and Sacrobosco offered a simple explanation of what an eclipse is and the conditions required for a lunar eclipse to take place; on the other extreme, Ptolemy's complex mathematical account in the *Almagest* was criticized and emended in equally complex and protracted calculations in Geber's *De astronomia* and in the *Almagesti minor*.[42] Grosseteste set out a more detailed and ordered geometrical account than Alfraganus and Sacrobosco without entering into the sharp disagreements concerning numerical values that formed the battle ground between Ptolemy and his later critics. The result is an account of lunar eclipses that demonstrates how the conditions under which the moon is eclipsed issue necessarily from what has been established concerning lunar and solar movements so far (see Ch. 4, §§57–9). The basic definition of an eclipse is assumed and alluded to, but not explicitly established. Instead, Grosseteste points out that it follows from the preceding account that the eccentric of the moon only intersects with

---

[41] Panti, *Moti, virtù e motori celesti*, 118.
[42] Cf. Sacrobosco, *De sphera*, 115–16; Alfraganus, *Aggregatio* c. 27; Ptolemy, *Almagest*, VI.6–7; Geber, *De astronomia* IIII, 45–52; *Almagesti minor* IV, ed. Zepeda, 276–329.

270   THE SCIENTIFIC WORKS OF ROBERT GROSSETESTE

the ecliptic in two points, called respectively the head and the tail of the dragon of the moon, and that it is only when the moon is within 12 degrees either side of these two points that a lunar eclipse can occur. This is because the definition of an eclipse requires the alignment of a light source, a non-transparent obstructing body, and the eclipsed body on a straight line, which in the specific case of the moon can only occur when the moon intersects with the orbit of the sun which lies under the ecliptic. A further consequence of this is that lunar eclipses are only possible around the time of the full moon, which is the only time that the sun and the moon are opposed. At this time only, then, can the shadow projected from the earth reach the moon as it passes the intersection with the ecliptic.

Grosseteste's account of how lunar eclipses follow from the basic geometric orientations of the sun and the moon is more pronounced and more tightly argued than parallel accounts in Alfraganus and Sacrobosco, who remain important sources for information but not for argumentative structure. Alfraganus's account of lunar eclipses is found many chapters after his description of the basic movement of the moon, and Sacrobosco provides a very brief account as part of a more generic explanation of the main aspects of planetary movement.[43] Grosseteste brings out more explicitly the geometrical reasons why lunar eclipses occur; again, the function of his exposition is to lead the reader to a deeper understanding of what is already known. The absence of the specific numerical values and complex calculations characterizing the discussions of Geber and the *Almagesti minor* produces this same effect, bringing focus and clarity to Grosseteste's demonstration of the causes of lunar eclipses from principles already established and known.

Again, a comparison with the *Commentary on Posterior Analytics* suggests that Grosseteste's argumentative structure is deliberately chosen to lead the reader to a fuller understanding. Understanding lunar eclipses means understanding the causes by which the earth is placed directly between the moon and the sun to block the passage of light from the latter to the former. Grosseteste's account in *On the Sphere* brings out precisely this sort of understanding, and anchors it in the wider context

---

[43] Alfraganus, *Aggregatio* c. 27; Sacrobosco, *De sphera*, 115–16.

STRUCTURE, SCOPE, AND SOURCES OF *ON THE SPHERE* 271

of the relative movements of the sun and the moon around the earth. That is to say, the account of the movements of the moon in *On the Sphere* lead directly to an explanation not only of the lunar eclipse in isolation, but of the specific conditions under which this occurs and the reason why these conditions only converge at certain times and not at others.

Grosseteste's decision to explain the conditions required for lunar eclipses with particular care compared to Alfraganus and Sacrobosco can be explained by the important role lunar eclipses play in Aristotle's *Posterior Analytics* and in Grosseteste's *Commentary*. As stated above, a demonstration properly speaking could only be derived from unchanging principles—ἐπιστήμη could only be had of what was always the case (Ch. 6, §3.1). Yet, Aristotle's oft-invoked example of lunar eclipses seems to fail this requirement, since it is clearly not true that eclipses are always the case, and so its causes cannot always be the case either. Grosseteste explicitly addressed this problem in the *Commentary*, book I, ch. 7:

> From what was demonstrated just now, that is, that demonstration is of what is incorruptible and not of what is corruptible, a puzzle arises: how can a demonstration be constructed concerning what is frequently the case and are not always the case, such as concerning eclipse, because neither a particular eclipse nor the universal eclipse is always the case, since the universal cannot be preserved except in some given particular.[44]

Grosseteste offered a clarification: 'Aristotle did not intend to say that an eclipse is always the case, but he intended to say that the conclusion in which an eclipse is demonstrated is a truthful proposition at any time whether there is an eclipse or not.'[45] He then proceeded to spell out what Aristotle had left implicit, in the form of a syllogism:

---

[44] Grosseteste, *Comm. Post An.* I.7, 143–4: 'Ex eo quod proximo demonstratum est, scilicet quod demonstratio est incorruptibilium et non est corruptibilium, emergit dubitation qualiter erigatur demonstratio super ea que frequenter sunt et non semper sunt, ut super eclipsim, quia neque eclipsis singularis neque eclipsis universalis semper est, quia non potest salvari universale nisi in aliquo individuorum suorum.'

[45] Grosseteste, *Comm. Post An.* I.7, 144.

272 THE SCIENTIFIC WORKS OF ROBERT GROSSETESTE

whenever the moon falls under the shadow of the earth, the moon is eclipsed,

and whenever the moon is placed opposite the sun by a diameter having latitude less than the quantity of the two semidiameters, that is, of the moon and of the [earth's] shadow, the moon falls under the shadow,

therefore whenever the moon is placed opposite the sun by a diameter having latitude less than the quantity of the two semidiameters, that is, of the moon and of the shadow, it is eclipsed.[46]

Both premises are true whether there is an actual eclipse or not, and therefore the demonstration is always true even if the phenomenon it demonstrates does not always occur. The middle term of this syllogism, that is, the moon falling under the shadow of the earth, is the definition of a lunar eclipse as elaborated in Grosseteste's example quoted above (Ch. 6, §4.3). Simply stating this definition as a middle term, however, would not provide the basis for a demonstration, since it would as yet be unclear how a lunar eclipse can follow from unchanging and necessary principles. Only by showing how the alignment of sun, moon, and earth necessary for the occurrence of a lunar eclipse follows from unchanging and necessary principles can a demonstration properly speaking be constructed. A *propter quid* explanation of a lunar eclipse requires a *propter quid* demonstration of the reason as to why the earth sometimes prevents sunlight from reaching the surface of the moon, and Grosseteste provides such a demonstration. This further strengthens the hypothesis that Grosseteste aimed to lead the reader to new understanding more than to new information.

Having provided a comprehensive account of lunar eclipses, Grosseteste turned, in the final part of *On the Sphere*, to explain the even more complex phenomenon of solar eclipses. Again, his explanation proceeds from what he has already established. One additional level

---

[46] Grosseteste, *Comm. Post An.* I.7, 144: 'quotienscumque luna cadit in umbram terre, luna eclipsatur, et quotienscumque luni opponitur soli per diametrum habens minorem latitudinem quam sit quantitas duorum semidiametrorum lune, scilicet, et umbre, luna cadit in umbram, ergo quotienscumque luna opponitur opponitur soli per diametrum habens latititudinem habens minorem latitudinem quam sit quantitas duorum semidiametrorum lune, scilicet, et umbre, luna eclipsatur, quelibet istarum propositionum in omni tempore est vera.'

STRUCTURE, SCOPE, AND SOURCES OF *ON THE SPHERE* 273

of complexity, however, must be added to account for the fact that solar eclipses occur more rarely than lunar eclipses, and that solar eclipses are relative to the observer on earth in ways that lunar eclipses are not. Owing to the relative size and position of the earth and the moon, the precise astronomical position of the moon in relation to the firmament differs from its apparent position relative to the firmament for an earthly observer, unless this observer is directly under the moon (see, Ch. 4, §§60–3). This difference between the real and apparent positions of the moon is called 'the diversity of aspect of the moon', and Grosseteste again provides a full geometrical account of the complex phenomena at hand anchored in reference points already established. Only when the geometrical account is complete does Grosseteste proceed to draw out the inferences explaining why the relevant conjunctions of the sun and the moon viewed from the earth are rare and relative to the observer.

The treatise ends somewhat abruptly at this point, without a concluding passage and in most manuscripts without an *explicit*. It cannot therefore be wholly certain that the treatise as it stands represents the finished form Grosseteste originally set out to realize. What the analysis of this chapter reveals is that while a wide range of influences and sources for *On the Sphere* can be identified, the treatise is far from wholly derivative, and the terminology and information shared with other texts is organized in an intelligible structure that lacks close parallel in the astronomical corpus. The final passages of *On the Sphere* complete an argumentative arc starting with the most remote and universal sphere of the firmament and concluding with what is the lowest celestial body, the moon, which gives the treatise as a whole the same *propter quid* explanatory structure as the individual demonstrations made within it. The treatise as a whole can be seen to provide a comprehensive understanding of the relative positions of the bodies of which the universe as a whole is constituted, their movements, and the observable variations over time. It does not lay claim to the discovery of new knowledge about the universe; on the contrary, it is a way of understanding what is already known. As such, it is a tightly structured and clear exposition of the foundational knowledge and understanding of the universe required for more profound explorations of natural philosophy.

274 THE SCIENTIFIC WORKS OF ROBERT GROSSETESTE

It should also be noted that the explanatory structure of Grosseteste's argument from basic principles and causes to observable phenomena is an intrinsic part of the treatise regardless of the origins for such an argumentative structure for Grosseteste at the time of writing. While it cannot be proved absolutely that Grosseteste had read the *Posterior Analytics* by the time he wrote *On the Sphere*, it is manifest that reading the latter work in the light of the former, alongside Grosseteste's *Commentary*, brings out a consistent and clear argumentative trajectory in *On the Sphere* that has so far escaped scholarly notice. As shown above there are some close overlaps between *On the Sphere* and the *Commentary*, and it seems more likely that these overlaps indicate a causal connection than that they are entirely coincidental. Given the comprehensiveness and depth of Grosseteste's *Commentary* it also seems unlikely, albeit not impossible, that he would reach such a level of understanding of Aristotle's difficult treatise in the period between the composition of *On the Sphere* and the *terminus ante quem* of the completion of the *Commentary*. While the inspirations for Grosseteste's argumentative structure remain open to debate, however, the argumentative structure itself is a crucial part of Grosseteste's contribution to the astronomical corpus as a whole.

7 The Scope of the Treatise

The assumption that Grosseteste's *On the Sphere* was an astronomy textbook, frequently expressed in previous scholarship, could easily produce the expectation that the treatise would present an overview of the entire range of astronomical knowledge. As stated above (Ch. 5, §4.2), however, *On the Sphere* does not contain a discussion of planetary movements beyond the movements of the sun and the moon, except a perfunctory mention of the apogees of all the planets in §53. There can be little doubt that Grosseteste knew of the intricacies of planetary motion; many of his sources discuss this in great detail, and there are also internal hints in the treatise itself that show that Grosseteste was well aware of this discussion (see above, Ch. 4, §53). While it remains possible that the treatise was left unfinished, arguments can also be advanced for the treatise as it stands to

STRUCTURE, SCOPE, AND SOURCES OF *ON THE SPHERE*   275

cover an intelligible and clearly defined scope. Two such arguments deserve particular notice here.

First, as the analysis presented above shows, Grosseteste's *On the Sphere* provides a way of systematizing and understanding knowledge already acquired. This function does not require a comprehensive summary of all available knowledge, but only a structure and method of understanding within which all available knowledge could be encompassed. Here, the Ptolemaic tradition marked out the sun and the moon as particularly important. In *Almagest* I.2, Ptolemy states that without a thorough comprehension of the movements of the sun and the moon and what follows from these movements, it is impossible to understand the account of the other planets.[47] If, as suggested above, Grosseteste's main purpose was to facilitate understanding of what was already known, his focus on the sun and the moon becomes explicable in light of the crucial role these two celestial bodies played within the Ptolemaic system as a whole. Once the matrix established by the movements of the sun and the moon had been understood, the learner would, if they wanted, be able to go on to map the other five planets onto this same system.

A second reason for limiting the scope of the treatise to the sun and the moon, compatible with the first, can be discerned by turning to the application of astronomy specifically for time reckoning. *On the Sphere* arguably provides sufficient understanding of the discipline of astronomy to prepare the reader for a subsequent study of computus. As mentioned above, Grosseteste emphasized the importance of astronomy for time reckoning in *On the Liberal Arts*, and like Sacrobosco he would later go on to write a computistical treatise himself. It is important to remember, therefore, that the reckoning of time, both in terms of its basic units of measurement and in its calendrical expression, was based on the cyclical movements of the sun and the moon. The Julian calendar, like its Gregorian successor, was based on the solar year, but the traditional dating of Easter with reference to the phases of the moon meant

---

[47] Ptolemy, *Almagestum seu Magnae Constructionis*, trans. Gerard of Cremona (Venice: Peter Lichtenstein, 1515): 'impossibile namque est comprehendere scientia stellarum, et que de earum scientia explanare volumus, ante horum [sc. soli et lune] scientie comprehensionem'; f. 2v.

276 THE SCIENTIFIC WORKS OF ROBERT GROSSETESTE

that the Christian calendar also had to incorporate the lunar cycle into its calculations. The scope of *On the Sphere*, therefore, covers the aspects of astronomy most crucial to accurate computistical work.

The specialized discipline of computus, like astronomy more generally, was in a state of transition in the early thirteenth century. On the one hand the inaccuracies of earlier calculations of the solar and lunar cycles led to increasing tension between calendrical and astronomical time; on the other, the increasing availability of Islamicate astronomy provided resources for significantly improving the foundations on which computus was based. Grosseteste's *Compotus* was an early contribution to calendrical reform based on more accurate astronomical knowledge, again paralleling a similar work by John of Sacrobosco.[48] As in *On the Sphere*, Grosseteste's *Compotus* 'expanded the boundaries of the discipline without shattering them.'[49] Philipp Nothaft has characterized the *Compotus* as not only pursuing reform of the ecclesiastical calendar, but also providing a presentation of

> the rules and concepts that made up the Latin computus in a form more streamlined, logically condensed, and mathematically abstract than usually found in works of this genre. Part and parcel of this approach was a decrease of the topics under discussion, which Grosseteste reduced to only the core requirements of ecclesiastical time-reckoning [...].[50]

The latter aim allowed Grosseteste to dispense with complicating themes like the zodiacal longitude of the sun and the moon, which had been treated in earlier computistical texts despite contributing little to the 'most basic objectives of the discipline'.[51] In other words, Grosseteste's *Compotus* followed the model of the earlier *On the Sphere* in presenting new learning in a way that prioritized logical and causal connections over breadth of information.

---

[48] See Jennifer Moreton, 'John of Sacrobosco and the Calendar', *Viator*, 25 (1994), 229–44.
[49] Grosseteste, *Compotus*, 10.   [50] Grosseteste, *Compotus*, 10.
[51] Grosseteste, *Compotus*, 10.

STRUCTURE, SCOPE, AND SOURCES OF *ON THE SPHERE*   277

The structure and scope of the *Compotus* are naturally very different to *On the Sphere*, as required by the differences in immediate aims; but the subject matter treated in the latter sets up the framework for the former. The movements of the sun and the moon, and how these movements are traced out against the two zodiacs and the fixed stars, with which *On the Sphere* concludes, becomes the starting point for an exploration of the complexities in calculating the solar and lunar cycles in order to bring calendrical time closer to astronomical reality. The other planets and the signs of the zodiac are only discussed to the extent that they impinge on the central concern of time measurement.

The scope of *On the Sphere*, then, is arguably streamlined to highlight the aspects of astronomy most of all needed for understanding the causal structure of the science according to its canonical texts. At the same time, it also provides a solid point of departure for the application of astronomy emphasized in *On the Liberal Arts*, that is, the reckoning of time. There is insufficient evidence to conclude that *On the Sphere* was ever planned specifically as a framework for a computistical treatise, but the lines of continuity running from Grosseteste's earliest work on the study of nature through to his magisterial astronomical and computistical works nevertheless suggest a unity of focus and an organic development of doctrine over an extended period of time. The theoretical possibility that Grosseteste planned to extend *On the Sphere* to include planetary motion must remain open; what can be concluded with much greater certainty is that the scope of the treatise as preserved has a completeness to it that emerges both from its internal logical structure and from Grosseteste's own explicit statements on, and practice of, the main applications of astronomy.

Starting from the *figura* and *situs* of the universe and placing the movement of celestial bodies within this framework, Grosseteste in this first half of *On the Sphere* produced a causal account consisting of *propter quid* demonstrations, through which the measurement of time could be understood, which was a central task of the science of astronomy. Both the treatise as a whole and each individual theme introduced proceed from geometrical and cosmological principles to conclusions about movement and time in line with what can be observed. In so doing, Grosseteste has also implicitly given demonstrations *propter quid* for the observable effects of the *situs* and *motus* of the universe. That is to say, the trajectory of Grosseteste's argument is consistently from principles derived from geometry, on which astronomy depends, through premises concerning the *situs* of the universe and its bodies, to conclusions about movement and time that can be corroborated by observation.

8

Trepidation or Precession:
The Turning Point in a Tradition

Within *On the Sphere* Grosseteste chooses to pay special attention to the measured drift, since antiquity, in the celestial coordinates of the stars. He compares two rival accounts of the causal pattern underlying these meas urements: the 'precession of the equinoxes' that is accepted today, and an alternative theory, which he attributes (in error) to Thābit, of 'trepidation', also known as 'access and recess'.[1] This chapter provides the mathematical astronomy necessary to interpret Grosseteste's discussion in the treatise, an example of his pattern of introducing and framing arguments in natural philosophy by geometry (Ch. 6). In what follows the two accounts and Grosseteste's response to both of them are summarized, contextualized, and analysed. Graphical representations of the predictions from the two rival theories clarify why trepidation rather than Ptolemaic precession could reasonably be preferred up to the middle of the thirteenth century, and also why soon afterwards it became rapidly discredited. Grosseteste himself gives an entirely different reason for disputing the precessional model, basing his argument on long term climate change, as outlined below.

1 The Precession of the Earth's Axis and the Alternative of Trepidation

A few simple astronomical fundamentals will be helpful in setting the scene to understand the two models of the slow changes in stellar

---

[1] The common expression found in medieval Latin sources is *accessio et recessio* or *accessus et recessus*, which is a literal translation of the Arabic *al-iqbāl wa'l-idbār*. The term 'trepidation' (*trepidatio*), although more popular in modern literature, was only introduced in the mid-fifteenth century. See C. Philipp E. Nothaft, 'An Alfonsine Universe: Nicolò Conti and Georg Peurbach on the Threefold Motion of the Fixed Stars', *Centaurus*, 61 (2019), 91–110.

280 THE SCIENTIFIC WORKS OF ROBERT GROSSETESTE

positions that Grosseteste discusses in *On the Sphere*. The two celestial great circles whose relative positions determine these motions are the *ecliptic* and the *celestial equator*. The first is the projection of the earth's orbit (and approximately, that of the other planets and the moon) onto the sky. The projection is therefore the path upon which the sun seems to move through the constellations over the course of a year, as observed from the earth (DS §17). In addition to its near-circular orbital motion around the sun, the earth also spins on its axis, which motion is responsible for the apparent rising and setting of the sun and stars in the course of each twenty-four hours. Importantly, the axis of the earth's rotation is tilted with respect to the perpendicular to its orbit (the ecliptic) by approximately 23.5°. This tilt is responsible for the seasons; it points the northern hemisphere towards the sun during the northern summer months, and away from the sun in its winter. As a result, the two coordinate systems of celestial latitude and longitude, one based on the earth's own latitude and longitude projected onto the celestial sphere, and the other on the plane and pole of the solar system, are tilted with respect to each other by 23.5°.

The points at which the two coordinate systems intersect are the equinoxes. They undergo a slow drift along the ecliptic as a result of a phenomenon known as precession (see Box 8.1 for more technical detail). As the earth's axis sweeps round the circle of precession so the point at which the celestial equator cuts the ecliptic also moves around its full circle at a steady rate and with the same period (Fig. 8.1(a)).

The same local motion of apparent drift in ecliptic longitude can be accounted for, over limited periods of time only, by the alternative model of trepidation, which places similar circles of precession, not as paths for the celestial pole, but on the ('fixed') ecliptic at the points at which the celestial equator cuts it: the equinoctial points of Aries and Libra. Attached to points on the surface of these two opposite circles is an apparent ecliptic, whose equinoctial points will precess at a rate such that the apparent drift in stellar ecliptic longitudes during the period from antiquity to 1200 is matched approximately to that of polar precession. The geometric scheme is illustrated in Fig. 8.1(b). This strange form of trepidation (which might equally be called 'equatorial precession' in distinction to the 'polar precession' model of Ptolemy) needs another constraint to determine the orientation of the mobile and fixed ecliptics.

Box 8.1 Milankovitch Cycles

It was not until the 1838 that Louis Agassiz and others realized that the geological record contained evidence that the temperature in the northern hemisphere had at one time been much lower than it is today. Over the next century data showing dramatic and cyclical changes in the earth's climate built up and in 1941 Milutin Milankovitch proposed that glaciation was driven by summer ice melting, rather than snow fall in winter, and that summer insolation (solar radiation received at the top of the atmosphere) at 65°N would be critical to the growth and decay of ice sheets, see footnote 39 in the present chapter. There are three principal astronomical effects that determine the insolation at such a latitude, all associated with the earth's orbit around the sun.

1. Eccentricity

It was Kepler who, in his 1609 book *Astronomia Nova* (*The New Astronomy*), showed that the orbits of the planets were not circles, but ellipses with the sun at one of the two foci of the ellipse (Fig. 8.B.1). An elliptical orbit is characterized by the sum of the distances to the two foci remaining a constant and the orbital eccentricity e is defined as $e = c/a$ (Fig. 8.B.1). Thus, unlike a circular orbit ($e = 0$), for $e \neq 0$ the distance of the earth from the sun varies through the orbit. Perihelion (closest distance) occurs currently in January, close to the winter solstice, though the date of perihelion regresses by about one full day every fifty-eight years. The intensity of the light from a point source falls off as the inverse square of distance, so the insolation at the perihelion is greater than at the aphelion.

The earth's motion is affected by the motion of all the other planets. These effects do not change the size of the orbit, i.e. the value of a, but they do change the value of the eccentricity e. The eccentricity is small ($e = 1.5$ per cent) but there is a periodic increase and decrease from almost zero to about three times its current value. The period of this cycle is about 100,000 years. In 1976, Hays, Imbrie, and Shackleton examined the $^{18}O/^{16}O$ ratio and relative abundance of small fossil sea

Continued

Box 8.1 Continued

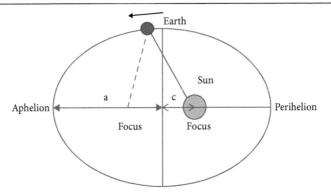

Fig. 8.B.1. Elliptical orbit of the earth around the sun.

creatures to plot the summer sea-surface temperature over geological time from deep-sea sediment cores. They found a dominant, 100,000-year variation in temperature, and by implication glaciation (Ice Ages), that has an average period close to, and in phase with, the earth's orbital eccentricity.

2. Obliquity of the Earth's Axis

Although the angle of tilt of the earth's rotational axis with respect to the perpendicular to its orbital plane, responsible for the seasons, is presently 23.4°, this is not fixed. Due primarily to the gravitational effect of the moon on the earth's equatorial bulge but also the effect of the other planets, this obliquity varies from 21.5° to 24.5° with a periodicity of about 41,000 years. It affects climate, as for lower values of obliquity, the difference in insolation between summer and winter is reduced and for decrease in obliquity there is a decrease in the solar radiation received at high latitudes, and an increase in the tropics. Hays, Imrie, and Shackleton ('Variations in the Earth's Orbit: Pacemaker of the Ice Ages'; see footnote 39 in the present chapter) found just such a climatic variation, implying glaciation, of the same period as the variation in the obliquity of the earth's axis and retaining a constant phase relationship with it.

3. Precession of the Earth's Axis

As Ptolemy had deduced, the earth's axis also precesses with respect to the fixed stars. Our current best measurement of the period of this motion is 25,765 years. There are no climatic effects specifically associated with the orientation of the earth's axis with respect to the fixed stars, but the orientation with respect to the earth's elliptical orbit does have impact. The direction of the earth's axis of rotation determines at which point in the earth's orbit the seasons will occur and thus precession will cause a particular season to shift from year to year. Therefore the earth's distance from the sun in, for example, the northern hemisphere summer shifts with time, as Grosseteste deduces as the consequence of precession in *On the Sphere*. The situation is complicated, however, by the precession of the perihelion itself, with respect to the fixed stars, with a longer period of 112,000 years. This too changes the date of the seasons and the two effects combine to explain the periodicities of 19,000 and 23,000 years found in the deep-sea cores of Hays et al. (see reference above in §2).

The so-called forcing mechanisms, which directly relate the astronomical variations with corresponding variations in climate, are still a matter of active debate. In particular it is unclear why the 100,000-year eccentricity cycle dominates the climate variation, rather than just modulating the shorter period cycles. Nevertheless, it is now generally accepted that the Ice Ages occurred as a result of changes in the earth's orbit and inclination of its polar axis. The 100,000-, 41,000-, 23,000-, and 19,000-year periodicities correlate extremely well with the periods of orbital eccentricity, obliquity, and precession. Although, as with his cosmogeny in *On Light*, Grosseteste did not have the correct physical mechanism, he was remarkably astute in realizing that changes in astronomical variables could have serious consequences for climatic phenomena.

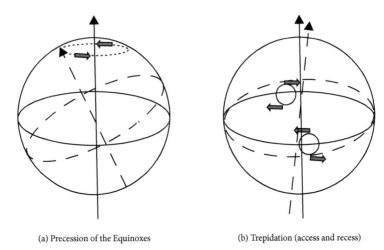

(a) Precession of the Equinoxes (b) Trepidation (access and recess)

Fig. 8.1. (a) The celestial sphere showing the ecliptic (horizontal solid curve) and the celestial equator (tilted dashed curve). The earth's axis (dashed line) precesses (dotted path) around the elliptic pole (vertical line with arrowhead) in a period of about 26,000 years. (b) Under the assumed motion of trepidation the apparent ecliptical system (dashed line and curve) oscillates with respect to the 'fixed' system by rotation on the extremities of two small circles fixed at the equinoctial points.

This is provided by the requirement that the declination of the solstices match in the two frames. In Fig. 8.1 the dashed (movable ecliptic) and solid (fixed ecliptic) circles coincide at the right and left of the figure at all times (the solstice points), irrespective of where the suspension points of the (dashed) movable ecliptic are along the circles of trepidation. The smaller diameter of these circles compared to the polar precession means that to match the drift rate, the period of trepidation is required to be shorter in proportion.

A comparison of the predictions for both ecliptic longitude and the declination ('celestial latitude') of a star with the celestial coordinates of Regulus (α Leonis, a star close to the celestial equator, and whose position has been recorded since antiquity) and for an entire precessional orbit, is given in Fig. 8.2 (for mathematical details on how celestial coordinates in one system are translated into another, as required by both mechanisms, see Box 8.2 on p. 303). The initial conditions are

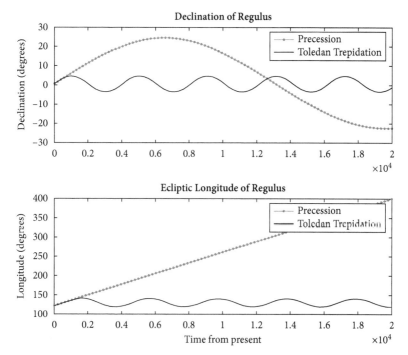

Fig. 8.2. Predictions of precession (dotted curves) and trepidation (smooth curves) for ecliptic latitude (a) and longitude (b) of a star close to the celestial equator (at the initial epoch for these plots), and over a period of 20,000 years.

chosen so that the two theories match as closely as possible at the (arbitrary) time $t=0$. The much shorter period of the trepidation oscillation is evident in the course of both coordinates. Note that for an interval of no more than 1,000 years, and not containing one of the turning points of the trepidation, the two models can be made to match relatively closely by virtue of equal gradients (but see §4 below for more detailed divergences).

Beyond the period for which the shorter period trepidation is an approximation for the longer period of precession, the two models diverge wildly. At these low celestial latitudes, the long period precession is almost linear (see lower panel of Fig. 8.2) while the trepidation is sinusoidal. The predictions for celestial latitude are close to sinusoidal in both cases but with radically differing amplitudes and periods, although these are of course arranged so that the initial rates of change are matched.

286   THE SCIENTIFIC WORKS OF ROBERT GROSSETESTE

2 Historical Background

The discovery that the longitude of certain stars changes in relation to the equinoxes was apparently made in the mid-second century BCE by the Greek astronomer Hipparchus, whose insights are quoted and expanded upon in Ptolemy's famous *Almagest*. Upon comparing Hipparchus's stellar observations with his own, Ptolemy concluded that the phenomenon was best explained by a slow eastward motion of the sphere of fixed stars about an axis that passes through the poles of the ecliptic. As the approximate speed of this motion, he accepted a relatively slow rate of 1° in a hundred years, which implied that the fixed stars would come full circle in relation to the vernal equinox after 36,000 years. The correct figure would have been closer to 26,000 years.[2]

Ptolemy's theory had important practical implications, in that it forced astronomers to choose between two different reference frames for measuring ecliptic longitudes. Such longitudes could be either sidereal (that is, as measured from a particular fixed star) or tropical (that is, as measured from the vernal equinox), but not both at the same time, since the alignment between the two reference frames kept changing at a linear rate. Ptolemy himself opted for a tropical coordinate system, which had the counterintuitive consequence of making the signs of the zodiac, now defined as 30°-segments of the ecliptic measured from the vernal point, independent of the star constellations that had originally given rise to them. This disparity between the signs and the constellations, whose mutual positions kept shifting over time, could make the theory of precession a difficult one for practising astrologers to accept. Evidence of its widespread rejection during the first two centuries after the *Almagest*'s completion has been discovered by Alexander Jones, whose

---

[2] See Ptolemy, *Almagest*, VII, 2–3, 13–34. For discussion, see Anton Pannekoek, 'Ptolemy's Precession', *Vistas in Astronomy*, 1 (1955), 60–6; Noel M. Swerdlow, 'Hipparchus's Determination of the Length of the Tropical Year and the Rate of Precession', *Archive for History of Exact Sciences*, 21 (1979–80), 291–309, at 300–6; Gerd Graßhoff, *The History of Ptolemy's Star Catalogue* (New York: Springer, 1990), 73–91; Bernard R. Goldstein and Alan C. Bowen, 'The Introduction of Dated Observations and Precise Measurement in Greek Astronomy', *Archive for History of Exact Sciences*, 43 (1991), 93–132, at 111–14; James Evans, *The History and Practice of Ancient Astronomy* (New York: Oxford University Press, 1998), 259–74; Olaf Pedersen, *A Survey of the* Almagest, rev. edn. (New York: Springer, 2011), 236–60.

TREPIDATION OR PRECESSION 287

study of late antique horoscopes indicates that many practitioners converted Ptolemy's tropical longitudes back into an older reference frame, using a supplementary calculation mentioned in Theon of Alexandria's fourth-century *Little Commentary* on Ptolemy's *Handy Tables*.[3] According to Theon, some astrologers adjusted planetary positions on the assumption that the solstitial points move at a rate of 1° over eighty years, but only until they have reached an elongation of 8°. Afterwards the solstices were supposed revert to their original starting points, the implication being that precession reverses direction every 640 years.[4]

This strange notion of a bi-directional motion of the equinoxes and solstices was to gain a new lease of life when astronomers active under the Abbasid Caliphate in the ninth century became interested in accounting for the way the tropical year and precession rate recorded by Ptolemy clashed very conspicuously with the parameters suggested to them by their own observations. A key example in this regard is the text accompanying the *Sabian Zīj*, a set of astronomical tables composed for the meridian of ar-Raqqa by al-Battānī (d. 929). Based on an equinox observation he made in 882, al-Battānī calculated the length of the tropical year as 365d 5h 46m 24s (where here and in the following, 'd', 'h', 'm', and 's' stand for days, hours, minutes, and seconds), which was a considerably shorter value than the 365d 5h 55m 12s Ptolemy had accepted in his *Almagest*.[5] He witnessed a roughly proportionate contraction when measuring the longitude of the stars Regulus (α Leonis) and β Scorpii, which he found to have undergone an increase of 11;50° over the 782 years that separated him from Ptolemy's predecessor Menelaus (*c.*98). This result revealed a significantly more rapid precession rate than Ptolemy had alleged: 1° in sixty-six years.[6]

---

[3] Alexander Jones, 'Ancient Rejection and Adoption of Ptolemy's Frame of Reference for Longitudes', in Alexander Jones (ed.), *Ptolemy in Perspective: Use and Criticism of His Work from Antiquity to the Nineteenth Century* (Dordrecht: Springer, 2010), 11–44.

[4] See *Le 'Petit commentaire' de Théon d'Alexandrie aux Tables faciles de Ptolémée*, ed. Anne Tihon (Vatican City: Biblioteca Apostolica Vaticana, 1978), 236–7, 319; and Otto Neugebauer, *A History of Ancient Mathematical Astronomy*, 3 vols (Berlin: Springer, 1975), ii. 631–4.

[5] Al-Battānī, *Opus astronomicum*, ch. 27, ed. Carlo Alfonso Nallino, 3 vols. (Milan: Hoepli, 1899–1907), i.42.

[6] Al-Battānī, *Opus astronomicum*, ch. 51, ed. Nallino, I, 124. It must be noted that the history of the precession parameter 1°/66y in Islamic astronomy goes back further than al-Battānī. See G. Ye. Kurtik, 'Precession Theory in Medieval Indian and Early Islamic Astronomy', in W. H. Abdi et al. (eds.), *Interaction between Indian and Central Asian Science and Technology*

288 THE SCIENTIFIC WORKS OF ROBERT GROSSETESTE

Al-Battānī's suggestion that precession had increased over time, from Ptolemy's 1°/100y to his 1°/66y, inspired subsequent theorists to develop a new breed of physical trepidation models, which were capable of varying continually both the rate of precession and the obliquity of the ecliptic.[7] They enjoyed their greatest influence and popularity in Muslim Iberia, where trepidation models were not just developed for speculative purposes, but used in actual computational practice. A well-known early example of this practical use of trepidation are the so-called *Toledan Tables*, which were the product of a group of astronomers assembled by Ṣāʿid al-Andalusī (d. 1070), a *qāḍi* based in the eponymous Iberian city. Translated into Latin at least twice during the first half of the twelfth century, these tables went on to leave a strong imprint on the practice of mathematical astronomy in Christian Europe, where they were still being copied and employed in the fourteenth century (Ch. 5, §2). Among the Toledan Tables' distinguishing characteristics was their use of a sidereal reference frame, which required calculators seeking tropical longitudes of the sun, moon, and five planets to take into account the 'accession and recession' of the sphere of fixed stars, as computed from special tables.[8]

Like the tables themselves, the treatise describing the underlying model survives only in a twelfth-century Latin translation, which is

---

in Mediaeval Times, 2 vols (New Delhi: Indian National Science Academy, 1990), i.94–110; S. Mohammad Mozaffari, 'A Medieval Bright Star Table: The Non-Ptolemaic Star Tables in the *Īlkhānī Zīj*', *Journal for the History of Astronomy*, 47 (2016), 294–316, at 303–7.

[7] The best study of the origin of these models is F. Jamil Ragep, 'Al-Battānī, Cosmology, and the Early History of Trepidation in Islam', in Josep Casulleras and Julio Samsó (eds.), *From Baghdad to Barcelona: Studies in the Islamic Exact Sciences in Honour of Prof. Juan Vernet*, Vol. 2 of 2 (Barcelona: Instituto 'Millás Vallicrosa' de Historia de la Ciencia Arabe, 1996), ii. 267–98. Further orientation on the history of precession/trepidation in medieval astronomy is provided in Jerzy Dobrzycki, 'The Theory of Precession in Medieval Astronomy' [originally published 1965], in Jarosław Włodarczyk and Richard L. Kremer (eds.), *Selected Papers on Medieval and Renaissance Astronomy* (Warsaw: Instytut Historii Nauki PAN, 2010), 15–60; John D. North, *Richard of Wallingford*, 3 vols (Oxford: Clarendon Press, 1976), iii. 155–8, 238–70; Raymond Mercier, 'Studies in the Medieval Conception of Precession', 2 pts., *Archives internationales d'histoire des sciences*, 26 (1976), 197–220, and 27 (1977), 33–71; Evans, *The History*, 274–80; Walcher of Malvern, *'De lunationibus' and 'De Dracone'*, ed. C. Philipp E. Nothaft (Turnhout: Brepols, 2017), 284–90; C. Philipp E. Nothaft, *Scandalous Error: Calendar Reform and Calendrical Astronomy in Medieval Europe* (Oxford: Oxford University Press, 2018), 99–106; Julio Samsó, *On Both Sides of the Strait of Gibraltar: Studies in the History of Medieval Astronomy in the Iberian Peninsula and the Maghrib* (Leiden: Brill, 2020), 579–654.

[8] G. J. Toomer, 'A Survey of the Toledan Tables', *Osiris*, 15 (1968), 5–174, at 118–22; Fritz S. Pedersen (ed.), *The Toledan Tables: A Review of the Manuscripts and the Textual Versions with an Edition*, 4 vols. (Copenhagen: Reitzel, 2002), i. 54–5; ii. 478–81, 686–7; iv. 1542–66.

TREPIDATION OR PRECESSION 289

known as *On the Motion of the Eighth Sphere*. With at least 110 manuscripts still extant, this translation was one of the most frequently copied technical astronomical works of the entire Middle Ages.[9] Its author postulates two reference points on a sidereally fixed 'mobile' ecliptic, designated as 0° Aries and 0° Libra, which undergo a slow circular motion on the surface of small circles (radius: 4;18,43°, where here and in the following this notation of three separated numbers signifies the degrees, minutes, and seconds of arc). The centres of these circles are fixed at the points of intersection between the celestial equator and a hypothetical mean ecliptic projected onto the ninth sphere. As they revolve on their respective circle, 0° Aries and 0° Libra become identical with their corresponding equinoctial point whenever they cross the celestial equator, which is by definition the moment when the motion of the eighth sphere reaches 0° (on 14 November 604) or 180° (as in Fig. 8.1(b)). Once it is at 90° or 270°, by contrast, the two reference points will have reached their maximum elongation from the respective equinox, which is assumed to be ±10;45°, such that the stars subsequently reverse their direction of precession. Over the whole cycle of 360°, which takes approximately 4,057 Julian years to complete, the actual rate of precession undergoes a continuous sinusoidal increase and decrease, making it possible to explain the apparent speeding up between the observations of Ptolemy and al-Battānī.[10]

---

[9] Charles Burnett and David Juste, 'A New Catalogue of Medieval Translations into Latin of Texts on Astronomy and Astrology', in Faith Wallis and Robert Wisnovsky (eds.), *Medieval Textual Cultures: Agents of Transmission, Translation and Transformation* (Berlin: de Gruyter, 2016), 63–76, at 69.

[10] For the Latin text, see José María Millás Vallicrosa, 'El "Liber de motu octave sphere" de Tābit ibn Qurra', *Al-Andalus*, 10 (1945), 89–108, repr. in José María Millás Vallicrosa, *Nuevos estudios sobre historia de la ciencia española* (Barcelona: CSIC, 1960), 191–209. Two different versions of the Latin text may also be found in Francis J. Carmody, *The Astronomical Works of Thabit b. Qurra* (Berkeley: University of California Press, 1960), 84–113, whose introduction is rather unreliable. An English translation, together with a very useful commentary, was published by Otto Neugebauer, 'Thâbit ben Qurra "On the Solar Year" and "On the Motion of the Eighth Sphere"', *Proceedings of the American Philosophical Society*, 106 (1962), 264–99, at 291–9. For technical descriptions of the model in question, see Bernard R. Goldstein, 'On the Theory of Trepidation According to Thābit b. Qurra and al-Zarqāllu and Its Implications for Homocentric Planetary Theory', *Centaurus*, 10 (1964–5), 232–47, at 232–8; John North, 'Medieval Star Catalogues and the Movement of the Eighth Sphere', *Archives internationales d'histoire des sciences*, 20 (1967), 71–83, at 78–81; Dobrzycki, 'The Theory', 20–30; Mercier, 'Studies', pt. 1, 209–20; pt. 2, 62–5; Julio Samsó, *Las ciencias de los antiguos en Al-Andalus*

290 THE SCIENTIFIC WORKS OF ROBERT GROSSETESTE

Even apart from the account given in his *On the Sphere* (see below), Grosseteste's familiarity with this model is suggested by the copy of *On the Motion of the Eighth Sphere* included in Oxford, Bodleian Library, MS Savile 21, fols. 153v–156r, which is very likely to be in his own hand and appears to date from *c.*1216 (Ch. 2, §1). This would put Grosseteste's copying *On the Motion of the Eighth Sphere* in close temporal proximity to his composing *On the Sphere*. The heading of this copy follows the common attribution of *On the Motion of the Eighth Sphere* to the Sabian mathematician and astronomer Thābit ibn Qurra, who was active in Baghdad in the second half of the ninth century. Although Thābit is known to have speculated about the trepidation of the eighth sphere,[11] there are on the whole excellent reasons to be sceptical of this attribution. Among these are some close links between *On the Motion of the Eighth Sphere* and the Toledan Tables, which suggest that its origins rather lie in eleventh-century al-Andalus.[12]

One possible candidate for having developed the model in question is Abū Ishāq Ibrāhīm al-Zarqālī (or al-Zarqālluh, d. 1100), known in Spain as Azarquiel, whom Latin astronomers, Grosseteste included, tended to regard as the principal author of the Toledan Tables.[13] His authentic works include a *Treatise on the Motion of the Fixed Stars* preserved only in Hebrew, in which he describes three different geometrical models.

(Madrid: Mapfre, 1992), 222–6; Samsó, *On Both Sides*, 580–3; C. Philipp E. Nothaft, 'Criticism of Trepidation Models and Advocacy of Uniform Precession in Medieval Latin Astronomy', *Archive for History of Exact Sciences*, 71 (2017), 211–44, at 213–15.

[11] Carmody, *The Astronomical Works*, 45–6, 84–5; Dobrzycki, 'The Theory', 28–30; Ragep, 'Al-Battānī', 282–3.

[12] These links are demonstrated in Raymond Mercier, 'Accession and Recession: Reconstruction of the Parameters', in Casulleras and Samsó, *From Baghdad to Barcelona*, Vol. I, 299–347. Further details relevant to the question of authorship are provided in Lutz Richter-Bernburg, 'Sā'id, the *Toledan Tables*, and Andalusī Science', in David A. King and George Saliba (eds.), *From Deferent to Equant: A Volume of Studies in the History of Science in the Ancient and Medieval Near East in Honor of E. S. Kennedy* (New York: New York Academy of Sciences, 1987), 373–401, at 385–9; F. Jamil Ragep, *Nasīr al-Dīn al-Tūsī's* Memoir on Astronomy (*al-Tadhkira fī 'ilm al-hay'a*), 2 vols (New York: Springer, 1993), ii. 400–8; Julio Samsó, 'Trepidation in Al-Andalus in the 11th Century', *Islamic Astronomy and Medieval Spain* (Aldershot: Variorum, 1994), 2–5; Samsó, *Las ciencias*, 220–2, 225, 239–40; Samsó, *On Both Sides*, 579–80, 583–6; Mercè Comes, 'Ibn al-Hā'im's Trepidation Model', *Suhayl*, 2 (2001), 291–408, at 293–6; Miguel Forcada, 'Saphaeae and Hay'āt: The Debate between Instrumentalism and Realism in Al-Andalus', *Medieval Encounters*, 23 (2017), 263–86, at 275–81.

[13] See the references to *Arzachel* in Grosseteste, *Compotus*, ch. 4, ll. 10, 16, 131, 136–7, 145, 151, 202; ch. 5, l. 17; ch. 7, l. 17.

TREPIDATION OR PRECESSION 291

The third and most successful of these shared some central features with the model known from the Latin *On the Motion of the Eighth Sphere*, but deviated from it by introducing a separate mechanism for varying the obliquity of the ecliptic.[14] Subsequent generations of astronomers active in al-Andalus and the Magheb continued on this path by fine-tuning Azarquiel's model and integrating versions of it into their astronomical tables.[15] Important as these developments may be for the wider history of astronomy, they have little bearing on the transmission of trepidation theories to Latin Europe, where Pseudo-Thābit's model appears to have been the only version of trepidation used in computational practice before the end of the thirteenth century.[16] Its popularity depended to a large degree on the wide dissemination of the Toledan Tables, which required astronomers to take into account the irregular motion of the eighth sphere whenever they sought to locate the time of the vernal equinox or perform some other calculation based on tropical coordinates.

This is not to say that Latin astronomers lacked any kind of opportunity to work with linear precession models in the tradition of Ptolemy's *Almagest*. Tables for the motion of the solar apogee based on an annual precession of 0;0,54°, which approximates al-Battanī's value of 1°/66y,

[14] Goldstein, 'On the Theory', 238–46; Julio Samsó, 'Sobre el modelo de Azarquiel para determinar la obliscuidad de la ecliptica', in *Homenaje al Prof. Darío Cabanelas Rodríguez*, 2 vols (Granada: Universidad de Granada, 1987), ii. 367–77, repr. as ch. 9 in Samsó, *Islamic Astronomy*; Samsó, *Las ciencias*, 227–39; Samsó, 'Trepidation in al-Andalus', 5–31; Samsó, *On Both Sides*, 586–610; Mercier, 'Accession', 327–30; Comes, 'Ibn al-Hā'im's Trepidation Model', 302–5.

[15] Mercè Comes, 'The Accession and Recession Theory in Al-Andalus and the North of Africa', in Casulleras and Samsó (eds.), *From Baghdad to Barcelona*, I, 349–64; Mercè Comes, 'Some New Maghribī Sources Dealing with Trepidation', in S. M. Razaullah Ansari (ed.), *Science and Technology in the Islamic World* (Turnhout: Brepols, 2002), 121–41; Julio Samsó, 'Astronomical Observations in the Maghrib in the Fourteenth and Fifteenth Centuries', *Science in Context*, 14 (2001), 165–78, at 169–74; Samsó, *On Both Sides*, 610–42; Montse Díaz-Fajardo, *La teoría de la trepidación en un astrónomo marroquí del siglo XV. Estudio y edición crítica del Kitāb al adwār fī tusyīr ul-unwār (parte primera) de Abu ʿAbd Allāh al-Baqqār* (Barcelona. Instituto 'Millás Vallicrosa' de Historia de la Ciencia Arabe, 2001).

[16] The only known minor exception are the tables for trepidation included in the *zīj al-Muqtabis* of Ibn al-Kammād (active in Cordova, *c*.1100), which survives in a Latin version made by John Dumpno in Palermo in 1260. See José Chabás and Bernard R. Goldstein, 'Andalusian Astronomy: al-Zīj al-Muqtabis of Ibn al-Kammâd', *Archive for History of Exact Sciences*, 48 (1994), 1–41, at 24–7; Chabás and Goldstein, 'Ibn al-Kammād's *Muqtabis* zij and the Astronomical Tradition of Indian Origin in the Iberian Peninsula', *Archive for History of Exact Sciences*, 69 (2015), 577–650, at 596–8; José Luis Mancha, 'On Ibn al-Kammād's Table for Trepidation', *Archive for History of Exact Sciences*, 52 (1998), 1–11; Samsó, *On Both Sides*, 610–17.

292 THE SCIENTIFIC WORKS OF ROBERT GROSSETESTE

survive in a number of Latin manuscripts copied in the twelfth to fourteenth centuries.[17] In most instances, these are related to the Tables of London, which are themselves a partial adaptation of the Tables of Pisa by the Jewish astronomer Abraham Ibn Ezra, who constructed them in the 1140s on the basis of an earlier set created by al-Ṣūfī (d. 986).[18] Against the prevailing Andalusian fashion of applying trepidation models to sidereal longitudes, Ibn Ezra strongly favoured a classical Ptolemaic approach of accepting linear precession and computing planetary longitudes in the ninth sphere.[19] Latin readers could find out more about his views from reading his *Book on the Principles behind Astronomical Tables* (*Liber de rationibus tabularum*) (1154), where Ibn Ezra attributed to the aforementioned al-Ṣūfī a precession rate of 1°/70y.[20] The latter value corresponds to an annual precession of $c.0;0,51,26°$, which comes remarkably close to the modern value of approximately 50 arcseconds per year. As with the aforementioned rate of $0;0,54°$, precession tables and calculations based on $0;0,51°$ per year are attested in a handful of twelfth- and thirteenth-century manuscripts.[21] None of these, however, can be plausibly linked to Robert Grosseteste, whose works show no familiarity with the astronomy of Abraham Ibn Ezra.

[17] MSS Cambridge, University Library, Add. 6866, f. 150v; London, British Library Arundel 377, f. 9v; Paris, Bibliothèque nationale de France lat. 16,208, f. 4v; Toledo, Archivo y Biblioteca Capitulares, 98–22, f. 104v; Oxford, Bodleian Library Savile 21, f. 91r; Oxford, Bodleian Library Savile 25, f. 203r; Oxford, Bodleian Library Ashmole 361, f. 27v. See also the tables for the motion of the fixed stars in MSS Cambridge, Fitzwilliam Museum McClean 165, f. 79v; Cambridge, Gonville & Caius College 456, p. 139, which appear to be derived from al-Battānī's work.

[18] Raymond Mercier, 'The Lost Zīj of al-Ṣūfī in the Twelfth-Century Tables for London and Pisa', *Studies on the Transmission of Medieval Mathematical Astronomy* (Aldershot: Variorum, 2004), ch. 8.

[19] Julio Samsó, '"Dixit Abraham Iudeus". Algunas observaciones sobre los textos astronómicos latinos de Abraham Ibn 'Ezra', *Iberia Judaica*, 4 (2012), 171–200, at 177–85, 193–6.

[20] José María Millás Vallicrosa (ed.), *El libro de los fundamentos de las Tablas astronómicas de R. Abraham Ibn 'Ezra* (Madrid: CSIC, 1947), 78, 83.

[21] See Pedersen, *The Toledan Tables*, III, 1226–7; *Artis cuiuslibet consummatio* (2.26), in Stephen K. Victor (ed.), *Practical Geometry in the High Middle Ages* (Philadelphia, PA: American Philosophical Society, 1979), 282–5; Nothaft, 'Criticism', 222–4; C. Philipp E. Nothaft, 'Henry Bate's *Tabule Machlinenses*: The Earliest Astronomical Tables by a Latin Author', *Annals of Science*, 75 (2018), 275–303, at 287–90.

TREPIDATION OR PRECESSION   293

3 *On the Sphere*

The discussion of precession and trepidation is found in DS §§48–53. Neither theory is referred to by name, but by the author to whom Grosseteste attributes it: Ptolemy is the proponent of precession and Thābit of trepidation. In DS §49 he records the Ptolemaic rate of precession as 1° of arc (of the 360 in a full circle) per century but has a novel way of refuting it (see §4 below). In DS §§50 and 51 he sets up the structure of the trepidation model by carefully describing the mobile ecliptic (or 'mobile zodiac') located in the eighth sphere, which is the sphere of fixed stars, and the fixed ecliptic located in the ninth sphere above it. Their mutual motion is defined, as discussed above, by the counter-turning of the one against the other on two small circles centred at the equinoctial points on the fixed ecliptic (the points of Aries and Libra). Grosseteste gives the diameter of these circles as 8;37° of arc ('with eight degrees 37 minutes being occupied'; DS §51), which is rounded from the 8;37,26° stated in *On the Motion of the Eighth Sphere*.[22] It follows that the sidereal heads of Aries and Libra located on the mobile ecliptic can be up to ±4;18,30° removed from the corresponding points of the fixed ecliptic. This will be the case at two instances of their circular path where the mobile heads of Aries and Libra come to lie on the fixed ecliptic itself, such that mobile Aries will be at 4;18,30° of fixed Aries ('the 19th minute of the 5th degree'; DS §51) or at 25;41,30° of fixed Pisces ('the 42nd minute of the 26th degree'; DS §51).

In DS §§52 and 53, he goes on to provide the other necessary condition to specify the orientation of the sphere of the mobile zodiac with respect to the fixed zodiac, namely that the mobile heads of Cancer and Capricorn (defined as +90° from the heads of Aries and Libra) always lie on the fixed ecliptic. Over the course of a full cycle of trepidation, the mobile head of Cancer (or Capricorn) will continually slide along the fixed ecliptic within a range of ±4;18,30° from the fixed head of Cancer (or Capricorn), meaning that, unlike Aries and Libra, the mobile and fixed heads of Cancer and Capricorn can sometimes occupy the same point.

For the angular motion of the mobile heads of Aries and Libra, DS §51 gives a rate of *c*.1;2° in twelve years, which amounts to *c*.0;5,10° per year. This numerical value comes from a table for the annual motion of the

---

[22] Millás Vallicrosa, 'El "Liber de motu octave sphere"', 99.

294 THE SCIENTIFIC WORKS OF ROBERT GROSSETESTE

eighth sphere contained in both *On the Motion of the Eighth Sphere* and the Toledan Tables, where the entry for twelve years is 1;1,59°.[23] Grosseteste leaves his readers in the dark about the fact that the years in question are Arabic lunar years of 354 11/30 days rather than Julian years of 365 1/4 days.

4 Climate Change and Trepidation

Despite his statement in *On the Sphere* DS §50, where it is claimed that Thābit 'found through certain experiments that the motion of the fixed stars is different' from what Ptolemy had proposed before him, there are no signs in his treatise that Grosseteste based his endorsement of trepidation on any detailed comparison of the two theories. What his text offers instead is an implicit argument from consequences, which appeals to the long term climatic effects produced by a linear precession model.

Grosseteste sets this argument up where he comments on the eccentricity of the solar path relative to the centre of the universe (DS §44).[24] According to Ptolemy's *Almagest*, 'the radius of the eccentre [the circle that describes the sun's path relative to the centre of the universe] is approximately 24 times the distance between the centres of the eccentre and the ecliptic'.[25] Ptolemy came to this conclusion after assigning an arbitrary value of 60 parts to the radius of the sun's eccentre and finding that there are 2;29,30 parts between its centre and the centre of the ecliptic. This value is very nearly 2;30 parts, or 1/24 of 60 parts.[26] Grosseteste refers to these parts as 'degrees' and doubles their number to obtain the portion of the eccentre's radius by which the sun at perigee draws closer to the earth than the sun at apogee.

Now, since the perigee is located in the sign of Sagittarius (DS §39), this means that the sun will come closest to the earth as it approaches the December solstice, at which time it is directly overhead those living at the

---

[23] Millás Vallicrosa, 'El "Liber de motu octave sphere"', 106. See also Pedersen, *The Toledan Tables*, IV, 1545.

[24] For more on Ptolemy's solar model, see Viggo M. Petersen and Olaf Schmidt, 'The Determination of the Longitude of the Apogee of the Orbit of the Sun According to Hipparchus and Ptolemy', *Centaurus*, 12 (1968), 73–96; Evans, *The History*, 205–43; Pedersen, *A Survey*, 122–58.

[25] Ptolemy, *Almagest*, III, 4, pp. 233–40. English translation from Toomer, *Ptolemy's Almagest*, 155.

[26] See Pedersen, *A Survey*, 146–7.

Tropic of Capricorn. Grosseteste writes that 'the cause of heat is doubled in their summer', by which he means summer in the southern hemisphere. He imagines that when the sun reaches Sagittarius, its great proximity will contribute further to the excessive heat it produces by being at or near zenith, such that its rays hit the earth below at a perpendicular angle. The converse is true for southern hemisphere winters, at which time the sun not only has a much lower altitude, but is close to the apogee and, hence, considerably more distant from the earth. Grosseteste suggests that this coincidence of factors is the cause of extreme climate in the southern hemisphere (or at least those parts close to the Tropic of Capricorn), to the extent that no human habitation is possible there. Things are different in the northern hemisphere, which is temperate and habitable precisely because the sun's altitude and position on its eccentre counterbalance each other.

The idea of factoring solar eccentricity into arguments about climate was a relatively new one in Grosseteste's time. Prior to the twelfth century, Latin Christian authors had usually based their descriptions of the earth's habitability on an ancient Parmenidean scheme transmitted by sources such as Macrobius's *Commentary on the Dream of Scipio* and Pliny's *Natural History*. It divided the world into five climatic zones, only two of which were temperate, while the other three were uninhabitable owing to excessive heat or cold.[27] In particular, the entire area between the two tropics was regarded to be an uninhabitable 'torrid zone', whose inhospitable nature was inferred from the fact that the sun in these regions stood directly overhead twice a year. According to Bede, for instance, the perpendicular angle of the sun's rays caused the land in these regions to be 'parched and burned up by flames [and] baked by being so close to the heat', making it impossible for humans to cross from one temperate zone to the other.[28]

---

[27] For further details, see Patrick Gautier Dalché, 'Guillaume de Conches, le modèle macrobien de la sphère et les antipodes. Antécédents et influence immédiate', in Barbara Obrist and Irène Caiazzo (eds.), *Guillaume de Conches. Philosophie et science au XIIe siècle* (Florence: SISMEL, 2011), 219–51; Alfred Hiatt, 'The Map of Macrobius before 1100', *Imago Mundi*, 59 (2007), 149–76.

[28] Bede, *De temporum ratione*, c. 34, ed. Charles W. Jones, CCSL 123B (Turnhout: Brepols, 1977), 389: 'Ipsa est aequinoctialis quam, quia semper sol aut praesens aut hinc vel inde vicinius illustrat, nimirum subiecta terrarum exusta flammis et cremata cominus vapore torrentur.'

296 THE SCIENTIFIC WORKS OF ROBERT GROSSETESTE

Although Grosseteste's acknowledges this five-zone model in *On the Sphere* (DS §16), he modifies its climatic aspects by drawing on some alternative cosmographical ideas that had entered Latin Europe only during the past century through contact with Arabic and Greek sources.[29] One of the early disseminators of these new ways of looking at the earth's habitability was Petrus Alfonsi, who in his *Dialogue against the Jews* (*c.*1110) argued that the equatorial region was exceptionally temperate and therefore populated, as seen from the fact that the city of Arim, which marked the prime meridian and 'middle of the world', was located below the celestial equator (see Ch. 5, §4.3).[30] A different story, however, could be told about regions closer to the southern tropic, which were affected adversely by their proximity to the sun. According to Petrus, this proximity was such that the sun 'renders the earth unfruitful for all things and altogether sterile, and this is why it is uninhabitable'.[31] In 1175, an English computist named Magister Cunestabulus seized on this idea to argue that the existence of the Antipodes could be safely excluded on natural grounds.[32] His scientific justification of this stance is worth citing in full given its similarities to *On the Sphere* (DS §44):

It is not a very safe thing for believers to opine that anybody can live beyond the equinoctial, where the sea seethes like a kitchen pot, and further beyond. But even according to the astronomical truth the depression of the Sun in Sagittarius and its elevation in Gemini are so great that there is necessarily extreme cold there when we have

[29] On the background, see Patrick Gautier Dalché, 'La renouvellement de la perception et de la représentation de l'espace au XIIe siècle', in García de Cortázar and José Angel (eds.), *Renovación intelectual del Occidente Europeo (siglo XII)* (Pamplona: Gobierno de Navarra, 1998), 169–217, repr. in Patrick Gautier Dalché, *L'espace géographique au Moyen Âge* (Florence: SISMEL, 2013), 293–344, at 327–9; Patrick Gautier Dalché, 'Géographie Arabe et Géographie Latine au XIIe siècle', *Medieval Encounters*, 19 (2013), 408–33, at 411–21; Gautier Dalché, 'Un débat scientifique au Moyen Âge. L'habitation de la zone torride (jusqu'au XIIIe siècle)', *Topoi* Supplément, 15 (2017), 145–81.

[30] Petrus Alfonsi, *Dialogus contra Iudaeos*, I.66, ed. and trans. (German), Peter Stotz (Florence: SISMEL, 2018), 38.

[31] Alfonsi, *Dialogus contra Iudaeos*, I. 72, p. 42: 'Unde cum sol ad sex meridianae plagae signa, quae sunt a libra ad arietem, descenderit, quia tunc terrae propior est, vicinitate sua calore terram exurens omnium rerum infecundam et omnino sterilem reddit, ideoque inhabitabilis existit'. See also Hermann of Carinthia, *De essentiis*, bk. II, ed. Charles Burnett (Leiden: Brill, 1982), 220–2.

[32] See C. Philipp E. Nothaft, 'A Reluctant Innovator: Graeco-Arabic Astronomy in the *Computus* of Magister Cunestabulus (1175)', *Early Science and Medicine*, 22 (2017), 24–54, at 31–8.

summer and a most immoderate heat when we have winter. The [amount by which] the Sun is more depressed in Sagittarius than in Gemini and closer to the centre of the Earth is the 12th part of the line that separates the centre of the Earth from the centre of the Sun at mean distance. This 12th part is more than one hundred times the Earth's radius. For this reason it seems to me not just a possible opinion, but much rather necessary to believe that nobody can live there because of the extreme climate. Ptolemy also suggests that it is true for all [observers] that the Sun, whether it is more elevated or more depressed, still increases its heat as it ascends towards their zenith and diminishes it as it descends, because the ray of the Sun that falls down from a right angle imposes the heat and makes it last, whereas the one coming from an oblique angle decays and passes the heat further on. This is the reason why in our regions, even though the Sun is more remote [from us] in summer than in winter, it nevertheless generates greater heat. But whatever lies beyond the equator is not just heated, but burned up completely during the summer, due to both the proximity [of the Sun] and the verticality [of its rays].[33]

While *On the Sphere* DS §44 can be shown, therefore, to be rooted in pre-existing discussions of the habitability of the southern hemisphere, Grosseteste appears to be entering completely new territory in DS §49, by claiming that Ptolemy's model of precession causes not just the fixed stars, but also the apogees of all planets, to shift their position relative to

---

[33] Cunestabulus, *Compotus*, ch. 12, ed. Alfred Lohr, *Opera de computo saeculi duodecimi* (Turnhout: Brepols, 2015), 75: 'Ultra aequinoctialem, ubi mare fervet ut caccabus, atque ulterius quempiam habitare non valde tutum est fidelibus opinari. Sed et secundum veritatem astronomicam tanta est depressio solis in Sagittario et elevatio in Geminis, ut necesse sit ibi esse frigus acutissimum, dum nobis est aestas, et, dum nobis hiems, calorem immoderatissimum. In XIIa parte lineae, quae est a centro terrae usque ad centrum solis in remotione media, depressior est sol in Sagittario quam in Geminis et centro terrae vicinior. Quae XIIa semidiametro terrae plus quam centupla est. Quapropter mihi videtur non opinabile, immo necessarium arbitrandum propter nimiam aeris intemperantiam neminem illic posse habitare. Suadet etiam Ptolomaeus solem sive elevatiorem sive depressiorem, omnibus tamen, quando illis ascendit ad punctum capitalem, calorem intendere et remittere, cum descendit, eo quod tardius solis directius descendens calorem infigat et remanere faciat, ex obliquo veniens prolabatur et calorem ad ulteriora transmittat. Unde apud nos, licet sol remotior sit in aestate quam in hieme, magis tamen calefacit. Quod autem ultra aequinoctialem est, in sua aestate tum propter vicinitatem tum propter directiorem non modo calefacit, sed exurit'. The first scholar to draw attention to this passage was Cecilia Panti, *Moti, virtù e motori*, 86.

298 THE SCIENTIFIC WORKS OF ROBERT GROSSETESTE

the equinoxes. The passage reveals that Grosseteste's understanding of precession was guided not such much by his reading of the *Almagest* as by his familiarity with the trepidation model of *On the Motion of the Eighth Sphere* and the Toledan Tables, which allowed the solar apogee to partake in the accession and recession of the eighth sphere. It was not so with precession as originally conceived by Ptolemy, who treated the perigee as fixed in the sign of Sagittarius (at 5;30°).[34] Evidence against this notion was provided by a treatise *On the Solar Year* (*De anno solis*), another work falsely attributed to Thābit ibn Qurra, which had already been cited by Cunestabulus in 1175.[35] Originally written in Arabic in the ninth century, *On the Solar Year* documented astronomical observations made in 830–832 in Baghdad by a group of astronomers known as the Banū Mūsā. Their measurements of solstices and equinoxes demonstrated that the sun's apogee was no different than the other planetary apogees in being subject to precession.[36]

Whether Grosseteste knew of *On the Solar Year* or not, he drew the same conclusion with regard to the mobility of the sun's line of apsides. He also considered the consequence of this mobility, which was that the perigee could be expected eventually to arrive in the northern hemisphere, rendering humanity's present habitat unsuitable for life. To be sure, this process was bound to be very slow. On Ptolemy's precession rate of 1°/100y, it would take 3,000 years for the perigee to traverse a single sign of the zodiac and 18,000 years to swap places with the apogee.

It is important to note that in his analysis, Grosseteste does not use the standard *quaestio* format that he deploys to demolish prevailing theories of the structure and origins of comets in *On Comets*. He argues in *On the Sphere* that if Ptolemy's view of continuous precession is correct, the apogee of the sun and the stars at present in the

---

[34] Ptolemy, *Almagest*, III, 4, p. 233.   [35] Nothaft, 'A Reluctant Innovator', 39–41, 45–6.
[36] The Latin text of *De anno solis* was edited by Carmody, *The Astronomical Works*, 63–79. For an English translation, see Neugebauer, 'Thâbit ben Qurra', 265–89. The Arabic text, together with a French translation and a discussion of the work's authorship, may be found in Thābit ibn Qurra, *Oeuvres d'astronomie*, ed. Régis Morelon (Paris: Les Belles Lettres, 1987), pp. xlvi–liii, 27–67. Its content is discussed in Kristian Peder Moesgaard, 'Thâbit ibn Qurra between Ptolemy and Copernicus: An Analysis of Thâbit's Solar Theory', *Archive for History of Exact Sciences*, 12 (1974), 199–216; Régis Morelon, 'Eastern Arabic Astronomy between the Eighth and the Eleventh Centuries', in Roshdi Rashed (ed.), *Encyclopedia of the History of Arabic Science*, 3 vols. (London: Routledge, 1996), i, 20–57, at 26–31. On the Banū Mūsā, see Josep Casulleras, 'Banū Mūsā', in Thomas Hockey et al. (eds.), *The Biographical Encyclopaedia of Astronomers* (New York: Springer, 2007), 92–4.

TREPIDATION OR PRECESSION   299

northern hemisphere would end up in the southern hemisphere, and in consequence the presently inhabited region would become uninhabitable. Here, however, Grosseteste does not go on to state that this implies the linear precession model is therefore wrong, leaving a conundrum. From the way Grosseteste transitions from DS §49 to the trepidation model introduced in DS §50, it is nevertheless clear that he regarded the migration of the solar perigee as an unacceptable aspect of any linear precession model.

There can have been no major issue with the movement of the apogee/perigee from creation to the thirteenth century as Grosseteste, with his contemporaries, will have taken the age of the universe to have been only a few thousand years, a figure commensurate with the Biblical account of the number of generations since the creation.[37] A drift of a few tens of degrees would not have led to dramatic climate change. The distant future would, however, present the problematic coincidence for the northern hemisphere of the coincidence of perihelion with summer. That the consequences of the double-effect were overestimated by Grosseteste is not relevant to the thought that drove him towards the trepidation theory of Pseudo-Thābit, which assumed a bounded precession of ±10;45°. By consequence, it was impossible for the sun's perigee ever to leave the sign of Sagittarius. Trepidation could hence be regarded as an insurance against the kind of major global climate change precession held in store. In thinking through these implications, Grosseteste became the first in a longer line of thirteenth-century thinkers who pointed out a connection between climatic change, and the theories of trepidation and precession.[38] An understanding in detail of how climate change depends directly on astronomical changes would, however, have to wait until the twentieth century before the full impact of changes in the earth's orbit and axis of rotation would be understood (see Box 8.1).[39]

---

[37] In the *Hexaemeron*, Grosseteste notes the biblical reference to the six ages of the world, the first two being ten generations, the next three being fourteen each, and the sixth age being since the coming of Christ: Robert Grosseteste, *Hexaemeron*, 8.XXXI.1, ed. Richard C. Dales and Servus Gieben (Oxford: Oxford University Press, 1982), 255.

[38] For further information, see C. Philipp E. Nothaft, 'Climate, Astrology and the Age of the World in Thirteenth-Century Thought: Giles of Lessines and Roger Bacon on the Precession of the Solar Apogee', *Journal of the Warburg and Courtauld Institutes*, 77 (2014), 35–60.

[39] J. D. Hays, John Imbrie, and N. J. Shackleton, 'Variations in the Earth's Orbit: Pacemaker of the Ice Ages', *Science*, 194 (1976), 1121–32; D. Paillard, 'Climate and the Orbital Parameters of the Earth', *Comptes Rendus Geoscience*, 342 (2010), 273–85.

300   THE SCIENTIFIC WORKS OF ROBERT GROSSETESTE

5 A Comparison of Historical Observations with Calculations

Even though Grosseteste cited no observational evidence in support of trepidation, it is unlikely that he would have devoted as much attention to Pseudo-Thābit's theory as he did in *On the Sphere* had he regarded it as empirically unsound. Indeed, his reference to reliable experiences (*certa experimenta*) in DS §51 may reflect a certain degree of confidence in the theory's predictive success, that is, its ability to predict changes in stellar longitude brought about by the motion of the sphere of fixed stars. In order to test this ability, an astronomer in Grosseteste's time would have had to compare the current position of a given star with observations of the same star recorded by earlier astronomers, ideally many centuries in the past. However, the pool of relevant historical data that would have been at the disposal of a Latin scholar working in the early thirteenth century is rather thin, being confined for the most part to the works of Ptolemy and al-Battānī. Since the trepidation model attributed to Thābit had been created specifically to accommodate the findings of both these astronomers, a test based on them was condemned to be circular.

Table 8.1 gives a better sense of the success of the Toledan model. It records four historical observations of the star Regulus (α Leonis), from Hipparchus in the second century BCE to Tycho Brahe in the sixteenth century CE.[40] To these a modern (calculated) value for the longitude of Regulus in 1220 CE has been added, chosen for being close to Grosseteste's time of writing *On the Sphere*. In the two rightmost

Table 8.1 Historical Longitude Observations of α Leonis

Year	Longitude α Leo	Observ. diff.	Calc. diff.
−128/7 (Hipparchus)	119;50°	—	—
139 (Ptolemy)	122;30°	+2;40°	+2;39°
879 (al-Battānī)	134°	+11;30°	+11;30°
1220 (modern)	139°	+5°	+4;20°
1585 (Brahe)	144;5°	+5;5°	+1;59°

---

[40] These observations are drawn from the following sources: Ptolemy, *Almagest*, VII, 2, pp. 13–16; al-Battānī, *Opus astronomicum*, ch. 51, ed. Nallino, I, 124; Tycho Brahe, *Astronomiae instauratae progymnasmata*, pt. 2, in J. L. E. Dreyer (ed.), *Tychonis Brahe Dani Scripta Astronomica*, 15 vols. (Copenhagen: Glydendal, 1913–29), ii. 254.

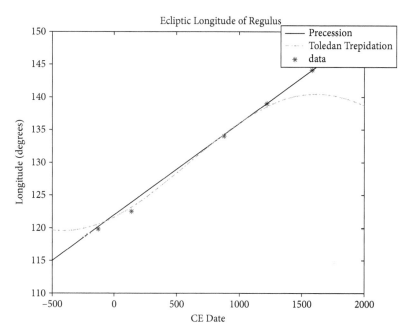

Fig. 8.3. Comparison of the predictions of Toledan trepidation and modern precession for the ecliptic longitude of Regulus from 500 BCE to 2000 CE together with measured pre-modern positions (see table).

columns the increase in longitude implied by these successive observations is compared with the change in the so-called equation of eighth sphere predicted by the tables attached to Pseudo-Thābit's model.

If these data are presented in graphical form, the result gives an immediate impression of the reason that trepidation could be considered a competitive alternative to precession as late as 1220. Fig. 8.3 gives the predictions of both models for the ecliptic longitude of Regulus against the observational data included in Table 8.1.

The first notable impression is the remarkably close agreement of the two models from 100 BCE until 1200 CE, and their rapid departure outside this, historically critical, interval. Grosseteste was writing at the very end of this period, over which naked-eye measurements of celestial coordinates were subject to errors not much greater than the divergences in the two models. The longitude of Regulus attributed by Ptolemy to his predecessor Hipparchus, for instance, falls approximately −0;34° behind the modern calculated value for the epoch of his observation. Tycho

302 THE SCIENTIFIC WORKS OF ROBERT GROSSETESTE

Brahe was to lower this error margin to <0;1°. A more dramatic deviation from the modern value is exhibited only by Ptolemy's observation for 139 CE, which falls short by $c.-1;34°$. Significantly, this error matches almost exactly the difference between the two models and is much closer to the prediction of trepidation than to the (correct) value given by precession.

It may be seen that an accurate measurement of Regulus's longitude in 1220 would have yielded a result that still fell very close to the curve described by the trepidation model (the very last of the points in Fig. 8.3 to do so). Whether or not Grosseteste ever made any stellar observations himself, it is unlikely that the results would have given him any grounds for arguing against Pseudo-Thābit.[41] The situation would have looked significantly different a century later owing to the way the trepidation model employed by the Toledan Tables falsely predicted an increasing slowing down of the rate of precessions. By $c.1300$, observable and computed longitudes would have fallen roughly 1° apart, an error great enough to be detected by naked-eye observations. It is therefore no accident that Latin astronomers of the late thirteenth and early fourteenth century were discussing alternatives to the Toledan model, in particular a return to a simple linear precession model. Dissatisfaction with this model appears to have been one of the main factors in pushing Latin astronomers towards adopting the Alfonsine Tables, as happened in Paris around 1320. These tables maintained the theory of trepidation, but improved predictive accuracy by combining it with linear precession.[42]

Grosseteste's discussion of Toledan Trepidation in *On the Sphere* opens up an inheritance of sources and ideas from antiquity to his contemporaries and anticipates a critical period in the evaluation of two competing models for the slow observed changes in celestial coordinates. He wrote in the very last few decades for which the standard trepidation model could compete with precession in the light of observations. Grosseteste's own contribution represents a characteristic combination of transmission of and commentary on his sources, together with original thinking of his own. His discussion of the climactic consequences of

---

[41] A text on the motion of the fixed stars written $c.1301$ by Walter Odington credits *Robertus Lincolniensis* with having found the longitude of Aldebaran (α Tau) to be 28;40°. It may be worth noting that this is the longitude the star would have had in $c.1203$. See MS Cambridge, University Library, Ii.I.13, fol. 180r.

[42] For more on these developments in Latin astronomy, see Nothaft, 'Criticism', 216–35.

astronomical alignment and distance of the earth with respect to the sun finds resonance in contemporary research into long term climate cycles (see Box 8.1 on Milankovitch Cycles). At the same time, it falls well within the astrological and theological thinking of Grosseteste's own time, insofar as it draws out the earthly effects of heavenly movements which is the concluding theme of *On the Sphere*.

Box 8.2 Notes on Calculation of Right Ascension and Declination under Precession and Trepidation

The easiest way to proceed is to work from the ecliptic coordinates of a star (for example, Regulus), then to transform into the Right Ascension (RA) and Declination of any celestial coordinate system rotated relative to the ecliptic pole. In the case of precession, for example, the time-dependent second coordinate system is tilted from the ecliptic system by a 'theta' angle of 23.5° and a 'phi' angle that precesses on a period of approximately 26,000 years. The underlying calculation is to transform between spherical angular coordinates of two systems tilted from each other (see Fig. 8.B.2).

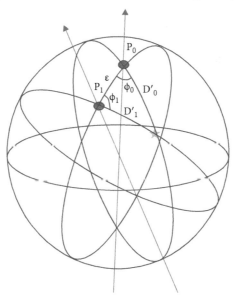

Fig. 8.B.2. Spherical trigonometric construction.

Continued

304   THE SCIENTIFIC WORKS OF ROBERT GROSSETESTE

Box 8.2 Continued

The poles of the systems are P_0 and P_1, the co-declinations of the star are D'_0 and D'_1, the angles needed to calculate the RAs in the two frames are ϕ_0 and ϕ_1. Standard spherical trigonometry can be used to find the unknown quantities with '1' suffices from the known quantities with '0' suffices:

$$\cos D'_1 = \cos \epsilon \cos D'_0 + \sin \epsilon \cos \phi_0$$

and

$$\frac{\sin \phi_1}{\sin D'_0} = \frac{\sin \phi_0}{D'_1}$$

The current celestial coordinates of Regulus are $RA=10^h8^m22^s$ (242.1°) and Dec=11.97°. The strategy is to convert these to ecliptic coordinates first, finding $RA_e=269.1°$ and $Dec_e=0.98°$. Now these (fixed) ecliptic coordinates may be mapped onto the earth-centred celestial coordinates using

$$\phi_0 = \phi_{00} + 2\pi t/T$$

where T is the precession period of 26,000 years.

9

Kinematic Descriptions, Epicycles, and Modelling *On the Sphere*

This chapter employs mathematical and geometrical tools to analyse quantitatively the celestial phenomena described in Grosseteste's *On the Sphere*. In response to the analysis of Chapter 7, which draws attention to the particular way in which the contents of his treatise are framed by geometry, before expounding their natural philosophy, it uses more modern mathematical tools to re-capitulate that perspective. It is structured in two sections. The first presents a brief technical summary of the Ptolemaic theory of solar and lunar motion, described in his magisterial *Almagest*, available to astronomers of Christendom in Grosseteste's generation in a number of forms. A translation into Latin had been made from Greek in Sicily and from Arabic by Gerard of Cremona in Toledo.[1] The treatise circulated in whole and abridged versions, and also as discussed and used within other Islamicate astronomical treatises translated from Arabic, for example that by Alfraganus (Ch. 6, §2). The second gives an accessible overview of modern models of planetary motion, focusing on the *kinematics*, which is to say how their motions are described. The analysis shows that, mathematically, these can be expressed in a manner equivalent to models based on a geocentric description with epicycles, as in the context of the descriptions given in Grosseteste's *On the Sphere* of the sun's and moon's orbits around the earth. In this way, and as developed in this series of volumes, current scientific understanding is deployed to reveal more, rather than less, of the intellectual achievements of earlier periods, and to emphasize the

---

[1] David Juste, 'Ptolemy, *Almagesti* (tr. Sicily c. 1150)' (update: 04.03.2021), *Ptolemaeus Arabus et Latinus. Works*, http://ptolemaeus.badw.de/work/21 and David Juste, 'Ptolemy, *Almagesti* (tr. Gerard of Cremona)' (update: 07.05.2021), *Ptolemaeus Arabus et Latinus. Works*, http://ptolemaeus.badw.de/work/3 (accessed 13 July 2021).

306 THE SCIENTIFIC WORKS OF ROBERT GROSSETESTE

conceptual connections that exist between them and modern interpretation of the same phenomena, rather than to stress inadequacies of earlier accounts or discontinuities. The essential geometric components of the epicyclic (Ptolemaic) description that are analysed in this chapter are introduced by Grosseteste as follows:

> Moreover, the circle of the moon is eccentric as is the circle of the sun. Now, on the circumference of the eccentric is the centre of a little circle that the eccentric carries around. Moreover, the little circle and the eccentric are on one plane. Now, the centre of the body of the moon is always on the circumference of the little circle. The eccentric of the moon, therefore, is rotated about the diameter of the earth from east to west with a continuous, uniform movement; and so by this movement the centre of the eccentric will describe a circle around the centre of the earth, and the centre of the eccentric is always on the circumference of that circle. Now, the centre of the little circle is moved conversely from the west to the east in such a way that if a line is drawn from the centre of the earth through the centre of the little circle to the firmament, the end of the line would be moved with an equal movement. And this movement of the line is called the mean movement of the moon in the heavens. (DS §§54–5)

The chapter may be read alongside the online computational visualization of the description given in *On the Sphere*, the Virtual Celestial Model, which is available at https://ordered-universe.com/de-sphera-visualisation. An introduction to the online tool, and instructions for its use, can be found at the back of this volume (The Virtual Celestial Model of On the Sphere). In the spirit of both Ptolemy's and modern kinematic accounts of planetary motion, it presents a calculational tool, this time in contemporary computational and computer-visualized form, of the phenomena that Grosseteste's text treats. Remaining as faithful to the text as possible, this model brings to life the geometry and motions of the universe described, acting as an aid to visualization of the, sometimes complex, behaviours and enabling the reader to better understand this rich text. A comparable aid for the contemporary readers and students of *On the Sphere* would be the armillary sphere (Ch. 5, §3; Ch. 10).

1 The Ptolemaic System of the *Almagest*

The reception of the *Almagest* of Claudius Ptolemy of Alexandria in its various translations and versions into the universities and schools of late twelfth and thirteenth century Europe was as transformational to that intellectual world as it would have been to the predictive astronomy of the imperial age in which it was written (Ch. 5, §1 and Ch. 6, §1). No previous attempt at a calculational scheme accounting for the apparent motions of sun, moon, and planets, as seen from the earth, had come remotely within observational accuracy. While drawing on earlier ideas of Hipparchus (190–20 BCE) and others (of eccentrics and epicycles), deploying throughout the geometric tools of Euclid's *Elements* (*c.*300 BCE), and containing a catalogue of star positions and planetary tables drawn from half a millennium of observations, the *Almagest* permitted the calculation of celestial positions to unprecedented accuracy. For a detailed description, and a modern critical appraisal of its quantitative accuracy, the review by Pannekoek is recommended.[2]

The *Almagest* is structured into thirteen books, only the last six of which treat the planetary motions and retrograde motions, and which are ignored in Grosseteste's treatise. The first seven books cover the definitions of the celestial spheres, the motion of the sun, at great length the motion of the moon (books IV and V, devoted to this are together approximately four times the length of the later treatments of any of the planets, and over twice that of book III, devoted to the sun), eclipses and the precession of the fixed stars. In other words, books I–VII cover very similar territory to Grosseteste's *On the Sphere*.[3]

The challenges Ptolemy faced in the case of each celestial body he referred to as the two *anomalies* of longitudinal motion.[4] The first anomaly refers to a component of the non-constant apparent *mean* motion of a planet against the sky as referred to stars of the zodiac

---

[2] A. Pannekoek, 'The Planetary Theory of Ptolemy', *Popular Astronomy*, 55 (1947), 459–75.

[3] It is worth noting as well that this content has a close correlation with that of the epitomized version of the *Almagest* known as the *Almagesti minor* (see Ch. 6, §2).

[4] Ptolemy, *Almagest*, 209.

308　THE SCIENTIFIC WORKS OF ROBERT GROSSETESTE

(the ecliptic), against which all planetary paths lie. The second anomaly refers to a second component of motion, this time correlated to the relative position of the planet and the sun (this anomaly is not present in the case of the sun itself), and not correlated to the absolute direction in the sky, as for the first anomaly. In addition to this, Ptolemy was well aware of further anomalies of latitude, described by small angles of tilt, or 'inclinations', of the paths of each individual planet to that of the sun.[5] Precessions of the parameters of these paths, especially in the case of the moon and Mercury, added to the complex collection of observed inconstancies that appeared, especially over the lengthy periods of historical time for which he had access to observational data, in particular that of Hipparchus, however inaccurate.[6]

Ptolemy met these challenges within a geocentric kinematic model of the sun, moon, and planets, comprising the addition of multiple circular motions. Within this model, the moon and planets followed small circular orbits, the *epicycles*, the centres of which themselves followed large circular orbits whose centres were displaced by (relatively) small but significant distances from the earth. Ptolemy's central breakthrough, however (these structures were known to Hipparchus), was the proposal of the *equant*, a third point about which the angular speed of the planets was constant, and which accounted for the set of first anomalies.[7] Before describing in more detail the very complex case of the moon, treated summarily within the *On the Sphere*, it will be instructive to summarize the general theory for the other planets (from book IX of *Almagest*) since this is simpler.

1.1　Ptolemaic Planetary Theory

The general planetary scheme of the *Almagest*, parameterized individually for each of the five planets (apart from Mercury), is described in book IX, in several pages of text accompanying Fig. 9.1.[8]

---

[5] Ptolemy, *Almagest*, 535.　　[6] Ptolemy, *Almagest*, 13ff.
[7] Ptolemy, *Almagest*, 253; and see Pannekoek, 'Planetary Theory'.
[8] Ptolemy, *Almagest*, 254–94.

Fig. 9.1. The planetary scheme of the *Almagest*. The centre θ of the circular epicycle LM, traces a circular path (deferent) HBK around its centre Z. The earth is displaced from the centre of the deferent at E. At an equal distance from Z in the opposite direction lies the equant, D, about which the angular velocity of the epicycle centre is constant.

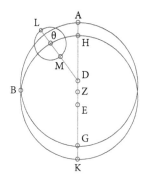

The planet is carried around a circle (small in the case of the outer planets but very large in the case of Venus), the *epicycle* (LM in Fig. 9.1) with a period equal to successive oppositions of the planet and the earth (that is, the moments when the sun and planet occupy the same place in the ecliptic). The centre of the epicycle (θ in Fig. 9.1) itself moves in the same sense (anti-clockwise when regarded from the north) as the epicycle, on a circular path (the *deferent* HBK in Fig. 9.1) and a period corresponding to the return of the planet to the same point. The crucial innovation of the *Almagest* is to allow that the angular speed of the epicycle centre around the observer at E vary (according to the 'first anomaly'). This is achieved by the combined move of displacing the centre of the epicyclic orbit, from the earth (Z and E respectively in Fig. 9.1), then constructing another point an equal distance from earth on the other side of the centre (D in Fig. 9.1). This is the *equant*. It is around the equant that the angular velocity (the rotation of the line Dθ in Fig. 9.1) is constant. To this point, the model accounts for the varying distance between the planets and the earth (with apogee at A and perigee at G of Fig. 9.1), the varying angular velocity of the planet, and also the phenomenon of *retrograde* motion, not discussed by Grosseteste, when close to opposition, planets appear to reverse their mean orbital motion from west to east along the ecliptic. Finally, Ptolemy accounts for precession by adding an angular rotation of the whole construction against the background of fixed stars, of 1 degree per century, again in the same sense of rotation as that of deferent and epicycle. The cases of Mercury (book IX), and the second of Ptolemy's lunar models

310   THE SCIENTIFIC WORKS OF ROBERT GROSSETESTE

below (§1.2) forces the introduction of a further complication that the centre of the deferent circle is itself carried on a small epicycle, centred on the earth, and with an opposite rotation to the deferent and planetary epicycle.

1.2 Ptolemaic Lunar Theory

Although the simpler planetary scheme above is not discussed in Grosseteste's *On the Sphere*, the more complex lunar theory does feature (as DS §§54–5). Ptolemy required several modifications to the planetary theory to account for lunar motion, which presented a further set of anomalies.[9] He made considerable use of historical eclipse observations, for at these moments one can be sure that both moon and sun are either at conjunction (solar eclipse) or opposition (lunar eclipse) and additionally that the moon is crossing the eclipse.

First, the moon's motion lies for one part of its orbit above, for the other below, the ecliptic. This required a considerable (five-degree) inclination of both lunar deferent and epicycle. Second, successive passages of the moon through the ecliptic plane were not at equal points. To account for this effect, the epicycle period was lengthened with respect to that of the deferent, so that the complexity of non-periodicity emerged from the two periodic motions. Third, the regular angular motion of the epicycle centre of the moon was not about its own centre, nor about a displaced equant, but about the earth itself, differing from all other planetary models. This point is specifically picked up by Grosseteste (DS §55).

The difficulties posed by the moon's motion and its several anomalies results in Ptolemy's presentation in the *Almagest* of two alternative models for Lunar motion. Book IV describes a model of the class described above, while book V introduces a new model with the same additional machinery as adopted in the case of Mercury (of a small circle carrying the centre of the eccentre, Z in the Fig. 9.1 in the opposite sense as the epicyclic motion). This is the model explained in detail by

---

[9] Ptolemy, *Almagest*, 366–400.

KINEMATIC DESCRIPTIONS, EPICYCLES, AND MODELLING   311

Alfraganus (a source with whom Grosseteste was familiar (Ch. 6, §2)) and by Grosseteste himself in *On the Sphere*, as discussed below (§2.4).

2 Modern Planetary Kinematics

Although Ptolemy's model addressed the projection of planetary kinematics onto the sky through a set of motions as to whose reality he seems to have been agnostic, modern kinematics similarly comprises that branch of physics which aims to provide a mathematical description of the motion of bodies relative to each other in three dimensional space, though without any necessary reference to the forces that cause the motion.[10] Importantly, classical mechanics describes these motions using any frame of reference (or viewpoint) preferred, and can transform between them without implying any special importance to one or the other. That is, it may equally be said, within kinematics rather than physics, that the sun orbits the earth, as the other way round. Such descriptions of relative motion are key in many branches of Physics and Engineering when describing complex motions.

The branch of physics concerned with the forces that cause motion is *dynamics*. A Newtonian model of dynamics, for example, would say that the sun and earth orbit their common centre of mass. Since the sun is so massive, this point is very close to its centre: a heliocentric model is an excellent approximation. However, this does not necessarily make such a model more 'correct' than its geocentric sibling, neither mathematically nor conceptually.

Comparisons of the Ptolemaic model and modern orbital models form part of the historiographical tradition.[11] As such, the aim of this chapter is not to give such detailed descriptions, beyond the summary in the

---

[10] On the celestial models assumed in Ptolemy's other works, for example, the astrological treatise and companion to the *Alamgest*, the *Tetrabiblos—Quadripartitus* in its twelfth-century Latin translation by Plato of Tivoli, which adopts different kinematic schemes, see Luigi Russo, *The Forgotten Revolution* (New York: Springer, 2003).

[11] Pierre Duhem, *Le système du monde. Histoire des doctrines cosmologiques de Platon à Copernic*, 10 vols. (Paris: Hermann, 1913–1959), i. 427–89 and ii. 185–89; Edward Grant, *Planets, Stars and Orbs: the Medieval Cosmos 1200–1687* (Cambridge: Cambridge University Press, 1994), 3–17.

312 THE SCIENTIFIC WORKS OF ROBERT GROSSETESTE

previous section, but to present a simple framework that conveys the key conceptual points, in the same spirit as discussion of the *sonativum* in the first volume of this series.[12] As noted above, the task is to present the ancient and medieval natural philosophy, not through a presentist lens of evaluation or judgement, but to allow the tools of science and mathematics now at hand to support an examination of the older achievements and descriptions in a clearer light. In particular, and in this case, simple models of the relative motions of the sun and earth and earth and moon, will show that constructions based on epicycles are exactly equivalent to a modern mathematical technique of asymptotic approximation to periodic functions known as the Fourier series, and can therefore be made arbitrarily accurate. Furthermore, it will become clear that only a small number of epicycles are required to achieve very high levels of accuracy for the sun-earth and earth-moon systems: here the ancient astronomer benefitted from these orbits being very close to circular. Finally, the point will be made that using epicyclic descriptions makes it easier to perform precise calculations of future motion based on historical observations than it would be to fit later Keplerian models to these observations. These latter points are well-established.[13] Nevertheless, whilst the equivalence of epicyclic orbits to Fourier has been noted, the comparison presented here is, to the authors' knowledge, new, and links modern and ancient astronomy in a manner accessible to students of both.[14]

2.1 Orbits

The first 'modern' kinematic description of planetary motion is given by Kepler's laws:

---

[12] *Knowing and Speaking*, ch. 14.

[13] Fred Hoyle, 'The Work of Nicholas Copernicus', *Proceedings of the Royal Society A*, 336 (1974), 105–14.

[14] Norwood Russel Hanson, 'The Mathematical Power of Epicyclical Astronomy', *Isis*, 51 (1960), 151–8; Giovanni Gallivotti, 'Quasi periodic motions from Hipparchus to Kolmogorov', *Atti della Accademia Nazionale dei Lincei. Classe di Scienze Fisiche, Matematiche e Naturali. Rendiconti Lincei. Matematica e Applicazioni*, 12 (2001), 125–52; Heinrich Saller, *Operational Symmetries: Basic Operations in Physics* (Cham: Springer, 2017), 159–61; Donald G. Saari, 'A Visit to the Newtonian N-body Problem via Elementary Complex Variables', *The American Mathematical Monthly*, 97 (1990), 105–19.

1) Each planet's orbit describes an ellipse with the sun at one of the two foci.[15]
2) A line segment joining the sun and the planet sweeps our equal areas in equal times.
3) The square of the orbital period of a planet is directly proportional to the cube of the semi-major axis of its orbit.

These laws are illustrated in Fig. 9.2. Box 9.1 gives more mathematical descriptions of these orbits. Whilst these appear simple, it must be realized that in practice the only observation the astronomer could reasonably make is of the path swept out by the planet in the sky: essentially the angle (geocentric longitude) between the planet and a reference. To express this path mathematically, deriving from Kepler's laws, requires a complicated series of manipulations, given later.

It would be essentially impossible for an astronomer to deduce the Keplerian construction by hand from a series of observations. The huge inductive leap that Kepler himself was able to make required at least his own genius, the much more accurate planetary measurements available in the early-seventeenth-century and the post-Copernican, early-modern cosmology. The solution based on epicycles is more elegant and simpler.

Fig. 9.2. Illustration of Kepler's laws, giving the key parameters of the elliptical planetary orbit. The semi-major axis is a, b is the semi-minor axis. Note that the sun is at one focus of the ellipse, the second is illustrated by a hollow circle, and the centre by ⋅. The planet orbits in an anticlockwise direction. See also Box 1.

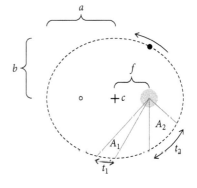

---

[15] Actually, the *common centre of mass* is at the focus.

314 THE SCIENTIFIC WORKS OF ROBERT GROSSETESTE

Box 9.1 Kepler's Laws

Kepler's laws may be illustrated as shown in Fig. 9.2. First, the ellipse, dotted line, has semi-major axis of length a (in this case $a = 4$) and semi-minor axis b (= 3.46); the ellipticity, e, is 0.5 ($e = 0$ for a circle). The first law states that the sun is at the focus of the ellipse, which is a distance f (= 1) from the centre, c, whilst the planet's orbit traces the dotted line.

Kepler's second law states that if the two times t_1 and t_2 are equal, then so will be the areas A_1 and A_2. This means that the planet moves more slowly when it is more distant from the sun, and more quickly when it is closer. It can be shown that this law is equivalent to saying that the angular momentum of the system is constant.

Finally, the third law states that when comparing planets in closer and more distant orbits, the period of the orbit, P (that is, the planetary year) is related to a by $P^2 \propto a^3$; this is a direct consequence of the (inverse-square) nature of the gravitational force between the sun and the planet.

2.2 Descriptions of Elliptical Orbits

Fig. 9.3 shows again the key features of an elliptical orbit. The sun is at one *focus* of the ellipse, which is a distance f from the centre. The ellipse has two axes, the semi-major (a) and semi-minor (b); these geometric properties are linked through the *eccentricity* (e), which is a measure of how non-circular, or elongated, the ellipse is: a circle has $e = 0$. The relationship between the properties is

$$e = \sqrt{1 - \frac{b^2}{a^2}} = \frac{f}{a}. \tag{1}$$

Fig. 9.3 also shows the heliocentric longitude, ϑ, and distance, r, which together describe the position of the planet relative to the sun; in this case, the heliocentric longitude and distance of the earth from the sun are the same as the geocentric longitude and distance of the sun from the

KINEMATIC DESCRIPTIONS, EPICYCLES, AND MODELLING 315

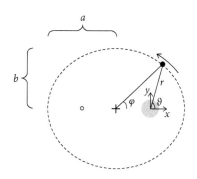

Fig. 9.3. The key parameters used to describe the position of a planet relative to the sun. Note that here, and in Figure 9.2, the ellipticities are highly exaggerated, real planetary orbits are much closer to circular.

earth: it is the geocentric longitude that that can be measured by an astronomer. Finally, another angle, φ, is defined, from the centre of the ellipse to the planet.

Box 9.2 describes the formal mathematical process of linking the movement of an orbiting body to a series of circles (both for a description based at the centre of the ellipse, and at one focus). Mathematically, it is shown that this is equivalent to describing the so-called Fourier series. This approach is very commonly used to efficiently describe periodic motions and is therefore well-suited to the analysis of orbits. The key points that bear on the planetary example of Fourier series are: (a) the complication of the sun's position at a focus, not at the centre, of the ellipse; (b) that the Fourier series describes a series of circles of decreasing size; and (c) that by adding more and more circles (formally *terms* of the series) the description can be made arbitrarily accurate, and can describe any periodic motion: realistically the sensible limit is determined by the accuracy of available astronomical data. A further point, which will be discussed later, is that it is more straightforward to implement the epicycle / Fourier series approach in calculations, than it is to derive or write out explicitly the equations of the ellipse. It is also, less obviously, easier to improve the accuracy of calculation, as additional terms can be added. This is, of course, the rationale for using this approach in the modern discipline of Engineering, and certainly also an advantage when all calculations had to be performed by hand. These methods can be applied to the orbits of the sun and moon, as described in *On the Sphere*.

316 THE SCIENTIFIC WORKS OF ROBERT GROSSETESTE

Box 9.2 Epicycles and the Fourier Series, Part 1

The ellipse may be expressed in two polar coordinate systems; both measure the angle in an anti-clockwise direction from the horizontal axis, but one is centred on the centre of the ellipse, and the other at the focus (Fig. 9.2). For convenience a Cartesian coordinate system will also be defined in which the focus is at the origin, so the centre of the ellipse is at $(-f, 0)$. The algebra is more simply performed using complex-exponential notation.

The aim here is to express the shape of the ellipse as the sum of a number of epicycles. This is exactly the same as expressing as a Fourier Series, which can be written as:

$$x(t) + iy(t) = c_0 + \sum_{n=1}^{N} c_n \exp(in\dot{\vartheta}t) + d_n \exp(-in\dot{\vartheta}t), \qquad (2)$$

where $\dot{\vartheta}$ represents the rate of change of angle, so $\dot{\vartheta}t$ is an angle that increases with increasing time; the (real) constants c_n and d_n are chosen to construct the target functions, and may be equivalently thought of as the radii of successive epicycles. Here, Euler's formula should be noted, that is

$$\exp(in\dot{\vartheta}t) = \cos(n\dot{\vartheta}t) + i\sin(n\dot{\vartheta}t), \qquad (3)$$

so that

$$x(t) + iy(t) = \exp(in\dot{\vartheta}t) = \cos(n\dot{\vartheta}t) + i\sin(n\dot{\vartheta}t) \qquad (4)$$

traces out a circle of radius 1 (see Fig. 9.B.1, 'one component').

Equation (2) describes how any periodic function, that is, a function which repeats every P seconds, can be described as the sum of a series of sin or cosine functions whose periods are P/n, where n is an integer. For example,

$$x = \cos\left(\frac{t}{P}\right) - \frac{1}{3}\cos\left(\frac{3t}{P}\right) + \frac{1}{5}\cos\left(\frac{5t}{P}\right) - \dots \frac{(-1)^{(2n/2)}}{(2n+1)}\cos\left(\frac{(2n+1)t}{P}\right) + \dots, \qquad (5)$$

describes a square wave, with increasing accuracy as the number of terms increases (Fig. 9.B.1). Expanding to two dimensions, the shape of a box can be described (Fig. 9.B.2) by combining square waves in the x and y directions. Indeed, this powerful tool allows us to describe periodic functions of any arbitrary shape, it is therefore used in many branches of Physics and Engineering.

KINEMATIC DESCRIPTIONS, EPICYCLES, AND MODELLING 317

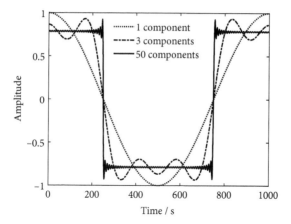

Fig. 9.B.1. Increasingly accurate approximations to a square wave can be made by adding cos functions of different amplitudes.

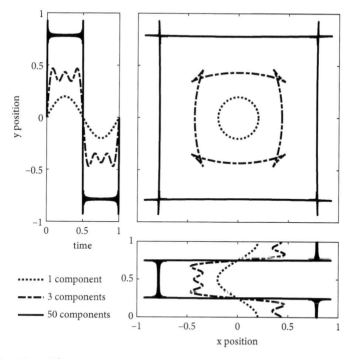

Fig. 9.B.2. The Fourier series can also be used to describe functions in two dimensions, by describing their projections onto the x and y axes separately. A one-term series (equation (4)) produces a circle, using increasingly accurate square wave approximations produces a square shape in two dimensions. Note, however, that the corners are not well represented.

318   THE SCIENTIFIC WORKS OF ROBERT GROSSETESTE

2.3 Epicycles for the Sun-Earth System

The mass of the sun is 2×10^{30} kg, whilst that of the earth is 6×10^{24} kg; this means that whilst the earth's orbit has a mean radius of 150 million km, the centre of mass of the sun-earth system, and hence the focus of the orbit, is only approximately 150 km from the centre of the sun itself: it might truly be said that the earth orbits the sun. Or, of course, it may also be said that the sun orbits the earth: the mathematical description is the same. The key parameters of this orbit are as outlined in Table 9.1.

Using these data, the circles described in equation (13) (Box 9.3), Fig. 9.4 can be calculated; because e is very small, only the first three terms are considered:

Offset: magnitude approximately $-3ae/2 = 3.75$ million km
Circle: radius approximately $a = 149.60$ million km
Circle: radius approximately $ae/2 = 1.25$ million km

Neglecting terms in e^2 or smaller is justified because these terms represent circles approximately fifty times smaller, i.e. of the order 0.075 million km. In *On the Sphere*, the sun's annual orbit, relative to the firmament, is described as a circle whose centre is offset by 'two and a half degrees [...] calculated from the diameter'.[16] Two and a half degrees is to be interpreted as $2.5/60 = 1/24 = 0.042$, which is then multiplied by the diameter to give the offset; however, the use of diameter is a mistake:

Table 9.1 Key Parameters of the Earth's Orbit

a	149.60 million km
e	0.0167
Period	365.256 days = 1 year

[16] The authors would like to thank Philipp Nothaft for very helpful discussion on this point.

Box 9.3 Epicycles and the Fourier Series, Part 2

The key step is now to realize that because the exponential functions in equation (2) are themselves represented by circles on the complex plane, the overall equation represents a series of circles added together. This is because the complex number $\exp(i\vartheta)$ traces a circle of unit radius as ϑ moves from 0 to 2π, as in equation (3) and Fig. 9.B.3. For example,

$$x(t) + iy(t) = \exp(i\dot{\vartheta}t) + \frac{1}{2}\exp(i2\dot{\vartheta}t) \qquad (6)$$

traces out the motion of an object moving in a circle of radius ½ whose centre is in turn moving in a circle of radius 1, with an orbit around the small circle taking half as long as around the larger (Fig. 9.B.3). The sum

$$x(t) + iy(t) = \frac{(a+b)}{2}\exp(i\dot{\varphi}t) + \frac{(a-b)}{2}\exp(-i\dot{\varphi}t) \qquad (7)$$

traces an ellipse (Fig. 9.B.4), with the angle φ measured around the centre (defined in Fig. 9.3). If it is so easy to describe an ellipse in terms of epicycles, one might ask why epicyclic orbital descriptions so complicated. It is because the bodies of interest are never at the centre: typically, we are interested in systems involving the sun (either explicitly as in the earth-sun system, or implicitly for the other planets), in which the sun is at a focus and the angle of interest is the heliocentric longitude, ϑ in Fig. 9.3.

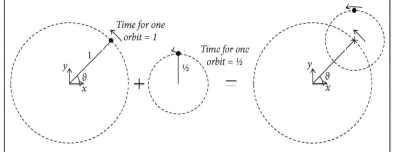

Fig. 9.B.3. A simple epicyclic orbit is the sum of two terms, each representing a circle. The terms can be added in either order: the centre of the larger circle could follow an orbit that traces out the smaller.

Continued

Box 9.3 Continued

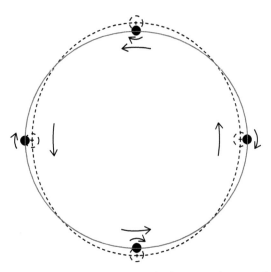

Fig. 9.B.4. An ellipse (grey line) is simply the sum of two circles orbiting in opposite directions (dotted lines), as long as they are referenced to the centre.

Box 9.4 Epicycles and the Fourier Series, Part 3

The standard equation for motion in an ellipse measured from a focus can be expressed as

$$x(t) + iy(t) = \frac{a(1-e^2)}{1+e\cos(\dot{\vartheta}t)} \exp(i\dot{\vartheta}t) \tag{8}$$

the $exp(i\dot{\vartheta}t)$ term describes motion in a circle, whilst the remainder of the equation forces the radius to be dependent on angle (and hence time). This equation can be expanded as follows:

$$x(t) + iy(t)$$

$$= \frac{a(1 - e^2)}{1 + \frac{e}{2}\left(exp(i\dot\vartheta t) + \exp(-i\dot\vartheta t)\right)} \exp(i\dot\vartheta t)$$

$$= a(1 - e^2)\left[1 - \frac{e}{2}\left(\exp(i\dot\vartheta t) + \exp(-i\dot\vartheta t)\right)\right.$$
$$\left. + \frac{e^2}{4}\left(\exp(i\dot\vartheta t) + \exp(-i\dot\vartheta t)\right)^2 ...\right]\exp(i\dot\vartheta t)$$

$$\approx a(1 - e^2)\left[\exp(i\dot\vartheta t) - \frac{e}{2}\exp(2i\dot\vartheta t) - \frac{e}{2} + \right.$$
$$\left. \frac{e^2}{4}\left(\exp(3i\dot\vartheta t) + 2\exp(i\dot\vartheta t) + \exp(-i\dot\vartheta t)\right)...\right]$$

$$= a(1 - e^2)\left[-\frac{e}{2} + \left(1 + \frac{e^2}{2}\right)\exp(i\dot\vartheta t) - \frac{e}{2}\exp(2i\dot\vartheta t) \right.$$
$$\left. + \frac{e^2}{4}\exp(3i\dot\vartheta t) + \frac{e^2}{4}\exp(-i\dot\vartheta t)...\right] \tag{9}$$

The motions represented by the terms in this equation are shown in Fig. 9.B.5, here $\dot\vartheta = 2\pi/P$, where P is the period of the orbit. Here, an approximation is made by assuming that e is much smaller than 1, as it is for planetary orbits, and then ignoring higher powers of e: this reflects the fact that the circles become smaller and smaller as the rotations become faster and faster; however, the key point is that the expression can be made arbitrarily accurate by increasing the number of circles. A further illustration is provided in Fig. 9.B.6, in which the approximations to an ellipse obtained by adding increasing numbers of epicycles are shown.

A final point needs to be made. The discussion so far implicitly assumes that the rate of change of angle is constant (i.e. $\dot\vartheta$ is constant). In fact, this is not the case: in order to meet the requirements of Kepler's second law, the angle must change more rapidly when the planet is close to the sun. This is discussed in Box 9.5.

Continued

Box 9.4 Continued

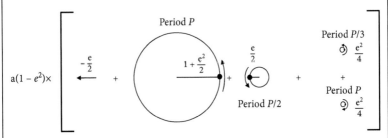

Fig. 9.B.5. Depiction of the circles described by the first five terms in the Fourier Series/Epicycle description of an ellipse in equation (9). Note that (1) The fixed displacement moves the centre of rotation approximately half-way towards the centre of the ellipse; (2) the circle P/2 is out of phase with the other circles, i.e. when $t = 0$ it moves in the negative x direction; (3) the fourth, smaller circle of period P moves in the opposite direction to the others (that is, retrograde).

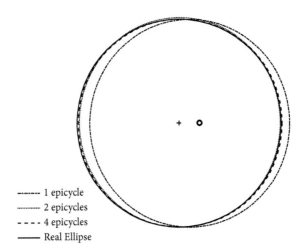

Fig. 9.B.6. Comparisons of the approximations to an ellipse obtained when using different numbers of epicycles: note that an excellent approximation is obtained with only two. In this case the eccentricity is 0.2, which is larger than most planetary orbits (only Mercury, 0.21 and Pluto, 0.25, exceed this). The focus of the ellipse is marked o, and the centre +.

KINEMATIC DESCRIPTIONS, EPICYCLES, AND MODELLING 323

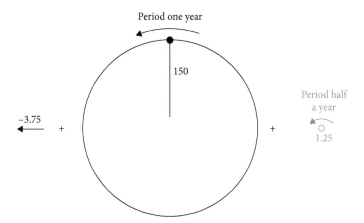

Fig. 9.4. Graphical representation of the first three terms in the epicycle approximation to the sun's orbit relative to the firmament. Distances are in million kilometres. Only two terms are included by Grosseteste. For presentation purposes, the scale of the large circle is reduced by a factor of 4 compared to the other graphics. This figure shows $-3a/2$, the value of $-a/24$, used in *On the Sphere*, is -6.25 million km.

the value in Ptolemy's *Almagest* is $1/24$ times the *radius*, which is equivalent to $5ae/2$ and is equal to 6.25 million km.[17]

Having established that this is a close approximation to the *mathematical description* of an elliptical orbit, the reader may still need convincing that it is a good approximation to the orbital motion observed: the actual position of the sun in the sky relative to the firmament, remembering that the sun and the firmament together also exhibit a daily rotation around the earth. This is addressed in Box 9.3, Fig. 9.B.7, for general orbits, which shows the exact motion produced by Kepler's laws and the motion obtained using different approximations. More relevant to this discussion, Fig. 9.5 compares three approximations, using the appropriate ellipticity for the sun. This demonstrates that,

---

[17] Ptolemy, *Almagest*, 223. Here, *a* has been used interchangeably with the mean radius of the sun's orbit. For most of the discussion this does not matter as we can normalize by the orbital radius, that is, only the ellipticity is important; however, in this case the offset is given as a function of the radius. It should be noted, though, that for the earth's orbit around the sun, the semi-major axis (a) is 149.598 million km, and the semi-minor axis is 149.577 million km: the mean radius is halfway between these, so *a* is a very reasonable approximation.

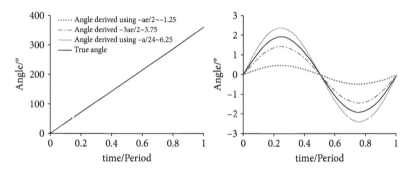

Fig. 9.5. Comparison of the position of the sun according to different models; $e = 0.0167$. Left: angle against time. Right: deviation from $2\pi t/P$ (i.e. a constant angular speed). Each model consists of an offset, of size given in the legend, and a single circle of radius a. The approximation using $-a/24$ is as good as that using $-3a/2$, which is the value in the Fourier series for the true angle as a function of time (i.e. equation (13)). Note that using an additional term in equation (13), would have resulted in a line indistinguishable, on this scale, from the true value. This could also have been achieved by using an offset of $-2ae$ (for the sun, $a/24 \approx 2.5ae$), to give a construction closely equivalent, for small e, to equation (14).

in terms of angular position, the error resulting from the use of $-a/24$ is almost exactly the same as that using the value of $-3ae/2$ derived from Kepler's laws, the only difference being that the peak angular deviation is slightly too large, instead of slightly too small. A more accurate approximation would have been obtained had $-2ae$ been used, which would have resulted in a construction equivalent to using Ptolemy's equant.

2.4 Epicycles for the Earth-Moon System

Because the moon has a relatively large mass compared to the earth, the common centre of mass, around which both bodies orbit, is offset by 4,640 km from the centre of the earth (for comparison the radius of the earth is approximately 6,370 km). Moreover, the greatest gravitational

KINEMATIC DESCRIPTIONS, EPICYCLES, AND MODELLING 325

Box 9.5 Refinement for Orbital Motion

Box 9.2 showed that the geometric shape of an ellipse can be approximated very well indeed for reasonable values of e using only an offset and two epicycles. However, the astronomer wishes to model and predict, not just the shape of the orbit, but the position of celestial bodies in the sky at any time. For the sun and moon, this means that the model must correctly describe the angular position, as observed from earth, as a function of time. The epicycles derived so far fall short, because they do not take into account the change in angular speed, $\dot{\vartheta}$, as the celestial body progresses around its orbit. This results from Kepler's second law, or equivalently from conservation of angular momentum: the product $r^2\dot{\vartheta}$, where r is the distance of the orbiting body from the focus, must be constant. This leads to the following equation:

$$\dot{\vartheta} = \frac{P}{2\pi} \frac{[1 + e\cos(\vartheta)]^2}{(1 + e^2)^{3/2}} \qquad (10)$$

The solution to this equation can only be expressed as an infinite series. This is discussed in Hoyle,* in which the solutions for the angle and radius of the orbit as functions of time are provided:

$$\vartheta = \frac{2\pi}{P}t + 2e\sin\left(\frac{2\pi}{P}t\right) + \frac{5}{4}e^2\sin\left(\frac{4\pi}{P}t\right) \\ - e^3\left[\frac{1}{4}\sin\left(\frac{2\pi}{P}t\right) - \frac{13}{12}\sin\left(\frac{6\pi}{P}t\right)\right] + \dots \qquad (11)$$

$$\frac{r}{a} = 1 - e\cos\left(\frac{2\pi}{P}t\right) + \frac{1}{2}e^2\left[1 - \cos\left(\frac{4\pi}{P}t\right)\right] \\ + \frac{3}{8}e^3\left[\cos\left(\frac{2\pi}{P}t\right) - \cos\left(\frac{6\pi}{P}t\right)\right] + \dots \qquad (12)$$

These are, again, effectively two Fourier series. In order to make them more comparable to the expression in Box 9.2, they can be combined to give a new Fourier series

Continued

Box 9.5 Continued

$$x(t) + iy(t) = a(1 - e^2)\left[-\frac{3e}{2} + \left(1 + \frac{e^2}{2}\right)\exp\left(i\frac{2\pi}{P}t\right) + \frac{e}{2}\exp\left(2i\frac{2\pi}{P}t\right)\right.$$
$$\left. - \frac{e^2}{8}\exp\left(2i\frac{2\pi}{P}t\right) - \frac{3e^2}{8}\exp\left(-i\frac{2\pi}{P}t\right)...\right]. \tag{13}$$

This is very similar indeed to the first three terms of the expression in equation (9): the main differences are that $-e/2$ has been replaced by $-3e/2$, whilst the third term is now added instead of subtracted. There are also differences in the fourth and fifth terms; however, these terms, which scale as e^2 are very small. Indeed, when ϑ is equal to $0°$ or $90°$ the first the approximations are the same. These equations are explored further in Fig. 9.B.7 which shows both the shape of the orbits described by equations (9) and (12), and also the angle of the orbiting body as observed from the focus. Here it is clear that the shapes are almost identical, but the angular position is much better approximated by the terms in equation (12). Hence, whilst epicycle descriptions of planetary orbits are equivalent to Fourier series, the result of Kepler's second law is that it is not possible to describe both the orbital shape and the planetary motion using the same series.

Further examining equation (13), it can also be shown that, neglecting terms in e^2, this series can be expressed as

$$x(t) + iy(t) = -2ae + a\left(1 + e\cos\left(\frac{2\pi}{P}t\right)\right)\exp\left(i\frac{2\pi}{P}t\right) + ... \tag{14}$$

The geometric interpretation of this is a little different, Fig. 9.B.8: it represents an orbiting body moving round a circle whose centre is a distance $-ae$ from the focus (that is, at the centre of the ellipse). However, it does not move at a constant speed around this circle, rather it moves at a constant angular speed around another point at location $-2ae$: this reproduces the construction of Ptolemy, in which the point at $-2ae$ is the equant.

KINEMATIC DESCRIPTIONS, EPICYCLES, AND MODELLING 327

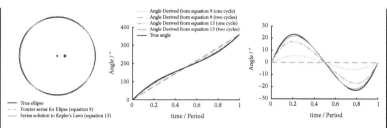

Fig. 9.B.7. Left: Two different approximations to an elliptical orbit, given by the first three terms (i.e. two cycles) of the ellipse Fourier series (equation (9)) and of the orbital motion Fourier series (equation (13)), compared to a true ellipse. The focus of the ellipse is marked o, and the centre +; the value of e is 0.2. Middle: angular position vs time obtained from the two series. Right: For clarity, the difference between the angular position and a straight line of gradient 2π shows the excellent agreement between equation (13) and the true angle. The true angle was derived by numerical integration of equation (10).

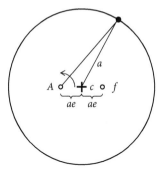

Fig. 9.B.8. Geometric interpretation of equation (14) for a body in orbit around a focus, f. In this approximation, the orbiting body moves in a circle of radius a around the centre, c, but the speed of the orbit varies so that a line from A to the orbiting body rotates at a constant rate. Viewed from f, the angular position as a function of time is the same as 'equation (13) (two cycles)' in Figure 9.4.

* F. Hoyle, *Nicolaus Copernicus* (London: Harper and Row, 1973), 13–14.

Table 9.2 A Simplified Lunar Orbit

a	384,748 km
e	0.055 (mean eccentricity)
Period	27.322 days = 1 sidereal month

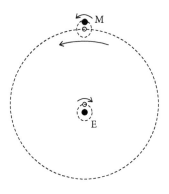

Fig. 9.6. Description of the moon's orbit as described in *On the Sphere*. The centre of the moon's epicycle traces a larger eccentric circle, the centre of which itself orbits the earth, but in the opposite direction.

influence on the moon is the sun, and this complicates its orbit. A simplified lunar orbit can be taken with the parameters shown in Table 9.2.

In *On the Sphere*, the description given is that the moon orbits in a small circle whose centre traces a larger circle; the centre of this larger circle also moves in circular motion about the earth, but in the opposite direction, this is illustrated in Fig. 9.6. This can be compared again to the terms in the Fourier expansion in equation (13) (see Fig. 9.7 for a graphical representation of the terms and their magnitudes). The small circle, whose centre traces a larger circle, is of course consistent with the first two circles in this expansion. Since previous discussion of the sun indicates that these terms would be sufficient to reproduce the apparent position of the moon very accurately, it is interesting to ask where the third term comes from: it turns out that this is a result of the sun's influence on the two bodies.[18]

---

[18] Richard Fitzpatrick, *A Modern Almagest* (Austin: University of Texas, c.2005), http://farside.ph.utexas.edu/Books/Syntaxis/Almagest.pdf (accessed 24 February 2021).

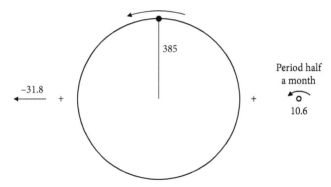

Fig. 9.7. Graphical representation of the first three terms in the epicycle approximation to the moon's orbit. Distances are in thousand kilometres and the large circle is scaled by a factor of two.

2.5 Implications

It can already be seen that any attempt to express the motion of the planets as a function of time must necessarily be an approximation: whether as a Fourier series/set of epicycles or as the series of sin and cosine functions in equations (11) and (12): these two approaches are really geometric and algebraic versions of the same asymptotic scheme of approximation. Furthermore, without first knowing Kepler's laws, it is difficult to see how these equations could be derived from observation alone, even if it were beneficial to do so. In any case, it has been shown that very precise descriptions of the observable solar and lunar positions can be obtained using only a small number of terms in these approximations, owing to the small ellipticities of these bodies. Although Grosseteste's *On the Sphere* deals only with the motions of the sun and moon, the same mathematics can, of course, be applied to the other planets. These orbits are discussed in more detail in Hoyle and in Carman and Recio.[19] This quickly becomes more complicated: equations

---

[19] Christián C. Carman and Gonzalo L. Recio, 'Ptolemaic Planetary Models and Kepler's Laws', *Archive for History of Exact Sciences*, 73 (2019), 39–124.

(10), (11), and (13) are expressed so as to give heliocentric positions. To convert these to observations made from the earth requires equations (10) and (11) to be combined, twice (describing the positions of the earth and planet relative to the sun), and then the difference taken. This is most straightforwardly done using an approach based on epicycles.

10
Illuminating *On the Sphere*

Medieval images of the study of astronomy sometimes show individuals or groups of scholars and students pointing to the sky or to a complex instrument such as an astrolabe or an armillary sphere (see Ch. 5, §3).[1] Examples of such images appear in British Library Harley MS 4350 (Lh3), a thirteenth-century collection of texts on astronomy and computus, which begins with Robert Grosseteste's *On the Sphere* (ff. 4–15). In the initial at the start of Grosseteste's text is a figure, presumably, given the context, to be identified with Grosseteste himself, holding and scrutinizing an armillary sphere (Fig. 10.1). This is followed, at the start of subsequent texts by different authors, by initials with similar figures writing numbers (f. 15v), holding an astrolabe (f. 31) and performing calendrical calculations (f. 68v). Yet despite these references to practical activities and specialist tools, the presence of such images in manuscripts is a reminder that astronomical knowledge was also transmitted through books.[2] Before the invention of the printing press, this required copying by hand, and the skill with which this was done, together with the scale and quality of the materials used, provides evidence for the investment of resources into a book, and therefore clues to the wealth and status of the patron and the volume's intended use.

Grosseteste's *On the Sphere* survives in over fifty manuscripts made between the thirteenth and fifteenth centuries (Ch. 3 §1).[3] These vary significantly in their size, contents, and decoration. Some appear to have been copied by scholars of limited means for their own study, while others were made by professional scribes and artists for individuals who

---

[1] See J. E. Murdoch, *Album of Science: Antiquity and the Middle Ages* (New York: Charles Scribner's Sons, 1984), esp. 179; L. Cleaver, *Education in Twelfth-Century Art and Architecture* (Woodbridge: Boydell, 2016), 179–97.

[2] See Murdoch, *Album of Science*, 3.

[3] See also K. W. Humphreys, *The Friars' Libraries* (London: British Library, 1990), 92.

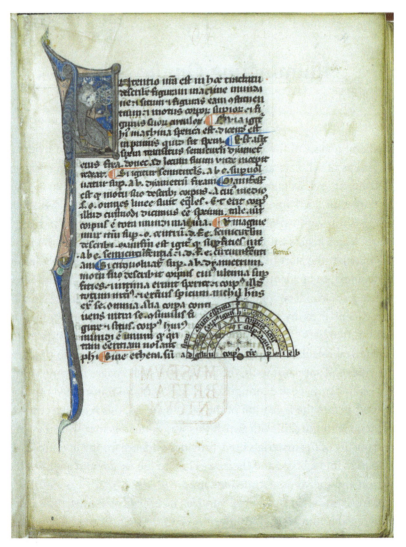

Fig. 10.1. British Library, Harley MS 4350, f. 4 © The British Library Board.

wanted to be able to display both their knowledge and status through their books.[4] In the course of copying the text decisions were made about

---

[4] For an example of the former, see British Library Cotton MS Otho D X (Lc), a fourteenth-century compilation of Grosseteste's works in which *On the Sphere* is copied without images and

ILLUMINATING *ON THE SPHERE* 333

the inclusion or omission of diagrams. These drawings were of a different status to decorated initials, since their function, in theory at least, was to make visual the content of the text in a way that would aid the reader's understanding. In this way such drawings have potential parallels with the use of scientific instruments. By contrast, initials served to mark the start of new sections and might add context as well as colour to a text. The decision to include or omit diagrams can indicate engagement with a text, as indeed with the current edition and translation, but can also provide insights into the dissemination and construction of knowledge in the light of multiple sources, as diagrams moved between texts. Moreover, the execution of the diagrams is an important reminder of the potential input of artists, as well as patrons and scribes, into the creation of some of these manuscripts and therefore into conceptions of the universe. The variations between images in different manuscripts reveal how their makers infused their own interpretation of Grosseteste's world view into their illustrations as they copied them.

Most of the surviving manuscripts containing *On the Sphere* are small and densely written, often with very narrow margins, making efficient use of the parchment. It is unusual for manuscripts to measure more than 25 × 17 cm and many are considerably smaller. For example, from the fourteenth century, Dublin, Trinity College MS 441 (Dt) measures 16.3 × 12.2 cm and Lincoln Cathedral MS A.7.9 (Lc) just 15.5 × 11 cm. In addition, the text can be compressed onto just four folios, as in two more fourteenth-century copies: Cambridge University Library MS Ii.1.13 (Cu3), ff. 36v–39v (21.5 × 15.1 cm); and British Library Harley MS 321 (Lh1), ff. 30–33v (23 × 16 cm); and in a late thirteenth- or early fourteenth-century manuscript: Erfurt, Universitäts- und Forschungsbibliothek, Bibliotheca Amploniana CA 4° 351 (Ea2), ff. 46–49v. Size is not necessarily an indication of economy or a lack of interest in the appearance of a manuscript (Ea2 is laid out in two columns and opens with a penwork initial), but most of the manuscripts of *On the Sphere* provide an intimate reading experience and therefore appear to have been made for an

---

compressed onto five folios of parchment (ff. 147–9). For the latter, see Verdun Bibliothèque municipale MS 25 (Vmu), a thirteenth-century presentation copy that contains five colourful diagrams set within the text block and decorated with gilding and playful figures (see Fig. 10.5).

334 THE SCIENTIFIC WORKS OF ROBERT GROSSETESTE

individual's study. Similarly, few copies feature painted initials or diagrams completed by a specialist artist, suggesting that most of the books were valued primarily for the content of the text rather than their attractiveness as an object.

In the second half of the thirteenth century some high-quality copies, complete with illuminated or elaborate penwork initials were produced in England and France. This suggests both that the importance of the text had been widely accepted, and that copies were being made for wealthy individuals and communities, but these remained the exception rather than the norm for copies of *On the Sphere*. In addition, most of the copies of *On the Sphere* appear in collections of works of multiple authors, usually focused on astronomy and computus, testifying to their reception as tools for the study of this subject. There is no evidence that anyone attempted to compile a volume of Grosseteste's complete works in the thirteenth century, although British Library Royal MS 6 E V (Lr) is a large (42 × 29 cm) and lavishly illuminated collection of Grosseteste's work from the end of the fourteenth century, testifying to enduring interest in his writings.[5] Moreover, the circumstances of the creation of the first manuscript copies of Grosseteste's work remain obscure, and it is unclear where *On the Sphere* was written (Ch. 1 §3). In Oxford in the first half of the thirteenth century, professional scribes and artists were available to meet the needs of students, scholars, and, increasingly, lay people. A now well-known artist, William de Brailes, was based in Catte Street, Oxford, in this period.[6] There is much less evidence for book production in Hereford at this time, despite the connection of many important scholars with the cathedral in the twelfth and thirteenth centuries.[7]

Unfortunately, it is impossible to know whether Grosseteste invested significant resources in the production of copies of *On the Sphere*.

---

[5] For collections of some of Grosseteste's works, see Harrison Thomson, *Writings*, 10–22.

[6] R. W. Hunt, 'The Library of Robert Grosseteste', in *Robert Grosseteste: Scholar and Bishop*, ed. D. A. Callus (Oxford: Oxford University Press, 1955), 128; G. Pollard, 'The University and the Book Trade in Medieval Oxford', *Miscellanea Medievalia*, 3 (1964), 336–44; C. Donovan, *The de Brailes Hours: Shaping the Book of Hours in Thirteenth-Century Oxford* (London: British Library, 1991), 13–15.

[7] See R. A. B. Mynors and R. M. Thomson, *Catalogue of the Manuscripts of Hereford Cathedral Library* (Cambridge: D. S. Brewer, 1993), xvii.

ILLUMINATING *ON THE SPHERE*   335

The scientific works carry no dedications to prospective patrons or clear indications of an intended audience. Grosseteste may have written drafts of the treatises on scraps of parchment or wax tablets, and then arranged for a neat copy to be made by payment or favour (perhaps by a student if the treatise emerged from a teaching environment).[8] After this point the dissemination of the treatise would rely on copying by or at the expense of other scholars and students.[9] William of Alnwick later claimed that Grosseteste 'wrote many scraps (*[s]c[h]edulas*) which are not all authoritative [...]. They are preserved in the Franciscan library at Oxford and I have seen them with my own eyes'.[10] These scraps may have included diagrams, as some were later bound into Bodleian Library MS Savile 21 (Os), a thirteenth-century collection of treatises on astronomy.[11] Alnwick's words imply that the scraps were revered, but lacked the status of a completed text. Such fragments are very rare survivals from the Middle Ages but are nevertheless an important reminder of the role of pieces of parchment and notes in wax as a means of transmitting material as well as working out ideas.

1 The World Machine Diagram

Despite their varied appearance and the different resources invested in them, about half the surviving manuscripts of *On the Sphere* contain (or were intended to contain) one diagram, comprised of concentric semi-circles, which is closely related to Grosseteste's text (Fig. 10.1).[12]

[8] See C. Burnett, 'Give Him the White Cow: Notes and Note-Taking in Universities in the Twelfth and Thirteenth Centuries', *History of Universities*, 14 (1995–6), 1–30; A. Pelzer, 'Les versions latines d'ouvrages de morale conservés sous le nom d'Aristote en usage au XIIIe siècle', *Rev. néo-scolastique de philosophie*, 23 (1921), 398; Southern, *Grosseteste*, 38.

[9] See Hunt, 'The Library', 128–9; Pollard, 'The University and the Book Trade'.

[10] Vatican City, Vatican Library, Pal. Lat. 1805, f. 10v; A. Pelzer, 'Les versions latines', 398; Southern, *Grosseteste*, 38; Burnett, 'White Cow', 17–18. On the material preserved by the Oxford Franciscans, see Humphrey, *Friars' Libraries*, xix, 224–9.

[11] Hunt, 'The Library', 133–4; Burnett, 'White Cow', 18.

[12] A version of the diagram appears in: Bologna, Bib. Uni. Lat. 1845 (Bu); Cambridge, University Library MS Gg.6.3 (Cu2); MS Mm.3.11 (Cu6); Dublin, Trinity College MS 441 (Dt); Erfurt, Universitäts- und Forschungsbibliothek, Bibliotheca Amploniana CA 4° 351 (Ea2) and 355 (Ea3); London, British Library, Add. MS 27589 (La1); Egerton MS 847 (Le2); Harley MS 3735 (Lh2); Harley MS 4350 (Lh3); New Haven, Yale University, Medical Library MS 11 (Ny); New York, Columbia University, Smith Western Add. MS 1 (Ns); Oxford, Bodleian

336 THE SCIENTIFIC WORKS OF ROBERT GROSSETESTE

The relatively large number of copies with this diagram gives it the best claim of any of the diagrams found in copies of the treatise to have been part of Grosseteste's original plan for the work.[13] Moreover, it appears in copies of both the major groups of manuscripts identified by Cecilia Panti, suggesting that it was devised at an early date.[14] Although not every copyist thought the diagram necessary, many did invest the necessary labour and parchment (which was, in most cases, minimal) to include it.

The semi-circular diagram illustrates the opening section of the text of *On the Sphere*, in which Grosseteste describes the spherical form of the *machina mundi*, or world machine. He starts with a definition of a sphere based upon a two-dimensional semi-circle: 'Now a sphere is the passage of a semicircle with its diameter fixed until it returns to the place from which it started' (DS §2). With the sphere defined, the text goes on to describe a set of five nested spheres, each of which is bounded by a semi-circle described by three letters, such as ACB. Each letter marks a point on the semi-circle's circumference and is therefore equidistant to the central point, O. The letters are usually inscribed on either end of a diameter (through O) that provides a baseline, and midway around the arc (see Fig. 10.1). The text begins with the outermost semi-circle (ACB) and works inwards. The largest, outermost semi-circle represents the *primum mobile*. Between this line and the second largest semi-circle (DFE) are the seven planets and the fixed stars. The remaining three semi-circles create four further spheres, the innermost of which (bounded by NRP) is earth. The three spheres between earth and the planets contain the other three elements. Working inwards, between

---

Library MS Bodley 676 (Ob2); MS Digby 98 (Od3); MS Gough Linc. 13 (Og); MS Laud Misc. 644 (Ol); Paris, Bibliothèque nationale de France MS Lat. 7195 (Pb1); MS Lat. 7292 (Pb3); MS Lat. 7298 (Pb4); Princeton, University Library, Garrett MS 95 (Pg); Salamanca, Biblioteca Universitaria 111 (Su1); Utrecht, Bibliotheek der Rijksuniversiteit 722 (Ur); Vatican City, Vatican Library, Pal. Lat 1414 (Vp); Verdun, Bibliothèque municipale MS 25 (Vmu). Manuscripts consulted for this study are underlined. See also Appendix 2. In Oxford, Corpus Christi College MS 41 (Oc1), the lower part of the manuscript page upon which the text begins (f. 143v) has been excised and may once have contained the diagram. Space was apparently left for the diagram in Cambridge, Gonville & Caius College MS 137/77 (Cg1).

[13] B. Obrist, 'Démontrer, montrer et l'évidence visuelle. Les figures cosmologiques, de la fin de l'Antiquité à Guillaume de Conches et au début du XIIIe siècle', in E. C. Lutz, V. Jerjen, and C. Putzo (eds.), *Diagramm und Text. Diagrammatische Strukturen und die Dynamisierung von Wissen und Erfahrung* (Wiesbaden: Reichert Verlag, 2014), 45–78, at 76.

[14] See Panti, *Moti, virtù e motori*, 211–41.

ILLUMINATING *ON THE SPHERE* 337

DFE and GHI is fire (*ignis*), the space between GHI and KLM contains air (*aer*), and that between KLM and NRP is water (*aqua*). The diagrams often include brief labels to help clarify the contents of the spheres (see Figs 10.1, 10.2, 10.5), but Grosseteste offered no opinions on the relative proportions of the spheres, leaving the copyists to decide whether or not to space them evenly. In the text, the description of the structure is then followed by an explanation of why these bodies must be spherical.

The diagram of the world machine is usually, though not always, placed at or near the start of the text. In this position it provides the graphic framework of the *machina mundi* that Grosseteste goes on to populate. Grosseteste's language is very visual, although some of it is taken directly from his sources (Ch. 7). His stated aim is to 'describe the shape (*figuram*) of the world machine' (DS §1, using a term (*figura*) often used to refer to diagrams).[15] Moreover, Grosseteste invites his readers to imagine (*imaginemur*) the semi-circle DFE. This explicitly points to a need to visualize the material (and if the diagram did not originate with Grosseteste would explain why it became a popular element).[16] With the spheres laid out before them (albeit in semi-circular form) the reader is expected to be able to go beyond the limits of fixed imagery and mentally visualize the movement of the celestial bodies as the spheres of the world machine begin to turn.[17] That some readers were thinking about the larger form of the world machine is suggested by two copies of the diagram in which the spheres are rendered as full circles: in British Library Egerton MS 847 (Le2) the outermost sphere is drawn as a complete circle, and in Dt all the spheres are shown as complete circles. Some readers, at least, therefore, understood the text well enough to begin to expand the most common form of the diagram.

---

[15] K. Müller, *Visuelle Weltaneignung. Astronomische und kosmologische Diagramme in Handschriften des Mittelalters* (Göttingen. Vandenhoeck & Ruprecht, 2008), 186.

[16] See also Southern, *Grosseteste*, 143; Müller, *Visuelle Weltaneignung*, 184–7; Obrist, 'Démontrer, montrer', 76. For more on the use of mental images as a tool for scientific visualization in other medieval texts, see J. Franklin, 'Diagrammatic Reasoning and Modelling in the Imagination: The Secret Weapons of the Scientific Revolution', in G. Freeland and A. Corones (eds.), *1543 and All That: Image and Word, Change and Continuity in the Proto-Scientific Revolution* (London: Kluwer Academic, 2000), 53–115, at 92–6.

[17] The authors are grateful to Aylin Malcolm for discussion of spherical diagrams in the work of Sacrobosco. Her work on animating spheres is available via http://aylinmalcolm.com/sacrobosco/ (accessed 6 April 2020).

338 THE SCIENTIFIC WORKS OF ROBERT GROSSETESTE

The reference to the outermost sphere containing the planets also seems to have prompted some readers to develop the diagram by adding more divisions so that bodies in orbit around the earth are given distinct zones. Ea2, f. 46, includes fourteen semi-circular bands labelled (from the outermost, working inwards) as: the *primum mobile*, the stars in the firmament, Saturn, Jupiter, Mars, sun, Venus, Mercury, moon, [blank], fire, air, water, and earth. Vatican Library MS Pal. Lat. 1414 (Vp), f. 35, and British Library Harley MS 3735 (Lh2), f. 74, omit the blank band for a total of thirteen divisions and include drawings of stars, or in the case of Lh2 red dots, in the relevant zone (Fig. 10.2).[18] In all three cases only the letters given in the text are used to denote Grosseteste's spheres, but the subdivision of the outermost sphere using more semi-circles creates the impression of additional spheres. A further variation on this theme is found in Bibliothèque nationale de France MS Lat. 7298 (Pb4), f. 31, where the outermost sphere is labelled as the *primum mobile*, followed by bands (*orbis*) for the fixed stars (*stellarum fixarum*), Saturn, Jupiter, Mars, sun, Venus, Mercury, and the moon. In this case, all the letters referred to in Grosseteste's text are omitted, making the diagram of very little use as a visualization of his work, but providing the basic concept and additional information.

The variation in the nested semi-circles diagram to include the planets may, in part, have been inspired by knowledge of diagrams created in conjunction with other texts. Cambridge University Library MS Ff.6.13 (Cu1) provides an example of how both diagrams and text could be adapted by copyists. At the end of Grosseteste's *On the Sphere* in this thirteenth-century copy are excerpts from John of Sacrobosco's treatise on the same subject, together with a diagram of concentric circles (f. 18) (Fig. 10.3).[19] The innermost circle is labelled as earth and partially painted in yellow and brown. A second version of this diagram on f. 26v, placed before the rest of Sacrobosco's treatise, clarifies the significance of the colouring, as here green is used for both the sphere of water and to suggest water on the earth, which is painted yellow, while the brown area is labelled '*abiss*' suggesting that the designer of

---

[18] A further variation on a larger number of spheres appears in Cu2, f. 200v.
[19] See Harrison Thomson, *Writings*, 118–19; Thorndike, *The* Sphere *of Sacrobosco*, 64–6.

ILLUMINATING *ON THE SPHERE* 339

Fig. 10.2. British Library, Harley MS 3735, f. 74 © The British Library Board.

Fig. 10.3. Cambridge University Library MS Ff.6.13, f. 18. Reproduced by kind permission of the Syndics of Cambridge University Library.

ILLUMINATING *ON THE SPHERE* 341

this diagram was thinking in many dimensions. Around the earth, the bands are again labelled, working outwards, as: water, air, fire, moon, Mercury, Venus, sun (with the names of the zodiac), Mars, Jupiter, and Saturn. The penultimate band is blank, and the final, narrower, band is painted yellow. In the second version the zodiac is omitted, and the outermost sphere contains the firmament. This diagram, again in varied forms, appears in other copies of Sacrobosco's work, though in the case of Cu1 ideas from Grosseteste's text seem to have cross-fertilized with Sacrobosco's work in the creation of these two schemas, as Sacrobosco's diagram does not usually include spheres of the elements.[20] The more common form of Sacrobosco's diagram occurs in two other manuscripts that contain both Sacrobosco's and Grosseteste's treatises where additional spheres were introduced to Grosseteste's diagram: Lh2 (f. 17) and Pb4 (f. 20v). The transmission of diagrams dealing with similar subject matter in the same manuscripts may therefore help to explain the introduction of variations into some versions of both diagrams.

The idea of a diagram of concentric circles showing different celestial zones was not original to either Grosseteste or Sacrobosco. In the twelfth century, copies of William of Conches' *Dragmaticon Philosophiae* included a circular diagram in the context of the discussion of the movement of the stars and firmament.[21] In one version, found in Montpellier, Bibliothèque interuniversitaire Médecine H 145, f. 15, and Vatican Library MS Reg. Lat. 1222, f. 9v, this diagram had the same organization of the elements at the centre, although it presented the heavenly bodies in a slightly different order, with the moon as the innermost orbiting body, followed by the sun, and then Mercury, Venus, Mars, Jupiter, Saturn, and finally the firmament with the

---

[20] See, for example, British Library Egerton MS 844, f. 8; Harley MS 531, f. 4v; Harley MS 3647, f. 22v; Lh2, f. 17; Royal MS 12 C XVII, f. 17. See also, Murdoch, *Album of Science*, 335.

[21] William of Conches, *Dragmaticon Philosophiae*, ed. I. Ronca (Turnhoult: Brepols, 1997), 77; English translation in William of Conches, *A Dialogue on Natural Philosophy*, trans. I. Ronca and M. Curr (Notre Dame: University of Notre Dame Press, 1997). See also Müller, *Visuelle Weltaneignung*, 93–181; Obrist, 'Démontrer, montrer', pp. 60–1; E. Ramírez-Weaver, '"So That You Can Understand This Better": Art, Science, and Cosmology for Courtiers in William of Conches' *Dragmaticon Philosophiae*', in Lutz, Jerjen and Putzo, *Diagramm und Text*, 319–48.

342 THE SCIENTIFIC WORKS OF ROBERT GROSSETESTE

fixed stars.[22] An earlier, simpler, precursor can be found in some copies of Isidore of Seville's *On the Nature of Things* and other texts, where circular diagrams set out the order of the celestial bodies, sometimes with the periods of their orbits.[23] Although the understanding of the organization of the universe had moved on by the time of Sacrobosco's and Grosseteste's texts, therefore, the diagrams, like the text, built on existing works.

Different views about the importance of the diagram of nested semi-circles in Grosseteste's *De sphera* are suggested not only by the varied content, but also by the position of the schema on the page and the effort invested in its execution. In most of the manuscripts the diagram is set within the text block, sometimes extending into the margin (see Figs 10.1–10.2). In Dt the diagram is given a full page at the end of the text, despite available space on the end of the previous folio. This may have made it easier to execute (and may even suggest that it was included as an afterthought, or by a copyist working out the text through the process of making a diagram on a separate piece of parchment), but it is ultimately less convenient for the reader, who has to flip back and forth between text and diagram. In a further six copies (Pb4, British Library Add. MS 27589 (La1), Cambridge University Library MS Mm.3.11 (Cu6), Erfurt Universitäts- und Forschungsbibliothek, Bibliotheca Amploniana CA 4° 355 (Ea3), New York, Columbia University Smith Western Add. MS 1 (Ns), and Yale Medical Library MS 11 (Ny)) the diagram is entirely located in the margin, and in Cambridge University Library MS G.6.3 (Cu2) the diagram is largely in the lower margin. This solves problems of layout and ease of reference, but also serves to divorce the diagram from the text, presenting it as an additional tool, like a gloss, rather than an

---

[22] A similar diagram is present in Florence, Biblioteca Medicea Laurenziana, MS Ashburn 173 (Fa), f. 11v, see Müller, *Visuelle Weltaneignung*, 134–9. The order of the planets set down by Ptolemy (moon, Mercury, Venus, sun, Mars, Jupiter, Saturn), was generally accepted in medieval astronomy, where disagreement occurred it turned on where the sun was to be placed with respect to Venus and Mars, see B. R. Goldstein, 'Theory and Observation in Medieval Astronomy', *Isis*, 63 (1972), 39–47; E. Grant, 'The Medieval Cosmos: Its Structure and Operation', *Journal for the History of Astronomy*, 28 (1997), 147–67, esp. 154; E. Grant, *The Foundations of Modern Science in the Middle Ages* (Cambridge: Cambridge University Press, 1996), 138–9.

[23] Murdoch, *Album of Science*, 54; Müller, *Visuelle Weltaneignung*, 79–80; Obrist, 'Démontrer, montrer', 54–8, 91.

ILLUMINATING *ON THE SPHERE* 343

integral part of the work. In the case of Pb4 and Cu6 this is appropriate, given the variation in content from Grosseteste's text; this is not the case for the other examples. Nevertheless, the separation of the diagram from the text may help to explain the inclusion of a range of additional diagrams in some copies, a point which will be discussed in more detail below.

In the cases of La1, Cu6, Cu2, Dt, and Ea3 there may also be a correlation between the position of the diagram and the care that went into its execution, which may in turn be linked to limited investment in the overall appearance of the volumes. In Ea3 the lines have been executed in red and with the aid of a compass, but they are uneven. The makers of the diagrams in La1 and Dt (who were probably the scribes) also seem to have used a compass, though not very skilfully, whilst the diagrams in Cu6 and Cu2 appear to have been drawn free-hand. At the other extreme, the diagrams in three late-thirteenth-century manuscripts, Lh2, Bodleian Library MS Laud Misc. 644 (Ol), and Verdun Bibliothèque municipale MS 25 (Vmu), were carefully planned. The scribe left space to suit the form of the diagram, which was completed with neat, even lines. In Lh2 the diagram was executed in red and black (Fig. 10.2). In Ol the lines are in red with the letters in blue, and the diameter line has been decorated with penwork, rendering it an extension of the initial I with which the text opens (Fig. 10.4).[24] It seems to have been completed by the person who did the elaborate penwork. Both these manuscripts are larger than average measuring 30.5 × 21 cm and 33 × 21 cm respectively. However, Vmu is by far the most lavishly decorated surviving copy, but is small at 19 × 14 cm.

Vmu's version of the world machine diagram is fully painted and gilded, and executed by a skilled artist (Fig. 10.5).[25] Beneath the nested semi-circles is a panel of foliage that echoes the decoration of the initial on the left of the page, but bears no relationship to the content of the text,

---

[24] On this manuscript, see F. Saxl and H. Meier, *Catalogue of Astrological and Mythological Illuminated Manuscripts of the Latin Middle Ages*: Vol. 3, *Manuscripts in English Libraries* (London: Warburg Institute, University of London, 1953), 386–93; D. Blume et al., *Sternbilder des Mittelalters. Der gemalte Himmel zwischen Wissenschaft und Phantasie. Band I. 800–1200* (Berlin: Akademie Verlag, 2012), 403–9.

[25] The authors would like to thank Michaël George for his help with Vmu.

Fig. 10.4. Oxford, Bodleian Library MS Laud Misc. 644, f. 143 © The Bodleian Libraries, The University of Oxford.

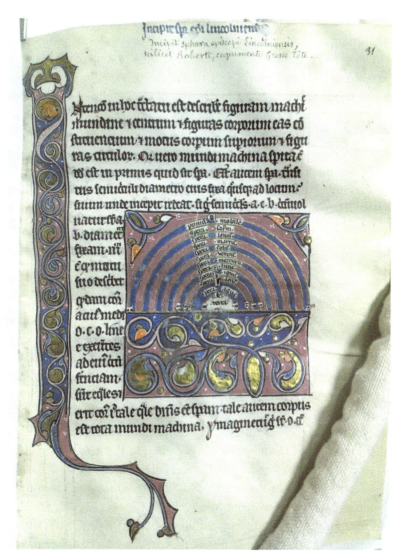

Fig. 10.5. Verdun, Bibliothèque municipal MS 25, f. 31. Reproduced by kind permission of Michaël George.

suggesting that one of the functions of this diagram was to make the volume more visually appealing. The diagram itself is another elaborated version, with twelve bands for the *primum mobile*, Saturn, Jupiter, Mars, sun, Venus, Mercury, moon, fire, air, water, and earth. Yet, whilst its

346 THE SCIENTIFIC WORKS OF ROBERT GROSSETESTE

scope and position find parallels in the copies now in the Vatican (Vp) and Erfurt (Ea2), the level of its decoration is unparalleled in the known copies of this text. A similar approach to the decoration of the diagrams is taken in Vatican Library MS Urb. Lat. 1428 (Vu), but in that manuscript there are no diagrams in the copy of Grosseteste's *On the Sphere*. In addition, Vmu includes four more painted and gilded diagrams on the folios with Grosseteste's *On the Sphere*, all set into the text block. In this manuscript, *On the Sphere* occupies seventeen folios, suggesting that there was no concern to save parchment. Vmu therefore stands out for the resources invested in its production, and this, together with its intimate scale, suggests that it was made for a very wealthy individual, interested in display as much as in the scientific content of the diagram.

2 Other Diagrams

While the world machine diagram occurs frequently in copies of Grosseteste's *On the Sphere*, the variation between each iteration demonstrates different levels of attention given to the image and, in some instances, different astronomical sources with which it cross-fertilized. This variety is further emphasized by those copies of the *On the Sphere* that include additional diagrams, between which there is little consistency. Vmu is relatively unusual in also including other diagrams in the main text area of *On the Sphere*. In addition to the nested semi-circles diagram it contains four more large and fully painted diagrams showing the eccentric orbit of the sun (f. 40v), the movements of the sun and moon that result in solar and lunar eclipses (ff. 44v, 45, with the lunar deferent and epicycle incorporated into the diagram on f. 44v), and the phenomenon of parallax, illustrating the consequences of different viewing positions (in a diagram which appears after the end of the text on f. 46v). In each case the artist has used the opportunity to display his range, including foliage, grotesques, and (on ff. 40v, 46v) rabbits. The grotesques in the lunar eclipse diagrams may have been inspired by the description 'the point through which the moon passes from the southern part of the ecliptic to the northern is called the head of the dragon, and the point opposite is called the tail of the dragon', as they are thin creatures comprised of just a head and tail, but the effect is to add visual

ILLUMINATING *ON THE SPHERE*   347

interest to the diagram rather than clarifying the information given in the text (DS §57).

Some of the additional diagrams in Vmu find parallels in schemas in other copies of Grosseteste's *On the Sphere*. Versions of the two eclipse diagrams appear, in less lavishly decorated form, in Vp (f. 87) and Ea2 (f. 50r–v), where they are located in otherwise blank space after the end of the text, and in Lh2 (ff. 80v, 81v), where they are set into the text block.[26] In Lh2 these diagrams also have references to dragons in the form of inscriptions that read *dracho lune*, and one of the diagrams in Ea2 also includes the label *draco*. Lh2 includes yet more diagrams, with an illustration of the movement of the sun through the zodiac (f. 78v), the climes (or climate zones) showing seven zones between areas that were uninhabitable because they were too hot or too cold (f. 79v) (a diagram also included in Ea2 and Ns, f. 135v), and the motion of the fixed stars (as described in §§51–4) (f. 80).[27] There is a logical progression within the diagrams in Lh2, as the final diagram on f. 82 combines information from the previous schemas, showing the movements of sun, moon, and zodiac, creating a solar eclipse on a small sphere labelled 'earth' and painted to evoke the bands of the climes. Like the final diagram in Vmu it includes an observer, labelled *oculus*, but here it is marked as a dot on the earth's surface. In addition to these diagrams that recur in a small number of manuscripts, another schema, an uninhabited zodiac ring, was part of Bibliothèque nationale MS lat. 7292 (Pb3) and Utrecht, Bibliotheek der Rijksuniversiteit MS 722 (Ur).[28] The repetition of diagrams is, in part, a result of the need for a model to copy to create a new manuscript, but overall the variations in the selection of diagrams suggest that this was an element that was carefully considered by the patron, who might also be the copyist.

The recurrence of diagrams in the surviving copies of *On the Sphere* testifies not only to the copying tradition, but also to the value of such visualizations in astronomical study and the potential for these

---

[26] On broader similarities between these manuscripts, see Panti, *Moti, virtù e motori celesti*, 216–17.

[27] Consultation of Ns was made possible by Adam Harris Levine, to whom the authors are grateful.

[28] See Panti, *Moti, virtù e motori celesti*, 222.

348   THE SCIENTIFIC WORKS OF ROBERT GROSSETESTE

schemas to circulate independently of their original texts.[29] In this context it is important to remember that Grosseteste's *On the Sphere* was usually copied and/or later bound with other texts. Moreover, other thirteenth-century thinkers were wrestling with similar problems, most notably Sacrobosco, whose own *On the Sphere* was probably produced before that by Grosseteste (Ch. 1 §1).[30] In a world in which diagrams might be placed in margins, either for ease of production or because they were conceived of as operating like a gloss, astronomical schema could easily be added to texts, though this supposes a widely-read designer.[31] Similarly, as we saw with the circular schemas of the spheres, some diagrams could travel between texts as ideas were adopted from earlier sources. For example, the climes diagram had a long history in copies of Isidore of Seville's *On the Nature of Things* (albeit usually with only five zones) and other texts.[32] Similarly, a circular diagram showing the eccentric orbit of the sun had appeared in copies of William of Conches' work, together with diagrams of eclipses and climate zones.[33]

Sacrobosco's version of *On the Sphere* seems to have usually contained more diagrams than Grosseteste's, though again there are many variations between the surviving manuscripts.[34] In addition to his diagram of the spheres, versions of the lunar eclipse diagram and climes appear in some copies of his text (although the eclipse diagram is more common than that of the climes).[35] As noted earlier in this volume, there has been disagreement between scholars as to whether Grosseteste's or Sacrobosco's treatise is the earlier, and the similar structure and use of shared text suggests that one was strongly influenced by the other (Ch. 1 §1; Ch. 5 §4.2; Ch. 7 §2). The diagrams do not resolve this issue, but they do point to a willingness to transfer schemas between texts as both works were copied (sometimes in the same manuscripts).

---

[29] For relationships between some of the manuscripts, see Ch. 3 §1; and Panti, *Moti, virtù e motori celesti*, 211–41.

[30] See Harrison Thomson, *Writings*, 115; Southern, *Grosseteste*, 145–6.

[31] See also Murdoch, *Album of Science*, 142.   [32] Murdoch, *Album of Science*, 340–2.

[33] Murdoch, *Album of Science*, 144–5. See for example Montpellier, Bibliothèque interuniversitaire Médecine H 145, ff. 17v, 20v, and Vatican MS Reg. Lat. 1222, ff. 11, 12v.

[34] Thorndike, *The* Sphere *of Sacrobosco and its Commentators*, 63–4. The authors are grateful to Arthur Hénaff for sharing his thoughts on the diagrams in Sacrobosco's *On the Sphere*.

[35] For example, the following folios contain diagrams of the climes and lunar eclipse respectively: Vatican Library MS Pal. Lat. 1400, ff. 22v, 24v; Pb4, ff. 28v, 29; Bibliothèque nationale de France MS Lat. 7421, ff. 30, 32v.

ILLUMINATING *ON THE SPHERE* 349

In the context of the transmission of diagrams between texts, a thirteenth-century manuscript in the Bibliothèque Mazarine, MS 3642, provides an intriguing piece of evidence. Within this collection of astronomical and astrological texts, which does not include Grosseteste's *On the Sphere*, are three leaves covered with diagrams (ff. 88–90), neatly executed in red and black, with the heading 'figures of circuits of the heavens, planets, elements, winds, eclipses, sun, and zodiac'.[36] In addition to covering the topics set out in the list, the diagrams include a world map and a consanguinity diagram setting the prohibited degrees of relationship for marriage. Such collections were not new in the thirteenth century (a twelfth-century collection survives at the end of a copy of Plato's *Timaeus* in Bodleian Library Digby MS 23, ff. 51v–54v, and a parallel is found in the two pages of diagrams after *On the Sphere* in Ea2, ff. 50r–v) but amongst the collected schemas in the Mazarine manuscript is the nested semi-circles diagram from Grosseteste's *On the Sphere* (Fig. 10.6). On the same page is a version of the Porphyrian tree setting out qualities of the elements, and circular diagrams, including three that deal with similar content to Grosseteste's spheres. One uses five concentric circles, with earth in the middle surrounded by water, air, fire, and finally *mundus* indicating that all the elements are within the subcelestial world. Another, like Sacrobosco's diagram, puts *infernus* at the centre, surrounded by earth, water, air, thick air in which are evil spirits, the ether in which are the good spirits, the fire between the angels and demons, and the Empyrean where the angels are situated.[37] These ideas are expanded further in a diagram at the top of the page, in which the outermost circle is labelled 'God'. A similar combination of themes is found in Digby MS 23, where diagrams of the elements, climes, and astronomy are accompanied by diagrams on musical harmony and the location of different spiritual creatures. On these pages, therefore, the collected diagrams bring together different ways of conceptualizing the universe, from the physical to the theological. The reader is left to

---

[36] 'De figuris circulorum celi, pl[anet]arum, elementorum, ventorum, eclipsis, solis, et zodiaci'. Paris, Bibliothèque Mazarine, MS 3642, f. 88.

[37] 'aer crassus in quo sunt cacodemones; Ethere primus(?) in quo sunt calodemones; ignis medius inter angelos et demones; celum empirium in quo angeli sunt'. Paris, Bibliothèque Mazarine, MS 3642, f. 89.

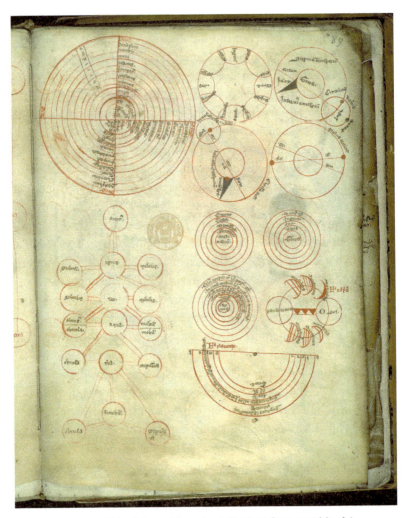

Fig. 10.6. Paris, Bibliothèque Mazarine MS 3642 f. 89 © Bibliothèque Mazarine.

meditate on all these diagrams as different ways of showing related ideas, without the aid or distraction of large amounts of text.

The inclusion of the nested semi-circles within this collection is a valuable indication that Grosseteste's *De sphera* was accepted as part of an authoritative corpus of astronomical texts in the thirteenth century.

Yet the question remains as to what role the image played within the collection. If the original function of the diagram, as demonstrated above, was to elucidate the geometrical opening of *On the Sphere* through the interaction of image and text, then here it achieves something quite different. The nested semi-circles are entirely divorced from the text, rendering the letters that label each point meaningless, that is, unless the copyist or reader was already familiar with *On the Sphere*. On this folio, the diagram serves less as an explanatory tool but instead as a reminder of Grosseteste's *On the Sphere* and its geometrical foundation. By depicting the image in isolation from the text, the Mazarine manuscript shows how the nested semi-circles, beyond being associated with the *On the Sphere*, came to stand for some of the ideas it contained. At the same time, in this collection the diagram is adapted slightly. The sphere beyond those of earth, water, and air is no longer simply labelled as fire, but instead contains an inscription, derived from Grosseteste's work, conflating that region with the fifth essence, ether, which in Grosseteste's text is actually outside the spheres of the elements.[38] Nevertheless, the presence of the diagram within this collection indicates Grosseteste's contribution to the different conceptions of the universe and suggests that some medieval viewers were able to recognize, or even understand, the image without the inclusion of the text.

The nested spheres diagram in the Mazarine manuscript is readily identifiable as the schema found in Grosseteste's work in part because semi-circular diagrams are unusual in astronomical texts. Other diagrams of the spheres usually show them as full circles, and it is striking that when the diagram was adapted in the context of Grosseteste's treatise, for example with the addition of more semi-circles for the planets, it continued to be presented in semi-circular form.[39] It therefore seems possible that, whether the world machine diagram was designed as part of the original text or whether it was a later inclusion that came to be

---

[38] 'Superficies quamque(?) nominat quinta ferva(?) sive(?) ethera, id est, corpus preter proprietates elementares'. Paris, Bibliothèque Mazarine, MS 3642, f. 89.

[39] Kislak Center for Special Collections, Rare Books and Manuscripts University of Pennsylvania LJS 26, f. 10, provides a rare example of a semi-circular diagram in a Sacrobosco manuscript. We are grateful to Aylin Malcolm for drawing our attention to this manuscript.

352 THE SCIENTIFIC WORKS OF ROBERT GROSSETESTE

widely copied, its unusual semi-circular form became associated with a conception of the world machine particular to Grosseteste. If so, the diagram may have rendered the text more immediately recognizable as Grosseteste's *On the Sphere* and, in doing so, differentiated it from other authors' works on the same subject.

3 Diagrams and Instruments

The collection of diagrams divorced from their usual explanatory texts in Bibliothèque Mazarine, MS 3642, together with the variations introduced into the diagrams in copies of Grosseteste's work, raises the question of how readers might have used such schemas, beyond the clarification of text. The cross-fertilization of diagrams suggests that they were powerful tools for conceptualizing information, but a question remains as to whether they also provide insights into more practical activities. One suggestion involves the extent to which the treatise might have been designed to be used in conjunction with, or as an alternative to, an armillary sphere. A fifteenth-century example in Oxford's History of Science Museum (Inv. 12765), thought to be one of the earliest to survive, shows a wooden sphere representing the earth at its centre surrounded by brass and gilt brass rings that indicate the motion of the sun and fixed stars.[40] The instrument thus displays the same basic structure of the universe as shown in Grosseteste's world machine diagram, but in three dimensions and with movable rings that could enact the movement of certain spheres. The link to an armillary sphere was explicitly made by the designer of the initial at the start of Grosseteste's text in Lh3 (Fig. 10.1).

The idea that the group of medieval *On the Sphere* texts were, in essence, descriptions of an armillary sphere is not new in itself and has also been discussed by those evaluating the images in early modern

---

[40] The armillary sphere is cautiously dated to *c*.1500, though John North suggested that it was made around 1425 based on the stellar longitudes marked on its ecliptic ring. F. Maddison, *Medieval Scientific Instruments and the Development of Navigational Institutions in the XVth and XVIth Centuries* (Coimbra: Junta de Investigações do Ultramar-Lisboa, 1969), caption to fig. 2. For another early model, see footnote 43 below.

ILLUMINATING *ON THE SPHERE* 353

editions, most commonly those of Sacrobosco. In fact, a full-page illustration of an armillary sphere became the standard frontispiece for Sacrobosco's text after Erhard Ratdolt, a German typesetter who worked in Venice, included it in his edition of 1482.[41] Adam Mosley has argued that in the sixteenth century instruments such as the armillary sphere were not only calculating devices but employed in pedagogical contexts to convey cosmological ideas. Mosley traced this attitude back to the medieval commentaries on Sacrobosco's *On the Sphere*, namely that of Michael Scot, from which he inferred that 'one taught a student the circles of the celestial sphere in virtue of asking them to study the instrument'.[42] In addition, a late thirteenth-century copy of Sacrobosco's work (British Library Harley MS 3647, f. 2?) opens with an initial enclosing an image of a teacher pointing to an armillary sphere, accompanied by students holding books. Although armillary spheres were popular in illustrations of astro-

---

[41] Ratdolt's edition came just ten years after the earliest printed edition in 1472. O. Gingerich, 'Sacrobosco Illustrated', in L. Nauta and A. Vanderjagt (eds.), *Between Demonstration and Imagination: Essays in the History of Science and Philosophy Presented to John D. North* (Leiden: Brill, 1999), 211–24, at 211.

[42] A. Mosley, 'Objects of Knowledge: Mathematics and Models in Sixteenth-Century Cosmology and Astronomy', in S. Kusukawa and I. Maclean (eds.), *Transmitting Knowledge: Words, Images, and Instruments in Early Modern Europe* (Oxford: Oxford University Press, 2006), 193–216 at 214. Mosley quotes Lynn Thorndike's transcription of Michael Scot's commentary (*The* Sphere *of Sacrobosco*, 248–9): 'Ideo volentibus machine mundane et corporum superiorum cognitionem habere primitus tractamus de spera materiali brevi et utili in qua figura machine mundane plenissime figuratur, ut habita notitia de ipsa in ipsam celestis spere specialiter notitiam intueamur. Nam, sicut dicit Plato *in Timeo*, Mundus iste sensibilis factus est ad similitudinem mundi archetypi, id est, mundi principalis. Mundus autem principalis dicitur esse mundus qui fuit in mente divina ab eterno ad cuius similitudinem factus est iste mundus sensibilis quem videmus. Hec autem spera materialis facta est ad similitudinem huius mundi sensibilis et ideo omnes demonstrationes que fiunt in spera materiali similiter debent intelligi in mundo sensibili, quia abstrahentium non est mendacium, ut vult philosophus *in libro Physicorum*'. The authors are grateful to Sigbjørn Sønnesyn for providing the following translation of the above passage: 'Therefore, wishing to acquire knowledge of the world machine and of the higher bodies, to begin with we discuss, briefly and profitably, the material sphere in which the shape of the world machine is fully modelled, so that having attained knowledge of this [sphere] we may observe a particular acquaintance with the celestial sphere. For, as Plato says in the *Timaeus*, "this world that we can perceive with our senses was made in the likeness of the archetypal world, that is, the principal world". The principle world, furthermore, is said to be the world that existed in the divine mind from eternity, in the likeness of which this perceptible world that we can see was made. This material sphere, however, was made in the likeness of this perceptible world, and therefore all demonstrations derived from the material sphere should also be held to pertain in the perceptible world, because 'there is no lie concerning abstractions', as the philosopher says in the book of *Physics*.' This use of instruments in the sixteenth century is discussed at length in J. Bennett, 'Knowing and Doing in the Sixteenth Century: What Were Instruments For?', *The British Journal for the History of Science*, 36 (2003), 129–50.

354 THE SCIENTIFIC WORKS OF ROBERT GROSSETESTE

nomical study, very few survive from the Middle Ages, making it difficult to assess how readily accessible they may have been in the thirteenth and fourteenth centuries.[43] In a world in which Grosseteste's text could be compressed onto four small folios, a copy of his text would seem likely to have been more affordable for an individual than their own instrument. Moreover, the correlation between the expense of added decoration in the form of figurative imagery in initials and depictions of teachers with equipment may not be insignificant. The individuals or communities for whom those copies were made might have been more likely to be able to afford such tools.

The debt of Sacrobosco's treatise to the armillary sphere is, however, explicitly suggested by diagrams in some of the early copies of his text. Midway through a thirteenth-century English copy of the text, which is part of a collection of texts that also includes Grosseteste's *On the Sphere* (Lh2, f. 23), is a drawing of the instrument (Fig. 10.7). From the main body of the image alone, it could be regarded as a standard diagram which although two-dimensional, explains the parts of the celestial spheres with clarity. Yet a handle sticks out of the *polus antarticus* towards the bottom right of the image, deliberately marking it as a drawing of an instrument.[44] In a visual culture in which certain diagrams were used as instruments, sometimes with the aid of mental manipulation, the presence of the handle does not necessarily interfere with the function of the image. Just as the user of an armillary sphere turned its rings around the earth, so could the viewer of the diagram imagine their rotation.[45] In some cases, drawings of instruments could be used as instruments themselves if,

---

[43] In Müller's discussion of the armillary sphere image in Lh3 and what it reveals about the teaching of astronomy at this time, she considers a model in the Whipple Museum of the History of Science (Wh. 0336) dated to the fifteenth century. While made much later, it is regarded as one of the earliest surviving examples of this type of instrument. Müller, *Visuelle Weltaneignung*, 199–202.

[44] See also London, British Library Egerton MS 844, f. 16; Harley MS 3647, f. 29; Royal MS 12 C XVII, f. 25v; Pb4, f. 26; Oxford, Bodleian Library MS Canon. Misc. 161, f. 20; and Paris, Bibliothèque Sainte-Geneviève MS 1043, f. 12. For more on these images in Sacrobosco's *On the Sphere* and their mental manipulation, see Müller, *Visuelle Weltaneignung*, 237–40.

[45] For mental manipulation and medieval images, particularly concerning the use of diagrams, see M. Carruthers, *The Craft of Thought: Meditation, Rhetoric and the Making of Images 400–1200* (Cambridge: Cambridge University Press, 1998), 77–81; M. Carruthers, 'Moving Images in the Mind's Eye', in J. Hamburger (ed.), *The Mind's Eye: Art and Theological Argument in the Middle Ages* (Princeton: Princeton University Press, 2006), 287–305, at 300.

ILLUMINATING *ON THE SPHERE* 355

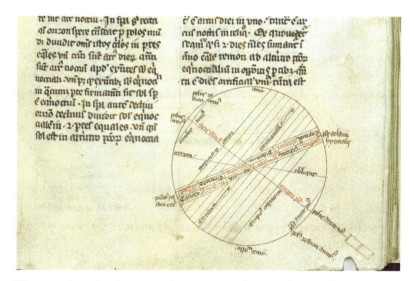

Fig. 10.7. British Library, Harley MS 3735, f. 23 © The British Library Board.

in Bert Hall's words, 'the image is the primary object of the viewer/reader's attention, while words of explication serve to illuminate details that might not be apparent at first glance'.[46] However, this is not the case for the images of Grosseteste's *On the Sphere* as their use, at least for the first-time reader, is clearly dependant on the accompanying text. The differentiation here between diagram and drawing of an instrument seems, therefore, purely referential to the use of an armillary sphere. This serves to emphasize the point that the medieval readers and copyists of Sacrobosco saw the armillary sphere to be central to the text, but it also raises a pertinent question as to why on the two-dimensional plane of the folio an instrument would be drawn where a diagram would suffice. The handle's inclusion implies that the instrument itself was embedded in the reading practices of the

---

[46] B. S. Hall, 'The Didactic and the Elegant: Some Thoughts on Scientific and Technological Illustrations in the Middle Ages and Renaissance', in B. S. Baigrie (ed.), *Picturing Knowledge: Historical and Philosophical Problems Concerning the Use of Art in Science* (Toronto: University of Toronto Press, 1996), 3–39, at 37. Crowther and Barker discuss the mental manipulation of images in early printed editions of Sacrobosco's *On the Sphere*, and consider how the placement of a crank on a particular type of diagram could infer the use of a similar apparatus in the lecture hall. K. M. Crowther and P. Barker, 'Training the Intelligent Eye: Understanding Illustrations in Early Modern Astronomy Texts', *Isis*, 104 (2013), 429–70, particularly 443–5.

356 THE SCIENTIFIC WORKS OF ROBERT GROSSETESTE

manuscript, perhaps functioning as a substitute when the instrument was unavailable.

A more practical reason for the use of three-dimensional, moveable models presents itself later in the text, where Grosseteste describes the movement of the solar body with such detailed precision that it is hard to believe he expected the viewer to retain an image of it in their mind's eye. He writes,

> Let us imagine (*imaginemur*) a line drawn from the 18th degree of Gemini through the centre of the earth to the degree of Sagittarius opposite to it. And from the centre of the earth on the same line let two and a half degrees be calculated from the diameter of the circle of the sun towards Gemini.   (DS §39)

The idea that the medieval reader would mentally picture a three-dimensional shape and its internal movement is wholly acceptable, but this detail pushes the boundaries of that concept. Whether Grosseteste expected the reader to measure, in their mind's eye, the 18th degree of Gemini, or 2.5 degrees from the diameter of the circle of the sun can be questioned. It could be argued that his point is still made if the medieval imaginer conjures a rough image of this description (see Ch. 4, Fig. 4.22). Nevertheless, the purpose of Grosseteste's precision and how he expected this to be demonstrated to the reader is still to be accounted for.

Grosseteste's description could be usefully pictured in two ways: through the manipulation of a three-dimensional instrument or by a diagram that incorporates the exact measurements he describes. Of the manuscripts surveyed, there is just one that records an attempt to reconstruct Grosseteste's model with these measurements: Lh2. At the base of the column upon which the above excerpt is written (f. 78v; Fig. 10.8) is an illustration of the movement of the sun through the ecliptic. The demarcations of the ecliptic ring divide each zodiac sign into ten parts and thus mark, albeit roughly, every three degrees. The red line that cuts through the centre of the image runs from approximately Gemini 18 (or just after) into Sagittarius. While other diagrams of this description show the sun and earth's positions relative to one another (see for example Ns, fol. 137v), the division of the zodiac band into 120 parts in Lh2

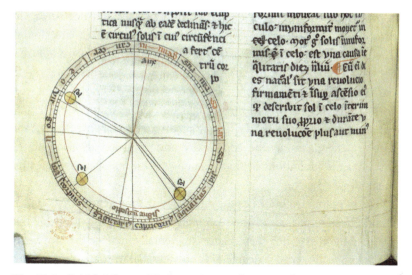

Fig. 10.8. British Library, Harley MS 3735, f. 78v © The British Library Board.

afforded the maker to include the numerical detail of the description and thus represent it with precision. Similar diagrams of the movement of the sun through the ecliptic can also be found in copies of the *Theory of the Movements of Planets* (*Theorica Motuum Planetarum*), a more advanced text on planetary motions, and their affinity to another type of astronomical instrument, the astrolabe, has been explored by Kathrin Müller.[47] While the author is unknown, it has been supposed that it was made to supplement Sacrobosco's *On the Sphere*.[48] Whether or not the

---

[47] Müller, *Visuelle Weltaneignung*, 266–71. The similar diagram in *Theorica Planetarum* illustrates a theory of the sun's motion that ends in this way: 'To find the mean motus of the Sun means to find a certain arc of the zodiac which has the same ratio to the whole zodiac as the arc of the eccentric circle described by the Sun has to the whole eccentric circle. This is found by the parallel line as seen in the following figure'. Pedersen, 'Anonymous. The Theory of the Planets', 452. The translation is based upon the text given in Kopenhagen, Det Kongelige Bibliotek MS Add. 447 2°, which includes the diagram mentioned above.

[48] Murdoch, *Album of Science*, 140. For discussion on the potential author of the *Theorica Planetarum* (which is inconclusive), see O. Pedersen, 'Origins of the "Theorica Planetarum"', *Journal for the History of Astronomy*, 12 (1981), 113–23, at 118–22; and O. Pedersen, 'Anonymous: The Theory of the Planets', in E. Grant (ed.), *A Source Book in Medieval Science* (Cambridge, MA: Harvard University Press, 1974), 451. Examples of the image include Cambridge University Library MS Ii.2.3, f. 79v (dated to 1276) and Kopenhagen, Det Kongelige Bibliothek MS Add. 447 2°, f. 49 (c.1300).

358 THE SCIENTIFIC WORKS OF ROBERT GROSSETESTE

copyist was working from diagrams from this text or vice versa, the drawing in Lh2 is distinct as it is tailored to Grosseteste's description. The Lh2 diagram is therefore exceptional as the only known example that shows an interest in the precise measurements Grosseteste offers. However, it still struggles to accurately capture his description. If it were correct, the red line would run straight into the sixth mark of Gemini, which marks exactly 18 degrees. The discrepancy between text and diagram demonstrates the complexity and precision of Grosseteste's text, and the limitations of books alone as a means of communicating these ideas.

The attempt also indicates the ambiguity with which copyists understood Grosseteste's use of the verb *imaginere*. While it is probable that many of them expected their readers to mentally imagine the movement of the sun to the nearest degree, it is possible that in some cases another visual prompt would have been made available. Looking to other texts from this period that describe the structure of the universe in a visual way, the question as to whether their authors ever intended the described picture to be physically made remains open to debate.[49] Nonetheless, what can be known is that some copyists of *On the Sphere*, such as that of British Library Harley MS 3647, tried to demonstrate Grosseteste's ideas through diagrams. This suggests that some read *imaginemur* as an instruction to create physical images, whether or not a diagram existed in the manuscript from which they copied. Such responses to the demand to *imaginemur* Grosseteste's descriptions by creating an image (whether or not the diagram was new or taken from other astronomical

---

[49] A notorious example of this ambiguity is observed in Hugh of St Victor's *De arca noe mystica* (*c*.1125–30). Hugh asks the reader to *pingo* (paint) the *pictura* of the Ark and describes certain components of it in cubits, to give the reader an idea of the proportional relationships between each of its parts. For example, 'The height of this pillar is as much less than the width of the Ark as three measures is to five, for the height of the Ark was thirty cubits and the width fifty', quoted in C. Rudolph, *The Mystic Ark: Hugh of Saint Victor, Art, and Thought in the Twelfth Century* (Cambridge: Cambridge University Press, 2014), 409. Hugh's description has been interpreted both as an instruction to physically construct the diagram and, alternatively, as a guide to mentally imagine the diagram. The debate as to Hugh's intention (summarized in Carruthers, *The Craft of Thought*, 243–6, n. 61) is ongoing and often centres around questions of the terms he uses.

ILLUMINATING *ON THE SPHERE* 359

sources) could be the reason for the variety of diagrams included in later copies of the manuscript. In turn, the creation of new images as the text was copied demonstrates the autonomy of the copyist in the dissemination of Grosseteste's ideas.

In the era before printing, once a text left the immediate orbit of its author the quality of both text and diagram was only as good as the level of skill and care offered by copyists. This had significant consequences for diagrams in scientific texts in the Middle Ages. Copies of Euclid's theorems circulated with diagrams that did not accurately illustrate them, and in the context of astronomy, images of the constellations differed dramatically in the number and location of the stars, or even omitted them entirely.[50] This suggests that the value for readers lay in the authority provided by the age of the text, and that the content was understood primarily through reading, and perhaps also that the ideas were not tested by readers through observation of the stars. With the exception of the ambitious diagram of the movement of the sun through the zodiac in Lh2, the variations introduced in the diagrams in copies of Grosseteste's *On the Sphere* were not obviously unhelpful or erroneous. Nevertheless, in the most lavishly decorated copies specialist artists did have the chance to put their own stamp on the appearance of the diagrams. For the maker of Ol (Fig. 10.4) the diagram of the world machine became an extension of the text through the integration of diagram and initial letter, but the orderliness of its nested semi-circles was also coupled with spiky penwork that extended into the margin, balancing the order of creation with the artist's creative freedom. Similarly, the maker of the diagram in Lh3 differentiated the spheres with geometric patterns, diluting the ink to add colour, whilst the artist of the initial combined the image of a scientist looking at an armillary sphere with a dragon that forms the body of the initial I (Fig. 10.1). Most dramatically, the artist of Vmu's interest in rabbits (which are not explicitly referred to in the text) and grotesque creatures, perhaps

---

[50] For Euclid, see E. A. Zaitsev, 'The Meaning of Early Medieval Geometry: From Euclid and Surveyors' Manuals to Christian Philosophy', *Isis*, 90 (1999), 522–53; Murdoch, *Album of Science*, 116. For the tradition of non-representation of stars (replaced by artistic representations of the constellations' references), see E. M. Ramírez-Weaver, *A Saving Science* (University Park, PA: Pennsylvania State University Press, 2017), 90.

360   THE SCIENTIFIC WORKS OF ROBERT GROSSETESTE

inspired by the dragons of the text, added humour as well as visual interest to his rendition of the movements of celestial bodies.

The surviving copies of Grosseteste's scientific works feature very few diagrams. *On the Sphere* is remarkable within the corpus of Grosseteste manuscripts for the frequent appearance of the diagram of nested semi-circles that may well have been part of Grosseteste's original design for the work. The value of being able to visualize the complex information given in the text probably explains the incorporation of additional diagrams into some copies of *On the Sphere*. This seems to have been part of a much wider phenomenon in which diagrams moved between texts and were sometimes collected without accompanying text. In part, this may be linked to the use of specialist artists, who were trained to copy existing imagery rather than demonstrate complex ideas. Through the different iterations the individual marks and interpretations of each copyist can be observed. Some of these are ornamental, with elaborate penwork or playful illuminations of animals, revealing the aesthetic sensibilities and humour of the artist. Others indicate the relationship between (and sometimes conflation of) Grosseteste's model and other authoritative studies of the spheres. However, manuscripts on astronomy made with the involvement of highly-skilled artists remained excep-tional. In addition, therefore, the transmission of diagrams may be linked to compilations of different texts in the course of wide reading by an individual patron, whether acting on their own behalf or designing a manuscript for an institutional library.

The evaluation of the copyist's visualization of the text raises the question of the role of the mental faculties of the reader in understand-ing, particularly in regard to the geometrical framework of the *machina mundi*. Visual prompts were significant to the activation of medieval astronomical thought, whether in the form of textual descriptions of visual phenomena, diagrams, or three-dimensional models. The elabor-ation of the diagrams in some copies of Grosseteste's work suggests an active engagement with the content both of his text and related works. Sacrobosco's work seems to have provided a source of additional dia-grams for some copies of Grosseteste's *On the Sphere*, but many of the diagrams in Sacrobosco's work were themselves adaptations of existing schemas, making it difficult to pinpoint their precise origins. In contrast, the semi-circular form of the diagram in Grosseteste's work was unusual, and may have become emblematic of his way of conceptualizing the spheres. In addition to the impact of existing manuscripts and the practicalities of book production, the surviving manuscripts provide tantalising hints about engagement with astronomical instruments, notably the armillary sphere. Grosseteste does not, however, mention the tool explicitly, and the fact that approximately half the surviving copies of *On the Sphere* include no diagrams serves as an important reminder that many readers in the later Middle Ages wrestled with the text alone. While there are indications that moveable models were sometimes used in the teaching of these concepts, it is the manuscripts themselves that consistently function as instruments of knowledge in the study and dissemination of the *On the Sphere*. Through them, the workings of the medieval imagination can be observed.

Concluding Reflections

On the Sphere is a complex treatise and one that should not be separated from the rest of the shorter scientific works as in some sense an elementary or basic textbook. Indeed, to attribute a singular purpose to the treatise is neither helpful nor is it responsive to the range of Grosseteste's interests on display. *On the Sphere* can be seen in continuity to the first of the shorter works, *On the Liberal Arts*. The high place accorded to astronomy in the latter is borne out in the former, and Grosseteste's framework for the arts can be seen to run through the later treatise. *On the Liberal Arts* sets out how the arts can correct and purge human error. The foundation of this is the relationship between *aspectus mentis*, that is, the sight of the mind, and *affectus mentis*, that is, the desire of the mind. *Aspectus* and *affectus* come as a pair; they are not divided. Both constrain and enhance each other, and if one overreaches then error and imperfection result; this applies to intellectual endeavour as much as anything else.[1] Grosseteste would later, in his *Hexaemeron*, the commentary on the six days of creation at the beginning of Genesis, point out that ancient authorities misunderstood that the universe had no beginning because of their failure to imagine properly. Their *affectus* was not sufficiently cleansed of error to allow their *aspectus* to grasp that eternity cannot be judged as simply another time or space but rather as something entirely different.[2] The same, in its own context, applies to *On the Sphere*. The geometry expounded within has to be imagined correctly for the treatise to work, that is, for the shape of the world machine and the movements and properties of its constituent parts to be fixed properly in the mind.

One of the ancient authorities that Grosseteste criticized for lacking imagination was Aristotle, and the long familiarity with his thinking so

---

[1] Grosseteste, *De artibus liberalibus*, §§1–2, 74; *Knowing and Speaking*, 97.
[2] Grosseteste, *Hexaemeron*, 1.VIII.5.

CONCLUDING REFLECTIONS 363

evident in later works is already apparent in *On the Sphere*. The treatise uses an Aristotelian methodology as a way to set out conclusions that inspire confidence. The purpose of demonstration is to be certain how things are known. Grosseteste applies the principles of demonstration throughout the treatise. This, and the extensive use of Euclid as the means to visualize the universe, indicate a shift in Grosseteste's intellectual methods from his earlier treatises. Astronomy is still presented as an art, but its demonstration is the product of a consistent engagement with Aristotle. This places into sharper relief other central aspects of the treatise, for example, familiarity with Ptolemy's astronomical system and its legacy in Islamicate thought. Grosseteste's use of al-Farghānī and *On the Motion of the Eighth Sphere* attributed to Thābit shows not only the range of reference but the critical approach taken to his sources and the distinctiveness of vision that results. The discussion of trepidation encapsulates this combination of familiarity and criticism. Grosseteste explicitly followed 'Thābit's' corrections to Ptolemy (DS §49–53) with a detailed account of Pseudo-Thābitian/Toledan Trepidation. Such a discussion is absent, for example, from Sacrobosco's similarly-named treatise. Futhermore, Grosseteste worked through the implications of trepidation in characteristically original fashion, forging a link to an argument on long term climate change and the habitability of the places on the earth.

In its range of sources and particular use of Aristotle, Euclid, and Islamicate astronomical writing, *On the Sphere* reveals the transformative effect for scholars of western Christendom of the translation into Latin of these works from the first third of the twelfth-century onwards. The scale of the change can be seen, for example, in the difference between Grosseteste's treatise and a mid-twelfth-century text by Bernardus Silvestris, the *Cosmographia*. An allegorical discussion of the universe closer to the more platonic instincts of writers like William of Conches or Thierry of Chartres, Bernardus's text is formed of two parts, the *Macrocosmos* and *Microcosmos*. At the beginning of the latter Bernardus addresses the reader:

> I would have you survey the heavens, inscribed with their manifold array of symbols, which I have set forth for learned eyes, like a book with its pages spread open, containing things to come in secret characters. I would have you regard the zones, and how, extending by fixed

364 THE SCIENTIFIC WORKS OF ROBERT GROSSETESTE

laws between the poles, they determine the climates of the underlying terrain. I would have you note the colures, and how, in their fourfold delineation they join to encircle the heavens, but never finish the extended journey which they had begun. I would have you consider the Zodiac, which a hidden plan has set atilt: hereby provision is made for the safety of the natural world, which would not endure perpetually if the Zodiac always conducted the blazing Sun in an unvarying course across the centre of the earth. I would have you gaze on the Galaxy, moderating the cold of the northern regions; for to regions lying so distant, the heat of the Sun does not bring its relief. And I would have you notice the line which corresponds to the two solstices, and likewise that which marks the prolongations of day and night at the time when they are equal.[3]

Grosseteste would have recognized the injunction to gaze upwards and learn the pathways of the celestial bodies, but his frame of reference was larger and more encompassing than anything available to Bernardus.

In a related vein the transformative interdisciplinary methodology of the Ordered Universe Project has brought a wider range of different perspectives to bear on the elucidation of Grosseteste's *On the Sphere*. As in the first volume of this series, the integration of background perspectives, methodologies, and reception from disciplines in the sciences as well as the humanities have assisted the authorial team in responding to Grosseteste on his own terms. Doing so has allowed, for example, a clarification of how stellar positional astronomy up to the thirteenth century still permitted, and even favoured, the hypothesis of trepidation over Ptolemaic precession. It would only be from the end of the

---

[3] Bernardus Silvestris, *Cosmographia*, in *Poetic Works*, ed. and trans. Winthrop Wetherbee (Cambridge, MA: Harvard University Press, 2015), 1–181 at 78–9: '"Caelum velim videas multiformi imaginum varietate descriptum, quod quasi librum, porrectis in planum paginis, eruditioribus oculis explicui, secretis futura litteris continentem. Zonas velim videas, quemadmodum intra polos certis legibus exporrectae, terras sibi subpositas afficiunt qualitate. Coluros velim videas, quemadmodum quadrifida lineatione caelum ambire convenient, sed continuationem quam coeperant non obsolvunt. Signiferum velim videas, quem ratio secretior obliquavit: rerum enim incolumitati provoisum est, quae perpetuo non duraret, si directo semper limite Solum Signifer excandentem terrae per medium reportaret. Galaxem velim videas, frigoribus hyperboreis temperantem; quia locis longe sepositis Solis calefaction remedium non ferebat. Lineam velim videas utrisque solstitiis respondentem, itemque illam quae diurni nocturnique temporis excrescentias ad momentum affigit".'

CONCLUDING REFLECTIONS 365

thirteenth century that the balance of the evidence would swing decisively the other way. The graphical traditions of contemporary physics have also enabled the construction of diagrams to accompany *On the Sphere* in continuity with the visualizing instincts of those medieval illustrators who felt it necessary to explain and mentally 'fix' the text in a similar way. The computational tools of the twenty-first century have been used to respond further to the imaginative invitations of the thirteenth. The Virtual Celestial model allows the text to be visualized in a different way entirely and for that experience to be shared. Many amongst the authors found this functionality extremely useful and all would recommend employing the model alongside a reading of the treatise.

What Grosseteste set up in *On the Sphere* remained important to him for the rest of his writing career. Throughout his life Grosseteste emphasized the influence of the celestial regions on life on earth in very concrete terms. In this sense the treatise builds the arena in which all of this later discussion will play out, from the motion of the super-celestial bodies, to the most convincing way to explain comets, or to define the horizon, which will be explored in subsequent volumes of this series. The influence of *On the Sphere* stretches beyond the scientific works. As argued in this volume Grosseteste was also highly involved in the theory and practice of pastoral care at the same time that he composed his astronomical treatise. Similarities in methodology have been noted, the desire both for natural phenomena and for pastoralia to not merely describe or proscribe but to explain the reason why, the *propter quid*. Specific examples from his scientific analyses feature regularly across his pastoral writings, not least in the *Dicta* collection. Grosseteste had other uses for Astronomy. The theme and metaphor of sphericity and its instantiation in the heavens as a reminder and illustration of the virtues to be adopted by priests and prelates occurs in several of the *Dicta*. *Dictum* 137 'God set the stars in the firmament' is an extended overlaying of the virtues and properties of the stars with the roles and responsibilities of Christian ministry. Grosseteste begins by stressing the spherical nature of stars:

And so a star is a spherical body, radiant, circularly mobile in a uniform movement, and therefore neither heavy nor light, unchangeably retaining the fixity of its ordered position. Moreover, a spherical

366 THE SCIENTIFIC WORKS OF ROBERT GROSSETESTE

object has no jutting corner by the sharpness of which it may hurt the touch, nor any sunken concavity where it may take in an accumulation of filth.[4]

In the same way priests are to be like the heavenly spheres: they should exhibit no acerbic sharpness nor contain concavities where sin may remain concealed, they should like the stars 'shine' both upwards to God and downwards to the world, and like sidereal motion they should continue uniform and steady, but circular in that they should begin and end with their Maker.

The heavens are for Grosseteste the great natural display of creation, and one that invites the contemplation of all human arts, that embodies the geometry of lines and circles that mathematics declares in perfection, but that also calls humankind to strive for, as an image of that perfection in mind and action. There is, however, in his later works, such as the *Hexaemeron* a more measured, and perhaps agnostic attitude to the practical limitations of the subject. His practical objection to astrology is well-known, that it was simply impossible to be accurate enough in observations of the stars and planets to make the detailed predictions claimed by its exponents.[5] In a broader vein considering the nature of the firmament and the heavens between it and the sphere of the moon Grosseteste was led to ponder how much could not be known and the incompatibility of the authorities who had thought on the matter:

> Since on this subject of the nature of the heavens, and of the movers of the heavens, and of the moving powers they have, so many philosophers and authorities have given so many and such uncertain opinions, what can I do except admit and bewail my own ignorance on the point?[6]

Grosseteste ends up characterizing the competing opinions of his authorities here, which include Ptolemy, as more fragile than cobwebs.[7]

---

[4] Grosseteste, *Dictum* 137: 'Cum itaque de celorum natura et motoribus celorum et de virtutibus eorum motivis tam diversimode et tam incerte sentiant philosophi et auctores tanti, quid possum ego nisi meam circa hec ignoranciam simul et dolere et fateri?'.

[5] Grosseteste, *Hexaemeron*, V.VIII–IX.

[6] Grosseteste, *Hexaemeron*, III.VII.1: 'Cum itaque de celorum natura et motoribus celorum et de virtutibus eorum motivis tam diversimode et tam incerte sentiant philosophi et auctores tanti, quid possum ego nisi meam circa hec ignoranciam simul et dolere et fateri?'. English translation from Grosseteste, *On the Six Days of Creation*, 107–8.

[7] Grosseteste, *Hexaemeron*, III.VIII.3.

The celestial realm, and its relation to the earthly, would continue to exercise Grosseteste until the end of his life. His final work was to translate Aristotle's *On the Heavens* from Greek into Latin. The problems that he saw later with respect to astronomy and cosmology are not expressed in *On the Sphere*. Here the voice is as confident as the treatise is complex and nuanced. Taking *On the Sphere* on its own terms and in the context of Grosseteste's other activities, resources, and intellectual interests allows its coherence, importance, and purpose to be better understood.

The Virtual Celestial Model of
On the Sphere

A difficulty for the modern reader, when reading *On the Sphere*, is to interpret the descriptions in the text. Whilst, on careful reflection these are invariably concise and self-consistent, a reader who is more familiar with the diagrammatic approach used in modern books, and preferred by many modern scientists, may find initial reading difficult. Moreover, whilst the medieval reader would be familiar with the layout of the cosmos, the motions of the moon and sun, and quite probably the descriptions of the Zodiac, the majority of modern readers are likely to experience light pollution to a level that prevents them from seeing all but the brightest celestial objects. For this reason, an online visualization of *On the Sphere* has been created which it is hoped will enable readers to better understand the motions described. The visualization is available here: <https://ordered-universe.com/de-sphera-visualisation/>.

This visualization is tied closely to the text; in this section, the reader will find an overview of the paragraphs covered, which are followed by a series of suggested observations that the reader may wish to make in order to familiarize themselves with the model. One consequence of the visualization is to bring to the fore the level of sophisticated thought that, by the early thirteenth century, had gone into understanding the motion of the sun and moon as observed from the earth's surface. At this point, it should be noted that the *On the Sphere*, as all medieval astronomy, is finally about the position of the stars and planets on the sky, not how they move in three dimensions.

1. Overview

This section uses the headings as seen on the website. Each part of the visualization gives an interactive display which highlights key points in the corresponding paragraphs of the *On the Sphere*, here those points are elucidated. This section, and the visualization itself, can be consulted helpfully alongside the diagrams embedded in the translation of the text.

Part 1: A Geometric Construction

Here, the constructions of *On the Sphere* §§2–16 are outlined. The first paragraphs (DS §§2–6) which are concerned with the formation of nested shells, starting from a semi-circle, are not dealt with explicitly; instead, the visualization starts by displaying the earth and the celestial sphere on which all the motions of all celestial bodies are projected. The circles described in the text, and also animated in the visualization, can be viewed as diagrams (Ch. 4). Within this scheme, the pole (DS §7) is represented using a line projecting from the North and South Poles of the earth. The positions of the *colori* (DS §11) and the equinoctial (DS §12) are shown. The zodiac (DS §13), appears as a band 12 degrees thick, angled at 23 degrees and 33 minutes relative to the poles (or equinoctial). Arranged around this band are the twelve signs of the zodiac (DS §14). The ecliptic, which describes the annual motion of the sun relative the firmament, passes along the zodiac. The tropics (DS §15) are also shown; here they are seen projected onto the celestial sphere, where they represent the most northerly and southerly points touched by the ecliptic, rather than their more usual modern depiction on the surface of the earth

Part 2: The Sun

The basic apparent motions of the sun are presented, and can be visualized as the days, months, and years progress. In one day, the sun

370 THE SCIENTIFIC WORKS OF ROBERT GROSSETESTE

and the firmament make a revolution around the pole, the sun also moves, although imperceptibly, by about 1 degree, relative to the firmament, advancing along the ecliptic. Over the course of a month, the sun's motion relative to the zodiac is more pronounced, by an entire zodiacal constellation, and as the year progresses the circle describing the sun's apparent daily revolution moves from its occupation of the winter tropic (Capricorn) at the northern hemisphere's winter solstice, past the equinoctial at the equinox and to the summer tropic (Cancer) for the summer solstice. It then reverses its annual motion in celestial latitude back to the tropic of Capricorn by the following winter; in consequence, therefore, the sun's apparent motion on the celestial sphere therefore describes a spiral (DS §§17–18). In addition to seeing these motions externally, the model provides an opportunity to observe virtually their effects from different terrestrial viewpoints (see 'Part 3: Seasons'). Another point to note here is the difference between the circular projection of the sun's orbit on the celestial sphere, and the orbit itself, which does not have to be a circle, and indeed, cannot be a circle centred on earth because summer and winter are not of even lengths. (DS §§39–41).

Part 3: Seasons

The model allows the viewer to stand virtually at any latitude on earth and observe the motions of the sun and cosmos. It is possible to stand on the equator, looking north or south, and see the poles on the horizon. Since the height of a person is infinitesimal compared to the radius of the earth, the horizon passes through the poles (DS §§19–20). The sun can be seen at its zenith twice a year for all the points between the tropics and once a year on the tropics themselves (DS §§23–5) and the lengths of the days in summer and winter at locations between the tropics and poles may be observed (DS §26). It is also possible to observe that when the sun is at zenith on the equator, day and night are equal everywhere (DS §§21–2). Switching between viewpoints on earth and the external viewpoint allows these observations to be understood from the point view of the observer, and in terms of the sun's position on the sky.

Part 4: The Zodiac

Here the model demonstrates the appearance and the movement of the zodiac (DS §32); of particular interest is the observation that although the constellations are evenly spaced around the zodiac, they take different lengths of time to rise and set, because the angle that the zodiac makes with the horizon changes. The closer this angle is to perpendicular, the faster the signs rise/set, owing to them having a shorter distance to travel in the direction perpendicular to the horizon. This effect is explained in *On the Sphere*, where diagrams can be found in the translation (DS §§33–7).

Part 5: Eccentricity

This section of the model corresponds to the subject of paragraphs §§38–40. In addition to the sun's daily motion with the firmament, its annual motion relative to the firmament can be observed, around a circle aligned with the zodiac. However, whilst the *daily* motion of the celestial system of *On the Sphere* is centred on the earth, the spatial orbit describing the *annual* motion is offset: the centre of the circle is co-located with the earth, but is offset towards the constellation of Gemini. This construction, introduced by Ptolemy, was necessary if, to an observer on earth, the angular motion of the sun relative to the firmament is to be non-uniform, while the sun progresses at a constant speed around its offset circle (eccentric).

Part 6: Trepidation

Here, the model allows the visualization of the description of trepidation of the zodiac, given in §§50–3. The model shows the fixed and mobile zodiacs, allowing visualization of the described motion of the mobile zodiac. In particular, the heads of Libra and Aries on the mobile zodiac trace circles, shown in the model, which causes the motion of the mobile Zodiac relative to the fixed. The model allows visualization of this complex portion of the treatise. Trepidation is discussed at greater length in Chapter 8.

372 THE SCIENTIFIC WORKS OF ROBERT GROSSETESTE

Parts 7 and 8: The Moon and Eclipses

The motion of the moon is described in §§54–5. The moon moves around a small circle (epicycle), the centre of which orbits a larger circle (deferent). In turn, the centre of this larger circle (equant) orbits the earth. The model allows visualization of the motion, and also of the overall motions of the moon and the sun with, and relative to, the zodiac. The plane of the moon's orbit relative to the orbit of the sun is also shown, together with the relative angle of its inclination. By showing the moon and the sun together, the model allows visualization of eclipses, which can only occur when the sun, earth, and moon align. However, because the relative sizes of the bodies are exaggerated, for the purposes of easy visualization, relative to the dimensions of the orbits, more angular alignments than in reality will lead to an eclipse.

2. Instructions for Using the Virtual Celestial Model

The description of the cosmos in *On the Sphere* can easily be visualized using the online tool associated with this book: https://ordered-universe. com/de-sphera-visualisation/. Navigate to this address in your preferred web browser to get started. The visualization is best navigated with a keyboard and mouse on a PC, although it also supports mobile browsers.

To begin, click on '**Part 1—A Geometric Construction**' in the menu on the left. To better view the diagram presented, it is possible to open and close the left-hand menu, by clicking the cross in the top left of the screen. Clicking 'color key' in the top right additionally toggles on and off the key and settings menu. The initial viewpoint is outside the firmament. The geometrical constructions on the celestial sphere described in the first part of the treatise can be visualized with reference to the key in the top right. The position of the zodiac can also be seen. Please note that the star constellation map in the background is not accurate and just represents the location of the firmament.

It is possible to view the model from different angles by holding down the left mouse button and dragging the mouse in the direction you want to view. The mouse wheel allows the user to zoom in and out. It is possible to reset to the original viewpoint by clicking 'reset camera' at the top right of the screen.

THE VIRTUAL CELESTIAL MODEL OF *ON THE SPHERE* 373

Clicking on 'Part 2—the Sun' in the left menu shows the basic daily and annual motions of the sun. The hour and day sliders in the right-hand menu can be used to adjust the hour of the day and day of the year. The motion over the course of a day can be viewed by clicking 'start day' in the right-hand menu, showing a single rotation of the firmament along with the sun. A helpful way to use this model is to step by day or month whilst keeping the same time of day: the sun stays in the same place, but the firmament moves.

Over the course of a year, the motions of the sun and firmament in the sky are visualized. It is easiest to observe this by clicking 'step year' in the menu at the right of the screen. This steps between the same time on each day over the course of a year, allowing the yearly motions to be viewed independently of the daily. The firmament, with the ecliptic attached, can be seen making a single rotation. The sun increases and decreases in declination as it follows the ecliptic's path between the tropics.

It is perhaps easiest to appreciate the descriptions in *On the Sphere* by observing from a terrestrial point of view. Clicking 'Part 3—Seasons' allows us to view the same celestial model from a point on earth, within it. To look around in this perspective, hold the left mouse button down with the mouse over the direction you want the camera to spin: further from the centre of the screen leads to a faster rotation. The latitude slider in the right-hand menu changes the location on earth, whilst clicking 'jump to Oxford' changes the latitude to that of Oxford, UK, a viewpoint that Grosseteste would have had for some of his life. It is possible to switch back and forth between the terrestrial and extra-terrestrial views with the 'viewpoint button'. It is interesting to view the motions of the sun again from this viewpoint by opening the 'motion' menu on the right. It is observed that, at latitudes of greater than +/- 66 degrees, within the Arctic or Antarctic circles, there are periods in the year of twenty-four-hour visibility of the sun, and of twenty-four-hour darkness.[1]

---

[1] Here, there is an inaccuracy in the model. In order to improve the user experience, the viewpoint is higher (that is, further from the surface of the earth), than the height of a real person. As a result, when standing on the Arctic or Antarctic circle, the length of the summer/winter day is not quite correct, in particular the sun does not graze the horizon on midsummer/midwinter, as it would for real human observer at the pole.

374 THE SCIENTIFIC WORKS OF ROBERT GROSSETESTE

Click 'Part 4—the Zodiac' to show the locations of the signs of the Zodiac and observe their movements in the same way as above, using the 'day' or 'hour' slider.

'Part 5—Eccentricity' displays the offset centre of the sun's presumed orbit of the earth within the model. Clicking the 'step year' button in the right-hand 'motion' menu displays the change in diameter of the sun's daily path of rotation around the earth; this path is displayed in brown and is shown at the same time of day over the course of a year. Adjusting the viewpoint by clicking 'viewpoint' in the right-hand 'view' menu shows how subtle this this effect appears when viewed from the earth. It is possible to see that the sun is closer to the earth during the southern hemisphere's summers than it is during northern hemisphere summers.

The trepidation of the equinoxes is visualized in 'Part 6—Trepidation'. Click 'Start Trepidation' to cycle through the 25,800-year trepidation cycle. The 'mobile', visible zodiac is shown as a semi-transparent band, with the 'fixed' zodiac behind. The mobile zodiac can be seen rotating with the firmament about their axis at the centre of the two turquoise circles shown.

'Part 7—The Moon' and 'Part 8—Eclipses' both display the same diagram, which is concerned with the motions of the moon. The epicycles and their axes are shown. As with the sun and zodiac in the previous diagrams, the moon's relative size and distance from the earth are exaggerated in order for the motions to be reasonably visualized: to ensure that they are not imperceptibly small on a computer monitor. The result of this is that eclipses occur far more regularly, and the shadows are of greater size, than in real life. For example, each year a solar eclipse can be observed in the model at zero hours and 264 days. Similarly, a lunar eclipse is shown at zero hours and forty-eight days.

The motion during one month can be cycled through to show the positioning of the moon and its epicycles, by clicking 'step month'. Again, to clarify the presentation of the movement of the moon, without the daily rotation of the cosmos, the same time each day is visualized. A smoother animation showing a whole month's motions can still be shown by clicking 'Start month' instead of 'step month'. Alternatively, the day and hour sliders can be manipulated manually in the right-hand menu.

APPENDIX 1

Commissio Cancellarii Universitatis Oxon'—Concerning the Commission to the Chancellor of the University of Oxford

As shown earlier in this volume (Ch. 1, §3.2) considerable store is placed within historiographical discussion of Grosseteste's career and, in particular his presumed association with Oxford, on the record of his title as master of the scholars (*magister scholarum*) rather the chancellor in the university, as mentioned by one of his later successors as bishop of Lincoln, Oliver Sutton, in 1294. Sutton's remarks on Grosseteste's occur in the context of the appointment of Master Roger of Weasenham as chancellor, following the resignation of Roger of Martival. Master Peter of Medbourne represented the university in reporting the election of Roger of Weasenham, prompting a rebuke from the bishop and his advisors, that the university should only nominate, not elect, their chancellors, since the gift of appointment lay with the bishop. It is in this discussion that mention of Grosseteste occurs. As Joseph Goering points out the context of 1294 is not that of the early part of the thirteenth century.[1] Sutton recalled the case of Grosseteste as an illustration of how episcopal authority had been exercised; the anecdote says nothing, per se, as to why the then bishop of Lincoln, Hugh of Wells, had offered and insisted upon a different title. Nevertheless, given the importance of this statement to the claims made for Grosseteste's 'Chancellorship' it seemed justifiable to include the whole document in this Appendix to place the remark in its fuller context.

Latin Text, taken from *The Rolls and Register of Bishop Oliver Sutton 1280–1299*: Vol. V, *Memoranda May 19, 1294 – May 18, 1296*, ed. Rosalind M. T. Hill, Lincoln Record Society, 60 (Printed for the Lincoln Record Society in Hereford: Hereford Times Ltd, 1965), 59–61; fols. 117–117v of the original.

Commissio Cancellarii Universitatis Oxon'

Memorandum quod tertio Idus Februarii anno domini M. CC. nonagessimo quarto apud Nettelham venit Magister Petrus de Medburn' juris canonici professor, exhibens episcopo quamdam litteram clausam magistrorum universitatis Oxon' sub hac forma:-

---

[1] Joseph W. Goering, 'Where and When did Grosseteste Study Theology?', in James McEvoy (ed.), *Robert Grosseteste: New Perspectives on his Thought and Scholarship* (Turnhout: Brepols, 1995), 49.

376 APPENDIX 1

Reverendo in Christo patri domino O. dei gracia Lincoln' episcopo, universitatis Oxon' cetus humilis magistrorum salute et obedienciam debitam ac devotam. Affectionis paterne dilectionem habere vos concedet circa querentes in agro studii sciencie margaritam que domum dei multipliciter convenustat in his que precipue non possunt absque denegatione gracie ac juris injuria denegari. Hinc est quod cum Magister Rogerus de Wesenham Archidiaconus Roff', sacre theologie professor, a nostra universitatis ad officium cancellari per cessionem venerabilis viri Magister Rogeri de Martivall' Archidiaconi Huntingd' vacantis concorditer sit electus, a vester paternitatis benevolencia ejus electionem more solito petimus confirmari. Talem ergo si placet vos exhibeatis in hac parte ut filiorum vestrorum sinceritas augeatur et paterne dilectionis caritas comprobetur. Valeat paternitas vestra per tempora longiora.

Qua quidem littera coram episcops perlecta, idem Magister Petrus petiit prout in littera per eum exhibita continetur electionem predictam more solito confirmari. Et cum queretur ab eo an hoc petendi potestatem haberet, exhibuit quoddam procuratorium sigillo communi universitatis predicte signatum sub continentia infrascripta:-

Inspecturis et audituris presentes litteras Christi fidelibus universis, cetus unanimis magistrorum in universitate Oxon' regentium salutem in omnium salvatore. Noverit universitas vestra quod cum per resignationem viri. Magistri Rogeri de Martivall' Archidiaconi Huntingdon' sacre theologie professoris nuper vacasset officium cancellarii prefate universitatis nostre, et nos ad electionem cancellari secundum consuetudinem hactenus usitatam procedentes Magistrum Rogerum de Wesenham Archidiaconum Roff' sacre theologie doctorem in nostrum cancellarium elegerimus, ad dictam electionem presentandam venerabili patri domino O. dei gracia Lincoln' episcopo, et ad petendam eam more solito confirmari, dilectum confratrem nostrum Magistrum Petrum de Medeburn' juris canonici professorem procuratorem nostrum constituimus per presentes. In cuius constitutionis testimonium sigillum nostre communitatis presentibus est appensum. Datum Oxon' die dominica proxime post festum purificationis beate virginis, anno domini M. CC. nonagesimo quarto.

Deinde dictum fuit eidem procuratori quod cancellarii pro tempore existentes non fuerunt electi sed tantummodo nominati. Et episcopus adjecit quod beatus Robertus quondam Lincoln' episcopus, qui hujusmodi officium gessit dum in universitate predicta regebat in principio creationis sue in episcopum dixit proximum predecessorum suum episcopum Lincoln' non permisisse quod idem Robertus vocaretur cancellarius, sed magister scholarum. At ipse Magister Petrus ad hoc non respondit, set petiit hujusmodi negotium more solito expediri. Cumque quereretur ab eo quare non venit personaliter nominatus sicut tenebatur, et sicut Magister Johannes de Monemuta nuper ejusdem cancellarius faciebat, respondit fatendo nominatum teneri personaliter venire ad episcopum pro officio cancellarii admittendo, set locorum distanciam, expensarum penuriam et temporis qualitatem pro causa non compareationis ipsius nominate allegans secum in hac parte agi petiit graciose. Episcopus vero causas allegatas attendens post deliberationem cum suis habitam aliqualem,

APPENDIX 1   377

commissionem sibi fieri precepit, qui statim facta extitit, et dicto Magistro Petro sub magno sigillo episcopi tradita sub haec forma:-

> Oliverus etc. magistris et scholaribus universitatis Oxon' salute, etc. Ad instanciam vestre devotionis officium cancellarii universitatis vestre Magistro Rogero de Wesenham, Archidiacono Roff', in theologia inter vos actualiter nunc regent, ad presens de gracia speciali committimus per presents donec aliud vobis super hoc dederimus in mandatis. Valete. Datum apud Nettelham III idus Februarii, anno domini M. CC. nonagesimo quarto.

English Translation
Concerning the Commission to the Chancellor of the University of Oxford

A memorandum that on the third of the Ides of February in the year of the lord 1294, Master Peter of Medburn, professor of canon law, came to Nettleham, displaying to the bishop a certain close letter of the masters of the university at Oxford in the following form:

> To the reverend in Christ, Father and Lord O[liver] by the grace of God bishop of Lincoln, the humble congregation of masters of the university of Oxford give salutation and their due and devout obedience. May you concede the possession of the love of your fatherly affection to those who seek, in the field of the study of knowledge, the pearl which makes the house of God beautiful in manifold ways, in those things in particular which cannot be denied without the denying of grace and an injury to justice. And so it is that, since Master Roger of Weasenham, Archdeacon of Rochester, professor of sacred theology has been unanimously elected by our university to the office of chancellor left vacant by the retirement of the venerable Master Roger of Martival, Archdeacon of Huntingdon, we request that his election be confirmed in the accustomed manner by your paternal benevolence. It if please you, may you therefore bestow this [benevolence] in this case so that the sincerity of your sons may be increased, and the charity of paternal love may be demonstrated. May your paternity be well for as long as possible.

This letter having been read in the presence of the bishop, the same Master Peter requested that, in accordance with the contents of the letter he had displayed, the aforementioned election be confirmed in the customary manner. And when it was inquired of him as to whether he had the power to make this request, he displayed a document from the procurators, bearing the seal of the community of the aforementioned university, containing what is written below:

> To all those faithful to Christ who are to see and hear this letter, the unanimous congregation of regent masters at the university of Oxford send their greetings in the saviour of all. Let it be known to everyone that, since the office of chancellor of our aforementioned university recently became vacant through the resignation of Master Roger of Martival, Archdeacon of Huntingdon and

378 APPENDIX 1

professor of sacred theology, and [since], proceeding to the election of a chancellor following the custom we have used until now, we have elected Master Roger of Weasenham, archdeacon of Rochester, doctor of sacred theology, as our chancellor, we the present have appointed our beloved fellow Master Peter of Medburn, professor of canon law and our procurator, to present this aforesaid election to the venerable father Lord O[liver], by the grace of God bishop of Lincoln, and to request that it be confirmed in the accustomed manner. As testimony to his appointment the seal of our community is attached the present letter. Given at Oxford on the first Sunday after the feast of the Purification of the Blessed Virgin, in the year of the Lord 1294.

After this the same procurator was told that chancellors in post at this time were not elected but only nominated. And the bishop added that the blessed Robert, formerly bishop of Lincoln, who had held such an office while he was teaching at the aforementioned university, said at the start of his episcopacy that his immediate predecessor as bishop of Lincoln had not allowed the same Robert to be called chancellor, but Master of the Scholars [*Magister Scholarum*]. But Master Peter did not respond to this but requested that the current matter be expedited in the accustomed manner. And when it was asked of him why the nominee had not come in person as he was required to do, as Master John of Monemuta, the recent chancellor of the same [university], had done, he responded by acknowledging that the nominee was required to come before the bishop in person in order to be admitted to the office of chancellor, but, citing the remoteness of the places, the monetary expense, and the nature of the times, as explanation for the failure on the part of the nominee to comply, and graciously requested that he [Peter] be permitted to act in his stead. Having deliberated with his advisors a little, the bishop, attending to the explanations cited, ordered that the commission should be made by himself, which was done at once, and handed over to the aforesaid Master Peter under the great seal of the bishop in the following form:

Oliver etc. greeting etc to the masters and students of Oxford. At the urgings of your devotion, we hereby commit by the present letter, by special grace, the office of chancellor of your university to Master Roger of Weasenham archdeacon of Rochester, now acting as regent of theology among you, until such time as we may give you a different mandate on this matter. Farewell. Given at Nettleham on the third of the Ides of February in the year of the Lord 1294.

APPENDIX 2

Manuscripts of *On the Sphere* with Diagrams

Note: it has not been possible to consult every known manuscript of *On the Sphere* and therefore this list makes no claim to completeness.

Shelfmark	Siglum	Date	Place of origin	Semicircular World Machine diagram	Other diagrams with *On the Sphere*
Bologna, Bib. Uni. Lat. 1845, ff. 101–6.	Bu	XIV	France	Yes	
Cambridge, University Library MS Ff.6.13, ff. 11–17.	Cu1	XIII		No	hybrid diagram of spheres informed by Sacrobosco, f. 18.
Cambridge, University Library MS Gg.6.3, ff. 200v–5v.	Cu2	XIV		Yes	No
Cambridge, University Library MS Ii.1.13, ff. 36v–39v.	Cu3	XIV		No	text immediately followed by circular diagram of the spheres with zodiac, f. 39v.
Cambridge, University Library MS Mm.3.11, ff. 144–50.	Cu6	XV		Yes	No

Continued

380 APPENDIX 2

Dublin, Trinity College MS 441, ff. 69–74v.	Dt	XIV	England	Yes	partial sketch of diagram to illustrate parallax in the margin of f. 74.
Erfurt, Universitäts- und Forschungsbibli- othek, Bibliotheca Amploniana CA 4° 351, ff. 46–50.	Ea2	c.1300	England	Yes	additional diagrams on ff. 50r–v including: lunar eclipse, the solar eccentric, solar eclipse, and climes.
Erfurt, Universitäts- und Forschungsbibli- othek, Bibliotheca Amploniana CA 4° 355, ff. 1–5	Ea3	c.1300	England	Yes	partial sketch of diagram in the margin of f. 5.
London, British Library, Add. MS 27589, ff. 69–76	La1	XIII		Yes	No
London, British Library, Egerton MS 847, ff. 59–62.	Le2	XV	England	Yes	No
London, British Library, Harley MS 3735, ff. 74–82v.	Lh2	1264– 93	France	Yes	movement of the sun through the zodiac (f. 78v), climes (f. 79v), trepidation (f. 80), lunar eclipse (f. 80v, 81v), combined diagram of movement of heavenly bodies (f. 82v).

Continued

APPENDIX 2   381

London, British Library, Harley MS 4350, ff. 4–15.	Lh3	1250–1300	France (?)	Yes	No
Milan, Bib. Naz. Braidense AD XII 53, ff. 18–28.	Mb	XIV–XV		No	
New Haven, Yale University, Medical Library MS 11, ff. 101v–106.	Ny	XIII	Germany	Yes	No
New York, Columbia University, Smith Western Add. MS 1, ff. 131v–8.	Ns	XIV	Flanders	Yes	diagram of climes, f. 135v.
Oxford, Bodleian Library MS Bodley 676, ff. 229–41.	Ob2	XIV		Yes	No
Oxford, Bodleian Library MS Digby 98, ff. 158–61.	Od3	*c.*1400		Yes	No
Oxford, Bodleian Library MS Gough Linc. 13, ff. 1–11.	Og	*c.*1760–70		Yes	No
Oxford, Bodleian Library MS Laud Misc. 644, ff. 143–7.	Ol	*c.*1273		Yes	No
Paris, Bibliothèque nationale de France MS Lat. 7195, ff. 67v–74.	Pb1	*c.*1300	England	Yes	No

Continued

382 APPENDIX 2

Paris, Bibliothèque nationale de France MS Lat. 7292, ff. 276v–80.	Pb3	*c*.1400		Yes	zodiac ring (f. 279), diagram labelled *thebit de motu octave sphere* (f. 279v).
Paris, Bibliothèque nationale de France MS Lat. 7298, ff. 31–6.	Pb4	XIV	France	Yes	No
Princeton, University Library, Garrett MS 95, ff. 111–20.	Pg	XV	England	Yes	
Salamanca, Biblioteca Universitaria 111, ff. 1–8.	Su1	*c*.1300		Yes	
Utrecht, Bibliotheek der Rijksuniversiteit 722, ff. 90–4.	Ur	XIV	Germany	Yes	zodiac ring (f. 93v); unfinished diagram (f. 94).
Vatican City, Vatican Library, Pal. Lat 1414, ff. 34–41.	Vp	*c*.1300	Netherlands	Yes	two diagrams immediately after text on f. 41r: lunar eclipse and parallax.
Verdun, Bibliothèque municipale MS 25, ff. 31–46.	Vmu	XIII	England	Yes	eccentric orbit of the sun (f. 40v); movements of sun and moon (f. 44v); lunar eclipse and phases of the moon (f. 45); parallax (f. 46v, after the end of the text).

Bibliography

Primary Sources: Ancient and Medieval

Abraham Ibn Ezra, *El libro de los fundamentos de las Tablas astronómicas de R. Abraham Ibn 'Ezra*, ed. José María Millás Vallicrosa (Madrid: CSIC, 1947).

Abū Ma'shar, *On Historical Astrology (On the Great Conjunctions)*, ed. and trans. Keiji Yamamoto and Charles Burnett, 2 vols. (Leiden, Brill, 2000).

Adam Marsh, *The Letters of Adam Marsh*, ed. and trans. Clifford H. Lawrence, 2 vols. (Oxford: Oxford University Press, 2006).

Alan of Lille, *Anticlaudianus*, ed. Robert Bossuat (Paris: Vrin, 1955).

Alan of Lille, *De planctu Naturae*, ed. Nicholas Häring, *Studi Medievali*, ser. 3, 19 (1978), 797–879.

Alexander Neckham, *Commentum super Martianum: Commentary on Martianus Capella's 'De Nuptiis Philogiae et Mercurii'*, ed. Christopher J. McDonough (Firenze: Sismel Edizioni del Gelluzzo, 2006).

Alexander Neckham, *Sacerdos ad altare*, ed. Christopher J. McDonough, CCCM 227 (Turnhout: Brepols, 2010), 196–7.

Alfraganus, *Compilatio Astronomica*, trans. John of Seville (Ferrara: Andrea Gallus, 1493).

Alfraganus, *Il 'libro dell'aggregazione delle stelle'*, trans. Gerard of Cremona, ed. Romeo Campani (Città di Castello: S. Lapi, 1910).

Almagesti minor, ed. Henry Zepeda, (Turnhout, Brepols, 2018).

Alpetragius (al-Biṭrūjī), *De motibus caelorum*, ed. Francis J. Carmody (Berkeley: University of California Press, 1952).

Annales de Dunstaplia, ed. Henry Richards Luard, *Annales monastici* (London: Longman, Green, Longman, Roberts and Green, 1864), iii. 3–408.

Anonymous, *The Berlin Commentary on Martianus Capella's De nuptiis Philologiae et Mercurii*, ed. Haijo Jan Westra (Leiden: Brill, 1994).

Aristotle, *Opera*, ed. August Immanuel Bekker, *Aristotelis Opera edidit Academia Regia Borussica, ex Recognitione Immanuelis Bekkeri*, 5 vols. (Berlin: Georg Reimer, 1831–70). English translations: *The Complete Works, The Complete Works of Aristotle, The Revised Oxford Translation*, Jonathan Barnes (ed.), 2 vols. (Princeton: Princeton University Press, 1995); *Posterior Analytics*, trans. and comm. Jonathan Barnes, 2nd ed. (Oxford: Oxford University Press, 1993).

Aristotle, *Posterior Analytics*, trans. James of Venice, in *Aristoteles Latinus* IV.1–4, ed. Lorenzo Minio-Paluello and Bernard G. Dod (Turnhout: Brepols, 1968).

Aristotle, *De caelo*, trans. Gerard of Cremona, in Albert the Great, *De caelo et mundo*, ed. Paul Hossfeld, *Alberti Magni Opera Omnia* (Münster i. W.: Aschendorff, 1971).

384 BIBLIOGRAPHY

Aristotle, *Physica*, ed. Fernard Bossier and J. Brams, *Aristoteles Latinus* VII.1, fasc. 2 (Turnhout: Brepols, 1990).

Artefius, *Clavis sapientiae*, in J.-J. Manget, Bibliotheca Chemica Curiosa, 2 vols. (Geneva: Sumpt. Chouet, G. De Tournes, Cramer, Perachon, Ritter, & S. De Tournes, 1702), 2. 2. 2, i. 503–9.

Artis cuiuslibet consummatio, in Stephen K. Victor (ed.), *Practical Geometry in the High Middle Ages* (Philadelphia, PA: American Philosophical Society 134, 1979), 282–5.

al Battānī, *De motu stellarum*, trans. Plato of Tivoli, in *Continentur in hoc libro. Rudimenta astronomica Alfragani. Item Albategnius astronomus peritissimus de motu stellarum* (Nuremberg: Petreius, 1537), fols. 88r–89r.

al-Battani, *Opus Astronomicum*, ed. and trans. C. A. Nallino *Al-Battani sive Albatenii Opus Astronomicum*, 3 vols. (Milan: Ulrico Hoepli, 1899–1907).

Bede, *De natura rerum*, ed. Charles W. Jones, CCSL 123A (Turnhout: Brepols, 1975).

Bede, *De temporum ratione*, ed. Charles W. Jones, CCSL 123B (Turnhout: Brepols, 1977).

Bernardus Silvestris, *Cosmographia*, ed. Peter Dronke (Leiden: Brill, 1978).

Bernardus Silvestris, *Cosmographia*, in *Poetic Works*, ed. and trans. Winthrop Wetherbee (Cambridge, MA: Harvard University Press, 2015), 1–181.

Boethius, *De institutione arithmetica*, ed. Henri Oosthout and Johann Schilling, CCSL, 94A (Turnhout: Brepols, 1999).

Calcidius, *On Plato's Timaeus*, ed. and trans. John Magee (Cambridge, MA: Harvard University Press, 2016).

The Cartulary of Haughmond Abbey, ed. Una Rees (Cardiff: University of Wales Press, 1985).

The Cartulary of Worcester Cathedral Priory (Register I), ed. Reginald Ralph Darlington, Pipe Roll Society Publications, 76 = ns 38 (1968).

Cassiodorus, *Institutiones*, ed. Roger A. B. Mynors (Oxford: Oxford University Press, 1937). English translation: *Institutions of Divine and Secular Learning and On the Soul*, trans. James W. Halporn (Liverpool: Liverpool University Press, 2004).

Chartularium Universitatis Parisiensis, Vol. i, ed. Heinrich Denifle and Émile Chatelain (Paris: Delalain, 1889).

Concilium Lateranense III a. 1179, in *Conciliorum Oecumenicorum Decreta*, ed. Josepho Alberigo, J. A. Dossetti, Periclīs-Petros Ioannou, Claudio Leonardi, and Paolo Prodi, 3rd ed. (Bologna: Istituto per le Scienze Religiose, 1973), 211–25. English translation: *Decrees of the Ecumenical Councils*, 2 vols., i: *Nicaea I to Lateran V*, ed. Norman P. Tanner (London: Sheed and Ward, 1990).

Concilium Lateranense IV a. 1215, in *Conciliorum Oecumenicorum Decreta*, eds. Josepho Alberigo, J. A. Dossetti, Periclīs-Petros Ioannou, Claudio Leonardi, and Paolo Prodi, 3rd ed. (Bologna: Istituto per le Scienze Religiose, 1973), 230–7. English translation: *Decrees of the Ecumenical Councils*, 2 vols., i: *Nicaea I to Lateran V*, ed. Norman P. Tanner (London: Sheed and Ward, 1990).

Cunestabulus, *Compotus*, ed. Alfred Lohr, *Opera de computo saeculi duodecimi* (Turnhout: Brepols, 2015).

Curia Regis Rolls, 4 and 5 Henry III (London: HMSO, 1952).

Daniel of Morley, '*Philosophia/Liber de naturis inferiorum et superiorum*', *Mittellateinisches Jahrbuch*, 14 (1979), 204–55.

BIBLIOGRAPHY 385

English Episcopal Acta VII: Hereford 1079–1234, ed. Julia Barrow (Oxford: Oxford University Press for the British Academy, 1993).

al-Farghani, *Differentie*, trans. John of Seville, ed. Francis J. Carmody (Berkeley: University of California Press, 1943).

Fasti Ecclesiae Anglicanae 1066–1300: Vol. 3, *Lincoln*, ed. Diana E. Greenway (London, 1977). *British History Online*: http://www.british-history.ac.uk/fasti-ecclesiae/1066-1300/vol3/pp30–32 (accessed 28 April 2020).

Fasti Ecclesiae Anglicanae 1066–1300: Vol. 4, *Salisbury*, ed. Diana E. Greenway (London, 1991).

Fasti Ecclesiae Anglicanae 1066–1300: Vol. 8, *Hereford*, ed. Julia Barrow (London, 2002).

Fasti Ecclesiae Anglicanae, compiled by John le Neve, corrected and continued by Thomas Duffus Hardy, 3 vols. (Oxford: Oxford University Press, 1854).

The Friars' Libraries, ed. Kenneth W. Humphreys (London: The British Library in association with The British Academy, 1990).

[Geber filius Affla Hispalensis] Jābir ibn Ḥayyān', *De astronomia* (Nüremberg: Iohannes Petreius, 1534).

Gerald of Wales, *Giraldi Cambrensis opera*, ed. John S. Brewer, 8 vols. (London: Longman, 1861–91).

Gerald of Wales, *Gemma ecclesiastica*, ed. Brewer in *Giraldi Cambrensis opera*, ii; English translation *The Jewel of the Church*, trans. John J. Hagen (Brill: Leiden, 1979).

Gerald of Wales, *Symbolum electorum, pars prima*, ed. Brewer, in *Giraldi Cambrensis opera*, i. 197–395.

Gerald of Wales, *Speculum duorum or A Mirror of Two Men*, ed. Yves Lefèvre and R. B. C. Huygens, trans. Brian Dawson, general ed. Michael Richter (Cardiff: University of Wales Press, 1974).

[Gerard of Cremona], *Theorica planetarum*, ed. Francis J. Carmody, *Theorica planetarum Gerardi* (Berkeley: University of California Press, 1942).

Gervase of Canterbury, *Mappa mundi*, in *The Historical Works of Gervase of Canterbury*, ed. William Stubbs, 2 vols. (London: Longman and Trübner, 1880).

Giles of Lessines, *De essentia, motu et significtione cometarum*, ed. Lynn Thorndike, *Latin Treatises on Comets Between 1238 and 1368 A.D.* (Chicago: The University of Chicago Press, 1950), 103–84.

Gregory Nazianzen, *Oration 2 'In Defence of his Flight to Pontus'*, 27 in Rufinus, *Orationum Gregorii Nazianzeni novem interpretatio*, ed. Augustus Engelbrecht, CSEL 46 (Vienna: Österreichische Akademie der Wissenschaften, 1910).

Historiæ Dunolmensis scriptores tres, Gaufridus de Coldingham, Robertus de Graystanes, et Willielmus de Chambre, ed. James Raine, Surtees Society 9 (London and Edinburgh: J. B. Nichols & Son, 1839).

Hugh of St Victor, *De sacramentis*, PL 176. English translation: Roy J. Deferrari, *Hugh of St Victor: On the Sacraments of the Christian Faith* (Cambridge, MA: Harvard University Press, 1951).

Ibn al-Haytham, *On the Configuration of the World*, ed. and trans. Y. Tzvi Langermann (London: Routledge, 2016).

386 BIBLIOGRAPHY

Ibn al-Muṯannā, *Commentary on the Astronomical Tables of al-Jwārizmī*, ed. Eduardo Millás Vendrell, *El comentario de Ibn al-Muṯannā a la Tablas Astronómicas de al-Jwārizmī. Estudio y edición crítica del texto latino, en la versión de Hugo Sanctallensis* (Madrid: CSIC, 1963), 131–2.

Ibn Sīnā, *Risāla fī ibṭāl aḥkām al-nujūm* (Avicenna's epistle on falsifying astrology) in *Opuscules d'Avicenne*, ed. Hilmi Zia Ülken (Istanbul: Ibrahim Horoz Basimevi, 1953), 49–67.

Isidore of Seville, *Etymologiarum siue Originum libri XX*, ed. Wallace M. Lindsay (Oxford: Oxford University Press, 1911).

Isidore of Seville, *De natura rerum*, ed. and trans. Jacques Fontaine (Paris: Institut d'études augustiniennes, 2002; repr. of Bordeaux 1960 edn).

Jābir ibn Ḥayyān' [Geber filius Affla Hispalensis], *De astronomia* (Nüremberg: Iohannes Petreius, 1534).

John of Sacrobosco, *De sphera*, ed. and trans. Lynn Thorndike, *The Sphere of Sacrobosco and Its Commentators* (Chicago: University of Chicago Press, 1949), 76–143.

Lanercost Chronicle, 1201–1346, ed. Joseph Stevenson (Edinburgh: Bannatyne Club, 1839).

Lucan, *De bello civili libri X. Editio altera.*, ed. David R. Shackleton-Bailey (Leipzig: Teubner, 1997).

Lucretius, *De rerum natura*, ed. Marcus Deufert (Berlin: De Gruyter, 2019).

Macrobius, *Commentarii in Somnium Scipionis*, ed. James Willis (Leipzig: Teubner, 1970).

Martianus Capella, *De nuptiis Philologiae et Mercurii*, ed. James Willis (Leipzig: Teubner, 1983).

Matthew Paris, *Chronica majora*, ed. Henry Richards Luard, 7 vols. (London: Longman, Green, Longman and Roberts, 1872–83).

Nasīr al-Dīn al-Ṭūsī', *Nasīr al-Dīn al-Ṭūsī's Memoir on Astronomy (al-Tadhkira fī 'ilm al-hay'a)*, ed. and trans. Jamil Ragep (Dordrecht: Springer Verlag, 1993).

Nicholas Trevet, F. *Nicholai Triveti, Annales ex regum Angliae, 1135–1307*, ed. Thomas Hog (London: English Historical Society, 1845).

Petrus Alfonsi, *Dialogus*, ed. and (German) trans. Peter Stotz (Florence: SISMEL, 2018). English translation: *Dialogue Against the Jews*, trans. Irven Resnick (Washington, DC: Catholic University of America Press, 2006).

Pliny, *Naturalis historia*, ed. Ludwig von Jan and Karl Friedrich Theodor Mayhoff, 5 vols. (Leipzig: Teubner, 1897–1933; repr. 1967–70).

Pseudo-Bede, *De mundi celestis terrestrisque constitutione. A Treatise on the Universe and the Soul*, ed. and trans. Charles Burnett (London: Warburg Institute, 1985).

Pseudo-Bede, *De mundi caelestis terrestrisque constitutione liber. La création du monde céleste et terrestre*, ed. Mylène Pradel-Baquerre, Cécile Biasi, and Amand Gévaudan (Paris: Classiques Garnier, 2016).

Pseudo-Thābit, *De motu octave sphere*, ed. Francis J. Carmody, *The Astronomical Works of Thabit b. Qurra* (Berkeley, Los Angeles: University of California Press, 1960), 102–7.

BIBLIOGRAPHY 387

Ptolemy, *Almagest*, ed. Johan L. Heiberg, *Claudii Ptolemaei opera quae exstant omnia*: Vol. I, *Syntaxis mathematica*, 2 vols. (Leipzig: Teubner, 1898–1903). Latin translation: Ptolemy, *Almagestum seu Magnae Constructionis*, trans. Gerard of Cremona (Venice: Peter Lichtenstein, 1515). English translation: *Ptolemy's Almagest*, trans. Gerald J. Toomer (London: Duckworth, 1984).

Ptolemy, *Tetrabiblos*, ed. and trans. Frank E. Robbins (Cambridge, MA: Harvard University Press, 1940).

Rabanus Maurus, *De universo libri viginti duo*, PL, 111 (Paris: Garnieri Fratres, 1864), cols. 9–614.

Raymond of Marseilles, *Tractatus astrolabii*, ed. with French translation in Marie-Thérèse d'Alverny, Charles Burnett, and Emmanuel Poulle, *Raymond de Marseille. Opera omnia* (Paris: CNRS Éditions, 2009).

Richard of Bardney, *Vita Roberti Grosthed*, ed. Henry Wharton, *Anglia Sacra*, 2 vols. (London: Richard Chiswel, 1691), ii. 325–41.

Robert Grosseteste, *Commentarius in VIII Libros Physicorum Aristotelis*, ed. Richard C. Dales (Boulder: University of Colorado Press, 1963).

Robert Grosseteste, *Commentarius in Posteriorum Analyticorum libros*, ed. Pietro Rossi (Florence: Leo S. Olschki, 1981).

Robert Grosseteste, *Compotus*, ed. and trans. C. Philipp E. Nothaft and Alfred Lohr (Oxford: Oxford University Press, 2019).

Robert Grosseteste, *De artibus liberalibus*, ed. and trans. Sigbjørn O. Sønnesyn, in Giles E. M. Gasper et al., *Knowing and Speaking: Robert Grosseteste's De artibus liberalibus* ('On the Liberal Arts') and *De generatione sonorum* ('On the Generation of Sounds') (Oxford: Oxford University Press, 2019), 74–95.

Robert Grosseteste, *De generatione sonorum*, ed. and trans. Sigbjørn O. Sønnesyn, in Giles E. M. Gasper et al., *Knowing and Speaking: Robert Grosseteste's De artibus liberalibus* ('On the Liberal Arts') and *De generatione sonorum* ('On the Generation of Sounds') (Oxford: Oxford University Press, 2019), 243–55.

Robert Grosseteste, *De intelligentiis*, ed. Ludiwg Baur, *Die philosophischen Werke des Robert Grosseteste, Bischofs von Lincoln* (Münster i. W.: Aschendorff, 1912), 112–19.

Robert Grosseteste, *De luce*, ed. Cecilia Panti, 'Robert Grosseteste's De luce: A Critical Edition', in *Robert Grosseteste and his Intellectual Milieu*, ed. John Flood, James R. Ginther, and Joseph W. Goering (Toronto: Pontifical Institute of Mediaeval Studies, 2013), 193–238.

Robert Grosseteste, *De modo confitendi*, ed. Joseph W., Goering, and Frank A. C. Mantello, 'The Early Penitential Writings of Robert Grosseteste', *Recherches de théologie ancienne et médiévale*, 54 (1987), 52–111.

Robert Grosseteste, *Dicta*, 'Grosseteste's Dicta: A Working Transcription,' ed. Joseph W. Goering and Edwin J. Westermann [download available at https://ordered-universe.com/dicta/].

Robert Grosseteste as Bishop of Lincoln, The Episcopal Rolls, 1235–1253, ed. Philippa M. Hoskin (Woodbridge: Boydell for The Lincoln Record Society, 2015).

Robert Grosseteste, *Epistolae*, ed. Henry Richards Luard (London: Longman, Green, Longman and Roberts, 1861). English translation: *The Letters of Robert*

388 BIBLIOGRAPHY

Grosseteste, Bishop of Lincoln, trans. Frank A. C. Mantello and Joseph W. Goering (Toronto: University of Toronto Press, 2010).

Robert Grosseteste, *Hexaëmeron*, ed. Richard C. Dales and Servus Gieben (Oxford: Oxford University Press, 1982). English translation: *On the Six Days of Creation*, trans. C. F. J. Martin (Oxford: Oxford University Press, 1996).

Robert of Flamborough, *Liber Poenitentialis*, ed. J. J. Francis Firth (Toronto: Pontifical Institute of Mediaeval Studies, 1971).

Roger Bacon, *Compendium studii philosophiae*, in *Roger Bacon, Opus tertium, Opus minus, Compendium studii philosophiae*,..., ed. J. S. Brewer, *Opera hactenus inedita Rogeri Baconi*, i (London: Longman, Green, Longman, and Roberts, 1859).

Roger Bacon, *Opus tertium*, ed. John S. Brewer, *Opera hactenus inedita Rogeri Baconi*, i (London: Longman, Green Longman and Roberts, 1859).

The Rolls and Register of Bishop Oliver Sutton 1280-1299: Vol. V, *Memoranda May 19, 1294 - May 18, 1296*, ed. Rosalind M. T. Hill, Lincoln Record Society, 60 (Printed for the Lincoln Record Society in Hereford: Hereford Times Ltd, 1965).

Seneca, *Naturalium questionum libri*, ed. Harry M. Hine (Leipzig: Teubner, 1996).

Snappe's Formulary extracts from a formulary attributed to John Snappe and other Records relating to Oxford University, ed. Herbert E. Salter, Oxford Historical Society, 1st ser. 80 (Oxford: Oxford University Press, 1923).

Statuta antiqua universitatis oxoniensis, ed. Strickland Gibson (Oxford: Oxford University Press, 1931).

Thābit [Thebit] b. Qurra, *The Astronomical Works of Thābit b. Qurra*, ed. Francis J. Carmody (Berkeley: University of California Press, 1960).

Thābit [Thebit] ibn Qurra, *Oeuvres d'astronomie*, ed. Régis Morelon (Paris: Les Belles Lettres, 1987).

Theon of Alexandria, *Le 'Petit commentaire' de Théon d'Alexandrie aux Tables faciles de Ptolémée*, ed. Anne Tihon (Vatican City: Biblioteca Apostolica Vaticana, 1978).

Thietmar of Merseburg, *Chronicon*, ed. Robert Holtzmann, MGH SS rer. Germ. N.S. 9 (Berlin: Weidmann, 1985).

The Toledan Tables, ed. Fritz.S. Pedersen, Historisk-filosofiske Skrifter 24, 4 vols. (Copenhagen: Det Kongelige Danske Videnskabernes Selskab, 2002).

Tycho Brahe, *Astronomiae instauratae progymnasmata*, pt. 2, ed. John. L. E. Dreyer, *Tychonis Brahe Dani Scripta Astronomica*, 15 vols. (Copenhagen: Glydendal, 1913-29).

Walcher of Malvern, *De lunationibus and De Dracone: Study, Edition, Translation, and Commentary* ed. and trans. C. Philipp E. Nothaft (Turnhout: Brepols, 2017).

William of Conches, *Dragmaticon philosophiae*, ed. Italo Ronca (Turnhoult: Brepols, 1997). English translation: *A Dialogue on Natural Philosophy*, trans. Italo Ronca and Matthew Curr (Notre Dame: University of Notre Dame Press, 1997).

William of Conches, *Glossae super Platonem*, ed. Édouard A. Jeauneau, CCCM, 203 (Turnhout: Brepols, 2006).

BIBLIOGRAPHY 389

Secondary Literature

Abdukhalimov, Bahrom, 'Ahmad al-Farghānī and his *Compendium of Astronomy*', *Journal of Islamic Studies*, 10 (1999), 142–58.

Ambler, S. T., *Bishops in the Political Community of England, 1213–1272* (Oxford: Oxford University Press, 2017), 64–5.

Andrée, Alexander, 'Editing the Glossa "ordinaria" on the Gospel of John: A Structural Approach', in Elisabet Göransson et al., eds. *The Arts of Editing Medieval Greek and Latin: A Casebook* (Toronto: Pontifical Institute of Mediaeval Studies, 2016), 1–20.

Angold, M., Baugh, J. G. C., Chibnall, Marjorie M., Cox, D. C. D., Price, T. W., Tomlinson, Margaret, and Trinder, B. S., 'Houses of Augustinian Canons: Priory of Chirbury', in A T Gaydon and R B Pugh (eds.), *A History of the County of Shropshire*, Vol. 2 (London: Victoria County History, 1973), 59–62. *British History Online*: http://www.british-history.ac.uk/vch/salop/vol2/pp59-62 (accessed 6 July 2020).

Arnaldi, Mario, 'Time Reckoning in the Latin World', in Anthony Turner (ed.), *A General History of Horology* (Oxford: Oxford University Press, 2022), 99–120.

Austin, Greta, 'Jurisprudence in the Service of Pastoral Care: The "Decretum" of Burchard of Worms', *Speculum*, 79 (2004), 929–59.

Aylmer, Gerald, Barrow, Julia, Caird, R., Lepine, D., and Tomlinson, H., 'Office Holders at Hereford Cathedral since 1300', in Gerald Aylmer and John Tiller (eds.), *Hereford Cathedral: A History* (London: Hambledon Press, 2000), appendix 2, 637–43.

Baldwin, John W., *Masters, Princes and Merchants: The Social Views of Peter the Chanter and His Circle*, 2 vols. (Princeton: Princeton University Press, 1970).

Barrow, Julia, 'The Canons and Citizens of Hereford c.1160–c.1240', *Midland History*, 24 (1999), 1–23.

Barrow, Julia, 'Athelstan to Aigueblanche, 1056–1268', in Gerald Aylmer and John Tiller (eds.), *Hereford Cathedral: A History* (London: Hambledon Press, 2000), 29–47.

Barrow, Julia, 'Foliot, Hugh (d. 1234), bishop of Hereford', *Oxford Dictionary of National Biography* (Oxford: Oxford University Press, 2004); online ed. 4 October 2007: https://doi.org/10.1093/ref:odnb/95044 (accessed 16 July 2021).

Barrow, Julia, *The Clergy in the Medieval World: Secular Clerics, their Families and Careers in North-Western Europe, c.800–c.1200* (Cambridge: Cambridge University Press, 2015).

Bartlett, Robert, *Gerald of Wales A Voice of the Middle Ages* (Oxford: Oxford University Press, 1982; repr. Stroud: Tempus, 2006).

Bartlett, Robert, 'Gerald of Wales (c.1146–1220x23)', *Oxford Dictionary of National Biography* (Oxford University Press, 2004); online edn., 28 Sept 2006: https://doi.org/10.1093/ref:odnb/10769 (accessed 26 April 2021).

Baur, Ludwig, *Die Philosophischen Werke des Robert Grosseteste, Bischofs von Lincoln*, (Münster: Aschendorff, 1912).

Beaujouan, Guy, 'The Transformation of the Quadrivium', in Robert L. Benson and Giles Constable (eds.), *Renaissance and Renewal in the Twelfth Century* (Cambridge, MA: Harvard University Press, 1982), 467–83.

390 BIBLIOGRAPHY

Bennett, Jim, 'Knowing and Doing in the Sixteenth Century: What Were Instruments For?', *The British Journal for the History of Science*, 36 (2003), 129–50.

Berggren, J. Len, 'Al-Bīrūnī on Plane Maps of the Sphere', *Journal for the History of Arabic Science*, 6 (1982), 47–112.

Bertola, Francesco, 'Tubi astronomici', in Filippomaria Pontani (ed.), *Certissima Signa: A Venice Conference on Greek and Latin Astronomical Texts* (Venice: Edizioni Ca' Foscari, 2017), 145–51.

Biller, Peter and Minnis, Alastair J. (ed.), *Handling Sin: Confession in the Middle Ages*, York Studies in Medieval Theology, 2 (Woodbridge: Boydell & Bewer, 1999).

Bisson, Thomas N., *The Medieval Crown of Aragon* (Oxford: Oxford University Press, 1986).

Bloch, David, 'James of Venice and the Posterior Analytics', *Cahiers de l'Institut du Moyen-Âge Grec Et Latin*, 78 (2008), 37–50.

Bloch, David, 'Robert Grosseteste's Conclusiones and the Commentary on the Posterior Analytics', *Vivarium*, 47 (2009), 1–23.

Bloch, David, 'Monstrosities and Twitterings: A Note on the Early Reception of the Posterior Analytics', *Cahiers de l'Institut du Moyen-Âge Grec Et Latin*, 79 (2010), 1–6.

Blume, Dieter et al., *Sternbilder des Mittelalters. Der gemalte Himmel zwischen Wissenschaft und Phantasie. Band I. 800–1200* (Berlin: Akademie Verlag, 2012), 403–9.

Borrelli, Arianna, *Aspects of the Astrolabe: 'Architectonica Ratio' in Tenth- and Eleventh-Century Europe*, Sudhoffs Archiv: Beihefte, 57 (Stuttgart: Steiner, 2008).

Boyle, Leonard E., 'Robert Grosseteste and the Pastoral Care', *Medieval and Renaissance Studies*, 8 (1979), 3–51.

Boyle, Leonard E., *Pastoral Care, Clerical Education, and Canon Law, 1200–1400*, ed. Peter Biller and Alastair Minnis (London: Variorum Reprints, 1981).

Brand, Paul A., *The Making of the Common Law* (London: Hambledon Press, 1992).

Bronstein, David, *Aristotle on Knowledge and Learning* (Oxford: Oxford University Press, 2016).

Brundage, James A., 'The Teaching and Study of Canon Law in the Law Schools', in Wilfried Hartmann and Kenneth Pennington (eds.), *The History of Medieval Canon Law in the Classical Period, 1140–1234, From Gratian to the Decretals of Pope Gregory IX* (Washington, DC: The Catholic University of America Press, 2008), 98–120.

Burnett, Charles S. F., 'Give Him the White Cow: Notes and Note-Taking in Universities in the Twelfth and Thirteenth Centuries', *History of Universities*, 14 (1995–6), 1–30.

Burnett, Charles S. F., *The Introduction of Arabic Learning into England* (London: The British Library, 1997).

Burnett, Charles S. F., 'King Ptolemy and Alchandreus the Philosopher: The Earliest Texts on the Astrolabe and Arabic Astrology at Fleury, Micy and Chartres', *Annals of Science*, 55 (1998), 329–68.

Burnett, Charles S. F., *Numerals and arithmetic in the Middle Ages* (Farnham: Ashgate, 2010).

BIBLIOGRAPHY 391

Burnett, Charles S. F., 'Petrus Alfonsi and Adelard of Bath Revisited', in Carmen Cardelle de Hartmann and Philip Roelli (eds.), *Petrus Alfonsi and his Dialogus: Background, Context, Reception* (Florence: SISMEL, 2014), 77–91.

Burnett, Charles S. F., 'Translation and Transmission of Greek and Islamic Science to Latin Christendom', in David Lindberg, Michael Shank, Charles S. F. Burnett, *The Cambridge History of Science*, Vol. 2 (Cambridge: Cambridge University Press, 2015) 341–64.

Burnett, Charles S. F., 'The Palaeography of Numerals', in F. T. Coulson and R. G. Babcock (eds.), *The Oxford Handbook of Latin Palaeography* (Oxford: Oxford University Press, 2020), 24–36.

Burnett, Charles S. F. and David Juste, 'A New Catalogue of Medieval Translations into Latin of Texts on Astronomy and Astrology', in Faith Wallis and Robert Wisnovsky (eds.), *Medieval Textual Cultures: Agents of Transmission, Translation and Transformation* (Berlin: de Gruyter, 2016), 63–76.

Burnyeat, Myles, 'Aristotle on Understanding Knowledge', in Enrico Berti, ed., *Aristotle on Science: 'The Posterior Analytics'* (Proceedings of the Eighth Symposium Aristotelicum) (Padua: Editrice Antenoire, 1981), 97–139, repr. in Myles Burnyeat, *Explorations in Ancient and Modern Philosophy*, Vol. 2 (Cambridge: Cambridge University Press, 2012), 115–44.

Callus, Daniel A., 'The Oxford Career of Robert Grosseteste', *Oxoniensia*, 10 (1945), 42–72.

Callus, Daniel A., 'Robert Grosseteste as Scholar', in Daniel A. Callus (ed.), *Robert Grosseteste, Scholar and Bishop* (Oxford: Oxford University Press, 1955), 1–69.

Carpenter, David, *The Minority of Henry III* (London: Methuen, 1990).

Carpenter, David, *Magna Carta* (London: Penguin, 2015).

Carpenter, David, *Henry III 1207–1258* (Newhaven: Yale University Press, 2020).

Carruthers, Mary, *The Craft of Thought: Meditation, Rhetoric and the Making of Images 400–1200* (Cambridge: Cambridge University Press, 1998), 77–81.

Carruthers, Mary, 'Moving Images in the Mind's Eye', in Jeffrey Hamburger (ed.), *The Mind's Eye: Art and Theological Argument in the Middle Ages* (Princeton: Princeton University Press, 2006), 287–305.

Casulleras, Josep, 'Banū Mūsā', in Thomas Hockey et al. (eds.), *The Biographical Encyclopaedia of Astronomers* (New York: Springer, 2007), 92–4.

Catlos, Brian, *Kingdoms of Faith: A New History of Islamic Spain* (London: Hurst & Company, 2018).

Catto, Jeremy, 'Franciscan Learning in England, 1450–1540', in James Clark (ed.), *The Religious Orders in Pre-Reformation England* (Woodbridge: Boydell, 2002), 97–104.

Chabás, José and Goldstein, Bernard R., 'Andalusian Astronomy: *al-Zīj al-Muqtabis* of Ibn al Kammâd', *Archivo for History of Exact Sciences*, 48 (1994), 1–41

Chabás, José and Goldstein, Bernard R., *A Survey of European Astronomical Tables in the Late Middle Ages* (Leiden: Brill, 2012).

Chabás, José 'Aspects of Arabic Influence on Astronomical Tables in Medieval Europe', *Suhayl*, 13 (2014), 23–40.

Chabás, José and Goldstein, Bernard R., 'Ibn al-Kammād's *Muqtabis* zij and the Astronomical Tradition of Indian Origin in the Iberian Peninsula', *Archive for History of Exact Sciences*, 69 (2015), 577–650.

392 BIBLIOGRAPHY

Cheney, Christopher R. and Cheney, Mary G., *The Letters of Pope Innocent III (1198–1216) Concerning England and Wales* (Oxford: Oxford University Press, 1967).

Cheney, Christopher R., *Pope Innocent III and England* (Stuttgart: Anton Hiersemann, 1976).

Cheney, Mary G., 'Master Geoffrey de Lucy, an Early Chancellor of the University of Oxford', *The English Historical Review*, 82 (1967), 750–63.

Church, Stephen, *King John: England, Magna Carta and the Making of a Tyrant* (London: Macmillan, 2015).

Clark, James G., 'Trevet, Nicholas (b. 1257x65, d. in or after 1334)', *Oxford Dictionary of National Biography* (Oxford: Oxford University Press, 2004); online ed. 23 Sept 2004: https://doi.org/10.1093/ref:odnb/27744 (accessed 23 February 2021).

Cleaver, Laura, *Education in Twelfth-Century Art and Architecture: Images of Learning in Europe, c. 1100–1220* (Woodbridge: Boydell, 2016).

Cobban, Alan B., *The Medieval English Universities: Oxford and Cambridge to c.1500* (Berkeley: The University of California Press, 1988).

Comes, Mercè, 'The Accession and Recession Theory in Al-Andalus and the North of Africa', in Josep Casulleras and Julio Samsó (eds.), *From Baghdad to Barcelona*, 2 vols. (Barcelon: Instituto 'Millás Vallicrosa" de Historia de la Ciencia Árabe, 1996), i. 349–64.

Comes, Mercè, 'Ibn al-Hā'im's Trepidation Model', *Suhayl*, 2 (2001), 291–408.

Comes, Mercè, 'Some New Maghribī Sources Dealing with Trepidation', in S. M. Razaullah Ansari (ed.), *Science and Technology in the Islamic World* (Turnhout: Brepols, 2002), 121–41.

Coxe, Henry O., *Laudian Manuscripts, Quarto Catalogues II*, repr. from the ed. of 1858–85, with corrections and additions, and an historical introduction by Richard W. Hunt (Oxford: Bodleian Library, 1973).

Crowther, Kathleen, McCray, Ashley Nicole, McNeill, Leila, Rodgers, Amy, and Stein, Blair. 'The Book Everybody Read: Vernacular Translations of Sacrobosco's *Sphere* in the Sixteenth Century', *Journal of the History of Astronomy*, 46 (2015), 4–28.

Crombie, Alastair C., *Robert Grosseteste and the Origins of Experimental Science 1100–1700* (Oxford: Oxford University Press, 1953).

Crook, David, *Records of the General Eyre* (London: HMSO, 1982).

Dales, Richard C., 'Robert Grosseteste's Scientific Works', *Isis*, 52 (1961), 381–402.

Dales, Richard C., 'The De-Animation of the Heavens in the Middle Ages', *Journal of the History of Ideas*, 41 (1980), 531–50.

Dekker, Elly, 'A Close Look at Two Astrolabes and Their Star Tables', in Menso Folkerts and Richard Lorch (eds.), *Sic Itur ad Astra: Studien zur Geschichte der Mathematik und Naturwissenschaften; Festschrift für den Arabisten Paul Kunitzsch zum 70. Geburtstag* (Wiesbaden: Harrassowitz, 2000), 177–215.

Díaz-Fajardo, Montse, *La teoría de la trepidación en un astrónomo marroquí del siglo XV. Estudio y edición crítica del Kitāb al-adwār fī tasyīr al-anwār (parte primera) de Abū 'Abd Allāh al-Baqqār* (Barcelona: Instituto 'Millás Vallicrosa' de Historia de la Ciencia Árabe, 2001).

BIBLIOGRAPHY 393

Dicks, D. R., 'Ancient Astronomical Instruments', *Journal of the British Astronomical Association*, 64 (1953–4), 77–85.

Dobrzycki, Jerzy, 'The Theory of Precession in Medieval Astronomy' [originally published 1965], in Jerzy Dobrzycki, *Selected Papers on Medieval and Renaissance Astronomy*, ed. Jarosław Włodarczyk and Richard L. Kremer (Warsaw: Instytut Historii Nauki PAN, 2010), 15–60.

Donovan, Claire, *The de Brailes Hours: Shaping the Book of Hours in Thirteenth-Century Oxford* (London: British Library, 1991).

Dugdale, William, *Monasticon Anglicanum*, ed. John Caley, Henry Ellis, and Bulkeley Bandinel, 6 vols. in 8 parts (London: T. G. March, 1846).

Duhem, Pierre, *Le système du monde. Histoire des doctrines cosmologiques de Platon à Copernic*, 10 vols. (Paris: Hermann, 1913–59).

Duke, Dennis W., 'Ptolemy's Instruments', in Alan C. Bowen and Francesca Rochberg (eds.), *Hellenistic Astronomy: The Science in Its Contexts* (Leiden: Brill, 2020), 246–58.

Dunn, Richard, 'Glossary', in Richard Dunn, Silke Ackermann, and Giorgio Strano (eds.), *Heaven and Earth United: Instruments in Astrological Contexts* (Leiden: Brill, 2018), 263–76.

Eastwood, Bruce, Review of 'Robert Grosseteste: The Growth of an English Mind in Medieval Europe by R. W. Southern', *Speculum*, 63 (1988), 233–7.

Eastwood, Bruce, *Ordering the Heavens: Roman Astronomy and Cosmology in the Carolingian Renaissance* (Leiden: Brill, 2007).

Ebbesen, Sten, 'The Posterior Analytics 1100–1400 in East and West', in Joël Biard (ed.), *Raison et démonstration. Les commentaires médiévaux sur les Seconds Analytiques* (Turnhout: Brepols, 2015).

El-Bizri, Nader, 'In Defence of the Sovereignty of Philosophy: al-Baghdādī's Critique of Ibn al-Haytham's Geometrisation of Place', *Arabic Sciences and Philosophy*, 17 (2007), 57–80.

Evan, James, *The History and Practice of Ancient Astronomy* (New York: Oxford University Press, 1998).

Evans, James, 'The Material Culture of Greek Astronomy', *Journal for the History of Astronomy*, 30 (1999), 237–307.

Faraday, Michael A., *Ludlow, 1085–1660: A Social, Economic, and Political History* (Chichester: Phillimore, 1991).

Ferruolo, Stephen C., *The Origins of the University: The Schools of Paris and their Critics, 1100–1215* (Stanford: Stanford University Press, 1985).

Fine, Gail, *The Possibility of Inquiry: Meno's Paradox from Socrates to Sexus* (Oxford: Oxford University Press, 2014).

Fitzpatrick, Richard, *A Modern Almagest* (Austin: University of Texas, c.2005): http://farside.ph.utexas.edu/Books/Syntaxis/Almagest.pdf (accessed 24 February 2021).

Flanagan, M. T., 'Lacy, Walter de (d. 1241), Magnate', *Oxford Dictionary of National Biography* (Oxford: Oxford University Press, 2004); online ed. 23 September 2004: https://doi.org/10.1093/ref:odnb/15864 (accessed 8 March 2020).

394 BIBLIOGRAPHY

Forcada, Miquel, 'Ibn Bājja's *Discourse on Cosmology (Kalām fī al-hay'a)* and the *"Revolt"* Against Ptolemy', *Zeitschrift für Geschichte der Arabisch-Islamischen Wissenschaft*, 20–21 (2012–14), 64–167.

Forcada, Miguel, '*Saphaeae* and *Hay'āt*: The Debate between Instrumentalism and Realism in Al-Andalus', *Medieval Encounters*, 23 (2017), 263–86.

Fourier, Jean-Baptiste Joseph, *Mémoire sur la propagation de la chaleur dans les corps solides* (Paris: Bernard, 1808), repr. in *Œuvres complètes de Ch. Fourier* (Paris: La Société pour la propagation et pour la réalisation de la théorie de Fourier, 1841, Tome 2).

Franklin, James, 'Diagrammatic Reasoning and Modelling in the Imagination: The Secret Weapons of the Scientific Revolution', in Guy Freeland and Anthony Corones (eds.), *1543 and All That: Image and Word, Change and Continuity in the Proto-Scientific Revolution* (London: Kluwer Academic, 2000), 53–115.

Friedman, Lee M., *Robert Grosseteste and the Jews* (Cambridge, MA: Harvard University Press, 1934).

Fuenzalida, José Antonio Valdivia, 'La contingence et la science. À propos de la réception des Seconds Analytiques au XIIIe siècle', *Scripta Mediaevalia*, 11 (2018), 43–79.

Gallivotti, Giovanni, 'Quasi Periodic Motions from Hipparchus to Kolmogorov', *Atti della Accademia Nazionale dei Lincei. Classe di Scienze Fisiche, Matematiche e Naturali. Rendiconti Lincei. Matematica e Applicazioni*, 12 (2001), 125–52.

Gasper, Giles E. M., 'Robert Grosseteste at Durham', *Mediaeval Studies*, 76 (2014), 297–303.

Gasper, Giles E. M., 'How to Teach the Franciscans: Robert Grosseteste and the Oxford Community of Franciscans c.1229–35', in Lydia Schumacher (ed.), *The Early English Franciscans* (Berlin: De Gruyter, 2021), 57–75.

Gasper, Giles E. M. et al., *Knowing and Speaking: Robert Grosseteste's De artibus liberalibus* ('On the Liberal Arts') and *De generatione sonorum* ('On the Generation of Sounds') (Oxford: Oxford University Press, 2019).

Gasper, Giles E. M., McLeish, Tom, and Smithson, Hannah E., 'Listening between the Lines: Medieval and Modern Science' *Palgrave Communications*, 2 (2016), 16062.

Gasper, Giles E. M. and Tanner, Brian K., '"The Moon Quivered Like a Snake': A Medieval Chronicler, Lunar Explosions, and a Puzzle for Modern Interpretation', *Endeavour*, 44 (2020), 100750: doi:10.1016/j.endeavour.2021.100750.

Gasper, Giles E. M., Tanner, Brian K., Sønnesyn, Sigbjørn O., and El-Bizri, Nader, 'Travelling Optics: Robert Grosseteste and the Optics behind the Rainbow', in Christian Etheridge and Michele Campopiano (eds.), *Medieval Science in the North: Travelling Wisdom 1000–1500* (Turnhout: Brepols, 2021), 25–60.

Gautier Dalché, Patrick, 'Guillaume de Conches, le modèle macrobien de la sphère et les antipodes. Antécédents et influence immédiate', in Barbara Obrist and Irène Caiazzo (eds.), *Guillaume de Conches. Philosophie et science au XIIe siècle* (Florence: SISMEL, 2011), 219–51.

Gautier Dalché, Patrick, 'Géographie Arabe et Géographie Latine au XIIe siècle', *Medieval Encounters*, 19 (2013), 408–33.

BIBLIOGRAPHY 395

Gautier Dalché, Patrick, 'La renouvellement de la perception et de la représentation de l'espace au XIIe siècle', in García de Cortázar and José Angel (eds.), *Renovación intelectual del Occidente Europeo (siglo XII)* (Pamplona: Gobierno de Navarra, 1998), 169–217, repr. in Patrick Gautier Dalché, *L'espace géographique au Moyen Âge* (Florence: SISMEL, 2013).

Gautier Dalché, Patrick, 'Le "tuyau" de Gerbert, ou la légende savante de l'astronomie. Origines, thèmes, échos contemporains (avec un appendice critique)', *Micrologus*, 21 (2013), 243–76.

Gautier Dalché, Patrick, 'Un débat scientifique au Moyen Âge. L'habitation de la zone torride (jusqu'au XIIIe siècle)', *Topoi* Supplément, 15 (2017), 145–71.

Gibson, Margaret, 'St Victor, Andrew of (c.1110–1175), Biblical Scholar and Abbot of Wigmore', *Oxford Dictionary of National Biography* (Oxford: Oxford University Press, 2004); online ed. 23 September 2004: https://doi.org/10.1093/ref:odnb/37116 (accessed 22 March 2020).

Gieben, Servus, 'Anecdota Lincolniensia. La preghiera mattutina del vescovo; La debolezza umana della sorrella Ivetta; L'eretica che non voleva bruciare', in P. Maranesi (ed.), *Negotium Fidei. Miscellanea di studi offerti a Mariano D'Alatri in occasionne del suo 80o compleanno* (Rome: Bravetta, 2002), 127–44.

Gingerich, Owen, 'Sacrobosco Illustrated', in Lodi Nauta and Arjo Vanderjagt (eds.), *Between Demonstration and Imagination: Essays in the History of Science and Philosophy Presented to John D. North* (Leiden: Brill, 1999), 211–24.

Ginther, James R., *Master of the Sacred Page: A Study of the Theology of Robert Grosseteste ca. 1229/30–1235* (Aldershot: Ashgate, 2004).

Giraud, Cédric and Mews, Constant, 'John of Salisbury and the Schools of the 12th Century', in Christophe Grellard and Frédérique Lachaud (eds.), *A Companion to John of Salisbury* (Leiden: Brill, 2014), 31–62.

Goering, Joseph W., *William de Montibus (c.1140–1213): The Schools and the Literature of Pastoral Care* (Toronto: Pontifical Institute of Mediaeval Studies, 1992).

Goering, Joseph W., 'Where and When did Grosseteste Study Theology?', in James McEvoy (ed.), *Robert Grosseteste: New Perspectives on his Thought and Scholarship* (Turnhout: Brepols, 1995), 17–51.

Goering, Joseph W., 'Robert Grosseteste and the Jews of Leicester', in Maura O'Carroll (ed.), *Robert Grosseteste and the Beginnings of a British Theological Tradition* (Rome: Istituto Storico dei Cappuccini, 2003), 181–200.

Goering, Joseph W., 'The Internal Forum and the Literature of Penance and Confession', *Traditio*, 59 (2004), 175–227, repr. and updated in W. Hartmann and K. Pennington (eds.), *The History of Courts and Procedure in Medieval Canon Law* (Washington, DC: The Catholic University of America Press, 2016), 379–428.

Goering, Joseph W., 'The Scholastic Turn (1100–1500): Penitential Theology and Law in the Schools', in Abigail Firey (ed.), *A New History of Penance* (Leiden: Brill, 2014), 219–38.

Goering, Joseph W. and Mantello, Frank A. C., 'The *Meditaciones* of Robert Grosseteste', *Journal of Theological Studies*, N.S. 36 (1985), 118–28.

Goering, Joseph W. and Mantello, Frank A. C., 'The Inter-conciliar Period 1179–1215 and the Beginnings of Pastoral Manuals', in Filippo Liotta (ed.),

396 BIBLIOGRAPHY

Miscellanea Rolando Bandinelli Papa Alessandro III (Siena: Accademia Senese degli Intronati, 1986), 45–5.

Goering, Joseph W. and Mantello, Frank A. C., 'The *Perambulauit Iudas…* (*Speculum confessionis*) Attributed to Robert Grosseteste', *Revue Bénédictine*, 96 (1986), 125–68.

Goering, Joseph W. and Mantello, Frank A. C., 'The Early Penitential Writings of Robert Grosseteste', *Recherches de théologie ancienne et médiévale*, 54 (1987), 52 111.

Goering, Joseph W. and Mantello, Frank A. C '*Notus in Iudea Deus*: Robert Grosseteste's Confessional Formula in Lambeth Palace MS 499', *Viator*, 18 (1987), 253–73.

Goldstein, Bernard R., 'On the Theory of Trepidation According to Thābit b. Qurra and al-Zarqāllu and Its Implications for Homocentric Planetary Theory', *Centaurus*, 10 (1964–5), 232–47.

Goldstein, Bernard R., 'Theory and Observation in Medieval Astronomy', *Isis*, 63 (1972), 39–47.

Goldstein, Bernard R., 'Casting Doubt on Ptolemy', *Science*, 199 (1978), 872.

Goldstein, Bernard R. and Bowen, Alan C., 'The Introduction of Dated Observations and Precise Measurement in Greek Astronomy', *Archive for History of Exact Sciences*, 43 (1991), 93–132.

Goldziher, Ignaz, 'The Attitude of Orthodox Islam Toward the "Ancient Sciences"', in Merlin L. Swartz (ed. and trans), *Studies on Islam* (Oxford: Oxford University Press, 1981), 185–215.

Grant, Edward, 'Celestial Orbs in the Latin Middle Ages', *Isis*, 78 (1987), 152–73.

Grant, Edward, *Planets, Stars and Orbs: the Medieval Cosmos 1200–1687* (Cambridge: Cambridge University Press, 1994).

Grant, Edward, *The Foundations of Modern Science in the Middle Ages* (Cambridge: Cambridge University Press, 1996).

Grant, Edward, 'The Medieval Cosmos: Its Structure and Operation', *Journal for the History of Astronomy*, 28 (1997), 147–67.

Graßhoff, Gerd, *The History of Ptolemy's Star Catalogue* (New York: Springer, 1990).

Hackett, Jeremiah, 'Robert Grosseteste and Roger Bacon on the *Posterior Analytics*', in Pia Antolic-Piper, Alexander Fidora, and Matthias Lutz-Bachmann (eds.), *Erkenntnis und Wissenschaft. Probleme der Epistemologie in der Philosophie des Mittelalters / Knowledge and Science: Problems of Epistemology in Medieval Philosophy* (Berlin: De Gruyter, 2004), 161–212.

Hackett, Michael Benedict, 'The University as Corporate Body', in Jeremy I. Catto (ed.), *The History of the University of Oxford*: Vol. 1, *The Early Oxford Schools* (Oxford: Oxford University Press, 1984), 37–98.

Hahn, Nan L., *Medieval Mensuration: 'Quadrans Vetus' and 'Geometrie Due Sunt Partes Principales'*, Transactions of the American Philosophical Society 72.8 (Philadelphia, PA: American Philosophical Society, 1982).

Hall, B. S., 'The Didactic and the Elegant: Some Thoughts on Scientific and Technological Illustrations in the Middle Ages and Renaissance', in Brian S. Baigrie (ed.), *Picturing Knowledge: Historical and Philosophical Problems Concerning the Use of Art in Science* (Toronto: University of Toronto Press, 1996), 3–39.

Hamel, Jürgen, *Studien zur 'Sphaera' des Johannes de Sacrobosco* (Leipzig: AVA, Akademische Verlagsanstalt, 2014).

Hanson, Norwood Russel, 'The Mathematical Power of Epicyclical Astronomy', *Isis*, 51 (1960), 151–8.

Harrison Thomson, S., *The Writings of Robert Grosseteste, Bishop of Lincoln 1235–1253* (Cambridge: Cambridge University Press, 1940).

Harvey, Joshua S. et al., 'A Thirteenth-Century Theory of Speech', *Journal of the Acoustic Society of America*, 146 (2019), 937–47.

Hays, James D., Imbrie, John, and Shackleton, Nick J., 'Variations in the Earth's Orbit: Pacemaker of the Ice Ages', *Science*, 194 (1976), 1121–32.

Healy-Varley, Margaret, 'Anselm's Afterlife and *De custodia interioris hominis*', in Giles E. M. Gasper and Ian Logan (eds.), *Saint Anselm of Canterbury and His Legacy* (Toronto: Pontifical Institute of Mediaeval Studies, 2012), 239–57.

Healy-Varley, Margaret, 'The *Admonitio morienti* and *Meditatio ad concitandum timorem* in Vernacular Compilations', in Margaret Healy-Varley, Giles E. M. Gasper, and George Younge (eds.), *Anselm of Canterbury: Communities, Contemporaries, and Criticism* (Leiden: Brill, 2021), 240–61.

Hiatt, Alfred, 'The Map of Macrobius before 1100', *Imago Mundi*, 59 (2007), 149–76.

Hill, Donald, R., 'Al-Bīrūnī's Mechanical Calendar', *Annals of Science*, 42 (1985), 139–63.

Hill, George F., *The Development of Arabic Numerals in Europe: Exhibited in Sixty-Four Tables* (Oxford: Oxford University Press, 1915).

Hill, R. M. T., 'Oliver Sutton, Bishop of Lincoln, and the University of Oxford', *Transactions of the Royal Historical Society*, 4th ser., 31 (1949), 1–16.

Holden, Brock, *Lords of the Central Marches: English Aristocracy and Frontier Society, 1087–1265* (Oxford: Oxford University Press, 2008).

Holland, Meridel, 'Robert Grosseteste's Greek Translations and College of Arms MS Arundel 9', in James McEvoy (ed.), *Robert Grosseteste: New Perspectives on His Thought and Scholarship* (Turnhout: Brepols, 1995), 121–47.

Holt, James, *Magna Carta*, 2nd ed. (Cambridge: Cambridge University Press, 1992, repr. 2015).

Honigmann, Ernst, *Die sieben Klimata* (Heidelberg: C. Winter, 1929).

Hoskin, Philippa M., 'Poor [Poore], Richard (d. 1237), bishop of Salisbury', *Oxford Dictionary of National Biography* (Oxford: Oxford University Press, 2004); 23 September 2004: https://doi.org/10.1093/ref:odnb/22525 (accessed 17 May 2020).

Hoskin, Philippa M., *Robert Grosseteste and the 13-Century Diocese of Lincoln* (Brill: Leiden, 2019).

Hoyle, Fred, 'The Work of Nicholas Copernicus', *Proceedings of the Royal Society A*, 336 (1974), 105–14.

Høyrup, Jens, 'Jordanus de Nemore, Thirteenth-Century Innovator: An Essay on Intellectual Context, Achievement and Failure', *Archive for the History of the Exact Sciences*, 38 (1988), 307–63.

398 BIBLIOGRAPHY

Hudson, John, *The Formation of the English Common Law: Law and Society in England from the Norman Conquest to Magna Carta* (London: Longman, 1996).

Humphreys, K. W., *The Friars' Libraries* (London: British Library, 1990).

Hunt, Richard W., 'The Library of Robert Grosseteste', in D. Callus (ed.), *Robert Grosseteste, Scholar and Bishop* (Oxford: Oxford University Press, 1955), 121–45, including 'Appendix A', 132–41.

Irwin, Terence H., *Aristotle's First Principles* (Oxford: Oxford University Press, 1988).

James, Montague Rhodes, *A Descriptive Catalogue of the Manuscripts in the Library of Gonville and Caius College*, 2 vols. (Cambridge: Cambridge University Press, 1907–8).

James, Montague Rhodes, *A Descriptive Catalogue of The Manuscripts in the Library of Corpus Christi College Cambridge*, 2 vols. (Cambridge, 1912).

Jacquemard, Catherine, Olivier Desbordes, and Alain Hairie, 'Du quadrant *vetustior* à l'*horologium viatorum* d'Hermann de Reichenau. Étude du manuscrit Vaticano BAV Ott. lat. 1631, f. 16–17^{v}', *Kentron*, 23 (2007), 79–125.

Jeauneau, Édouard, *Rethinking the School of Chartres*, trans. Claude Paul Desmarais from an unpublished original text in French (Toronto: University of Toronto Press, 2009).

Jones, Alexander, 'Ancient Rejection and Adoption of Ptolemy's Frame of Reference for Longitudes', in Alexander Jones (ed.), *Ptolemy in Perspective: Use and Criticism of His Work from Antiquity to the Nineteenth Century* (Dordrecht: Springer, 2010), 11–44.

Jonge, M. de, 'Robert Grosseteste and the Testaments of the Twelve Patriarchs', *The Journal of Theological Studies*, 42 (1991), 115–25.

Juste, David 'Neither Observation nor Astronomical Tables: An Alternative Way of Computing Planetary Longitudes in the Early Western Middle Ages', in Charles Burnett (ed.), *Studies in the History of the Exact Sciences in Honour of David Pingree* (Leiden: Brill, 2004), 181–222.

Juste, David, 'Hermann der Lahme und das Astrolab im Spiegel der neuesten Forschung', in Felix Heinzer and Thomas Zotz (eds.), *Hermann der Lahme. Reichenauer Mönch und Universalgelehrter des 11. Jahrhunderts* (Stuttgart: Kohlhammer, 2016), 273–84.

Juste, David, 'MS Paris, Bibliothèque nationale de France, lat. 7195' (update 20 April 2019), *Ptolemaeus Arabus et Latinus. Manuscripts*: http://ptolemaeus. badw.de/ms/99 (accessed 9 July 2021).

Juste, David, 'Ptolemy, *Almagesti* (tr. Gerard of Cremona)' (update 7 May 2021), *Ptolemaeus Arabus et Latinus. Works*: http://ptolemaeus.badw.de/work/3 (accessed 13 July 2021).

Juste, David, 'Ptolemy, *Almagesti* (tr. Sicily c. 1150)' (update 4 March 2021), *Ptolemaeus Arabus et Latinus. Works*: http://ptolemaeus.badw.de/work/21 (accessed 13 July 2021).

Juste, David, 'Ptolemy, *Preceptum canonis Ptolomei* (tr. before c. 1000)' (update 18 March 2021), *Ptolemaeus Arabus et Latinus. Works*: http://ptolemaeus.badw. de/work/52 (accessed 13 July 2021).

BIBLIOGRAPHY 399

Kennedy, Edward S., 'Al-Bīrūnī's Masudic Canon', *al-Abḥāth*, 24 (1971), 59–81.

Kennedy, Edward S. and Pingree, D., *The Astrological History of Masha'allah* (Cambridge, MA: Harvard University Press, 1971; reissued 2013).

Kennedy, Hugh, *Muslim Spain and Portugal: A Political History of al-Andalus* (Longman: London, 1996).

King, David A. and Samsó, J., 'Astronomical Handbooks and Tables from the Islamic World (750-1900): An Interim Report', *Suhayl*, 2 (2001), 9–105.

King, David A., *In Synchrony with the Heavens: Studies in Astronomical Timekeeping and Instrumentation in Medieval Islamic Civilization* (Leiden: Brill, 2005).

Kingsford, Charles L. and Kemp, Brian R., 'Poor [Pauper], Herbert (d. 1217), Bishop of Salisbury', *Oxford Dictionary of National Biography* (Oxford: Oxford University Press, 2004); 23 September 2004: https://doi.org/10.1093/ref:odnb/22524 (accessed 3 May 2020).

Knorr, Wilbur R., 'Sacrobosco's *Quadrans*: Date and Sources', *Journal for the History of Astronomy*, 28 (1997), 187–222.

Kunitzsch, Paul, *Typen von Sternverzeichnissen in astronomischen Handschriften des zehnten bis vierzehnten Jahrhunderst* (Wiesbaden: Harrassowitz, 1966), 39–46.

Kunitzsch, Paul, *Der Almagest. Die 'Syntaxis mathematica' des Claudius Ptolemäus in arabisch-lateinisch Überlieferung* (Wiesbaden: Otto Harrassowitz, 1974).

Kunitzsch, Paul, 'Glossar der arabischen Fachausdrücke in der mittelalterlichen europäischen Astrolabliteratur', *Nachrichten der Akademie der Wissenschaften in Göttingen, phil. hist. Kl., Jg. 1982*, 1 (1982), 459–571.

Kunitzsch, Paul, 'John of London and His Unknown Arabic Source', *Journal for the History of Astronomy*, 17 (1986), 52–7.

Kunitzsch, Paul, 'Gerard's Translations of Astronomical Texts, Especially the Almagest', in Pierluigi Pizzaiglio (ed.), *Gerardo da Cremona*, Annali della Biblioteca statale e libreria civica di Cremona 41 (Cremona: Libraria del convengno editrice, 1992), 71–84.

Kurtik, G. Ye., 'Precession Theory in Medieval Indian and Early Islamic Astronomy', in Wazir H. Abdi et al. (eds.), *Interaction between Indian and Central Asian Science and Technology in Mediaeval Times*, 2 vols. (New Delhi: Indian National Science Academy, 1990), i. 94–110.

Laird, Edgar, 'Robert Grosseteste, Ptolemy, and Christian Knowledge', in John Flood, James R. Ginther, and Joseph W. Goering (eds.), *Robert Grosseteste and His Intellectual Milieu* (Toronto: Pontifical Institute of Mediaeval Studies, 2013), 131–52.

Laird, W. Roy, 'Robert Grosseteste on the Subalternate Sciences', *Traditio*, 43 (1987), 147–69.

Landau, Peter, 'The Origins of Legal Science in England in the Twelfth Century: Lincoln, Oxford and the Career of Vacarius', in Martin Brett and Kathleen Cushing (eds.), *Readers, Texts and Compilers in the Earlier Middle Ages* (Aldershot: Ashgate, 2009), 165–82.

Lang, Helen S., *Aristotle's Physics and Its Medieval Varieties* (Albany: SUNY Press, 1992), 36–44.

Lang, Helen S., *The Order of Nature in Aristotle's Physics* (Cambridge: Cambridge University Press, 1998).

400 BIBLIOGRAPHY

Larson, Atria A., *Master of Penance: Gratian and the Development of Penitential Thought and Law in the Twelfth Century* (Washington, DC: Catholic University of America Press, 2014).

Lawrence-Mathers, A., *Medieval Meteorology; Forecasting the Weather from Aristotle to the Almanac* (Cambridge: Cambridge University Press, 2019), 40–65.

Lawrence, Clifford Hugh, 'The Origins of the Chancellorship at Oxford', *Oxoniensia*, 41 (1976), 316–23.

Lee, Richard A. Jr., *Science, the Singular, and the Question of Theology* (New York: Palgrave, 2002), 7–15.

Lewis, Neil, 'Robert Grosseteste's Notes on the Physics', in Evelyn A. Mackie and Joseph Goering (eds.), *Editing Robert Grosseteste* (Toronto: University of Toronto Press, 2003), 103–34.

Lewis, Neil, 'Robert Grosseteste', *The Stanford Encyclopedia of Philosophy* (Summer 2019 ed.), ed. Edward N. Zalta: https://plato.stanford.edu/archives/sum2019/entries/grosseteste/ (accessed 16 July 2021).

Lindberg, David C., 'The Theory of Pinhole Images from Antiquity to the Thirteenth Century', *Archive for History of Exact Sciences*, 5 (1968), 154–76.

Lindberg, David C. and Shank, Michael (eds.), *Cambridge History of Science*: Vol. 2, *Medieval Science* (Cambridge: Cambridge University Press, 2013), 341–64.

Lippiat, G. E. M., *Simon V of Montfort & Baronial Government 1195–1218* (Oxford: Oxford University Press, 2017).

Little, Andrew G., 'Review of S. Harrison Thomson The Writings of Robert Grosseteste, 1940', *The English Historical Review*, 56 (1941), 306–9.

Llanthony Prima (Priory), https://www.monasticwales.org/site/45 (accessed 14 Nov. 2022).

Lock, Peter, *The Franks in the Aegean: 1204–1500* (Longman: London, 1995).

Lorch, Richard P., 'The Astronomy of Jābir ibn Aflaḥ', *Centaurus* 19 (1975), 85–107.

Ludwig, Corinna, 'Die Karriere eines Bestsellers. Untersuchungen zur Entstehung und Rezeption der Sphaera des Johannes de Sacrobosco', *Concilium medii aevi*, 13 (2010), 153–85.

Luscombe, David, 'The Sense of Innovation in the Writings of Peter Abelard', in H. J. Schmidt (ed.), *Tradition, Innovation, Invention* (Berlin: De Gruyter, 2005), 181–94.

McEvoy, James, *The Philosophy of Robert Grosseteste* (Oxford: Oxford University Press, 1982).

McEvoy, James, 'The Chronology of Robert Grosseteste's Writings on Nature and Natural Philosophy', *Speculum*, 58 (1983), 614–55.

McEvoy, James, *Robert Grosseteste* (Oxford: Oxford University Press, 2000), 22–9.

McGinn, Bernard, 'The Role of the *Anima Mundi* as Mediator between the Divine and Created Realms in the Twelfth Century', in J. J. Collins and M. Fishbane (eds.), *Death, Ecstasy and other Worldly Journeys* (New York: State University of New York Press 1995), 285–316.

McKinlet, Richard, *Norfolk and Suffolk Surnames in the Middle Ages* (London, Chichester: Phillimore, 1975).

Macray, W. D., *Bodleian Library Quarto Catalogues IX: Digby Manuscripts*, repr. with addenda by R. W. Hunt and A. G. Watson (Oxford: Bodleian Library, 1999).

BIBLIOGRAPHY 401

Madan, Falconer, Craster, Herbert H. E., and Denholm-Young, N., *A Summary Catalogue of Western Manuscripts in the Bodleian Library at Oxford which have not hitherto been catalogued in the quarto series: With references to the Oriental and other manuscripts*, Vol. II, part ii (Oxford: Oxford University Press, 1937).

Maddison, Francis, *Medieval Scientific Instruments and the Development of Navigational Institutions in the XVth and XVIth Centuries* (Coimbra: Junta de Investigações do Ultramar-Lisboa, 1969).

Mancha, José Luis, 'Astronomical Use of Pinhole Images in William of Saint-Cloud's *Almanach planetarum* (1292)', *Archive for History of Exact Sciences*, 43 (1992), 275–98.

Mancha, José Luis, 'On Ibn al-Kammād's Table for Trepidation', *Archive for History of Exact Sciences*, 52 (1998), 1–11.

Mandelbrot, Benoit, 'How Long Is the Coast of Britain? Statistical Self-Similarity and Fractional Dimension', *Science*, 156 (1967), 636–8.

Mantello, Frank A. C., 'The Editions of Nicholas Trevet's *Annales sex regum Angliae*', *Revue d'histoire des textes*, 10 (1982 for 1980), 257–75.

Martin, Christopher J., 'The Development of Logic in the Twelfth Century', in Robert Pasnau and Christina Van Dyke (eds.), *The Cambridge History of Medieval Philosophy*, 2 vols. (Cambridge: Cambridge University Press, 2010), i. 129–45.

Mehren, August Ferdinand, 'Vues d'Avicenne sur l'astrologie et sur le rapport de la responsabilité humaine avec le destin', *Le Muséon*, III/3 (Juillet 1884), 1–38.

Mercier, Raymond, 'Studies in the Medieval Conception of Precession', 2 pts., *Archives internationales d'histoire des sciences*, 26 (1976), 197–220; 27 (1977), 33–71.

Mercier, Raymond, 'Meridians of Reference in Precopernican Tables', *Vistas in Astronomy*, 28 (1985), 23–7.

Mercier, Raymond, 'Astronomical Tables in the Twelfth Century', in Charles Burnett (ed.), *Adelard of Bath: An English Scientist and Arabist of the Early Twelfth Century* (London: Warburg Institute, 1987), 87–118.

Mercier, Raymond, 'Accession and Recession: Reconstruction of the Parameters', in Josep Casulleras and Julio Samsó (eds.), *From Baghdad to Barcelona: Studies in the Islamic Exact Sciences in Honour of Prof. Juan Vernet*, Anuari de Filologia, XIX, 2 vols. (Barcelona: Instituto 'Millás Vallicrosa' de Historia de la Ciencia Arabe, 1996), i. 299–347.

Mercier, Raymond, 'The Lost Zīj of al-Ṣūfī in the Twelfth-Century Tables for London and Pisa', in Raymond Mercier (ed.), *Studies on the Transmission of Medieval Mathematical Astronomy* (Aldershot: Variorum, 2004).

Michot, Yahya J., 'Ibn Taymiyya on Astrology: Annotated Translation of Three *Fatwas*', *Journal of Islamic Studies*, 11 (2000), 147–208.

Millás Vallicrosa, José María, 'La introducción del cuadrante con cursor en Europa', *Isis*, 17 (1932), 218–58.

Millás Vallicrosa, José María, 'El "Liber de motu octave sphere" de Tābit ibn Qurra', *Al-Andalus*, 10 (1945), 89–108, repr. in José María Millás Vallicrosa, *Nuevos estudios sobre historia de la ciencia española* (Barcelona: CSIC, 1960), 191–209.

402 BIBLIOGRAPHY

Minnis, Alastair J. and Nauta, Lodi, 'More Platonico loquitur: What Nicholas Trevet really did to William of Conches', in Alastair J. Minnis (ed.), *Chaucer's Boece and the Medieval Tradition of Boethius* (Suffolk: D. S. Brewer, 1993), 1–33.

Moesgaard, Kristian Peder, 'Thābit ibn Qurra between Ptolemy and Copernicus: An Analysis of Thābit's Solar Theory', *Archive for History of Exact Sciences*, 12 (1974), 199–216.

Morelon, Régis, 'Eastern Arabic Astronomy between the Eighth and the Eleventh Centuries', in Roshdi Rashed (ed.), *Encyclopedia of the History of Arabic Science*, 3 vols. (London: Routledge, 1996), i. 20–57.

Moreton, Jennifer, 'John of Sacrobosco and the Calendar', *Viator*, 25 (1994), 229–44.

Morini, Carla, '*Horologium* e *daegmael* nei manoscritti anglosassoni del computo', *Aevum*, 73 (1999), 273–93.

Mosley, Adam, 'Objects of Knowledge: Mathematics and Models in Sixteenth-Century Cosmology and Astronomy', in Sachiko Kusukawa and Ian Maclean (eds.), *Transmitting Knowledge: Words, Images, and Instruments in Early Modern Europe* (Oxford: Oxford University Press, 2006), 193–216.

Mozaffari, S. Mohammad, 'A Medieval Bright Star Table: The Non-Ptolemaic Star Tables in the *Īlkhānī Zīj*', *Journal for the History of Astronomy*, 47 (2016), 294–316.

Müller, Kathrin, *Visuelle Weltaneignung. Astronomische und kosmologische Diagramme in Handschriften des Mittelalters* (Göttingen: Vandenhoeck and Ruprecht, 2008).

Murdoch, John E., *Album of Science: Antiquity and the Middle Ages* (New York: Charles Scribner's Sons, 1984).

Murray, Alexander, 'Confession before 1215', *Transactions of the Royal Historical Society*, 6th ser., 3 (1993), 51–81, repr. in his *Conscience and Authority in the Medieval Church* (Oxford: Oxford University Press, 2015), 19–48.

Musatti, C. A., 'Alcune considerazioni sulla paternità del commento alla Sphaera di Giovanni Sacrobosco attribuito a Michele Scoto', in Pina Totaro and Luisa Valente (eds.), *Sphaera. Forma immagine e metafora tra Medioevo ed età moderna*, Lessico intellettuale europeo, 117 (Florence: Olschki, 2012), 145–65.

Musson, Anthony, *Medieval Law in Context: The Growth of Legal Consciousness from Magna Carta to the Peasants' Revolt* (Manchester: Manchester University Press, 2001).

Mylod, E. J., Perry, Guy, Smith, Thomas W., and Vandeburie, Jan (eds.) *The Fifth Crusade in Context* (London: Routledge, 2017).

Mynors, Roger A. B. and Thomson, Rodney M., *Catalogue of the Manuscripts of Hereford Cathedral Library* (Cambridge: Cambridge University Press, 1993).

Nasr, Seyyed Hossein, *An Introduction to Islamic Cosmological Doctrines* (Albany: SUNY Press, 1993).

Neugebauer, Otto, 'The Early History of the Astrolabe: Studies in Ancient Astronomy IX', *Isis*, 40 (1949), 240–56.

Neugebauer, Otto, 'Thâbit ben Qurra "On the Solar Year" and 'On the Motion of the Eighth Sphere', *Proceedings of the American Philosophical Society*, 106 (1960), 264–99.

BIBLIOGRAPHY 403

Neugebauer, Otto, *The Astronomical Tables of al-Khwarizmi* (Copenhagen: I kommission hos Munksgaard, 1962).

Neugebauer, Otto, *A History of Ancient Mathematical Astronomy*, 3 vols. (Berlin: Springer, 1975), ii. 631–4.

Newton, Robert R., *The Crime of Claudius Ptolemy* (Baltimore: Johns Hopkins University Press, 1977).

Nolte, Friedrich, *Die Armillarsphäre*, Abhandlungen zur Geschichte der Naturwissenschaften und der Medizin 2 (Erlangen: Mencke, 1922).

North, John D., 'Medieval Star Catalogues and the Movement of the Eighth Sphere', *Archives internationales d'histoire des sciences*, 20 (1967), 71–83.

North, John D., *Richard of Wallingford*, 3 vols. (Oxford: Clarendon Press, 1976).

North, John D., 'Astrology and the Fortunes of Churches', *Centaurus*, 24 (1980), 181–211.

Nothaft, C. Philipp E., 'Climate, Astrology and the Age of the World in Thirteenth-Century Thought: Giles of Lessines and Roger Bacon on the Precession of the Solar Apogee', *Journal of the Warburg and Courtauld Institutes*, 77 (2014), 35–60.

Nothaft, C. Philipp E., 'Bede's *horologium*: Observational Astronomy and the Problem of the Equinoxes in Early Medieval Europe (c.700–1100)', *English Historical Review*, 130 (2015), 1079–101.

Nothaft, C. Philipp E., 'A Reluctant Innovator: Graeco-Arabic Astronomy in the *Computus* of Magister Cunestabulus (1175)', *Early Science and Medicine*, 22 (2017), 24–54.

Nothaft, C. Philipp E., 'Criticism of Trepidation Models and Advocacy of Uniform Precession in Medieval Latin Astronomy', *Archive for History of Exact Sciences*, 71 (2017), 211–44.

Nothaft, C. Philipp E., 'Henry Bate's *Tabule Machlinenses*: The Earliest Astronomical Tables by a Latin Author', *Annals of Science*, 75 (2018), 275–303.

Nothaft, C. Philipp E., *Scandalous Error: Calendar Reform and Calendrical Astronomy in Medieval Europe* (Oxford: Oxford University Press, 2018).

Nothaft, C. Philipp E., 'An Alfonsine Universe: Nicolò Conti and Georg Peurbach on the Threefold Motion of the Fixed Stars', *Centaurus*, 61 (2019), 91–110.

Nothaft, C. Philipp E., 'An Overlooked Construction Manual for the *Quadrans Vetustissimus*', *Nuncius*, 34 (2019), 517–34.

Nothaft, C. Philipp E., 'The *Liber Theoreumacie* (1214) and the Early History of the *Quadrans Vetus*', *Journal for the History of Astronomy*, 51 (2020), 51–74.

Obrist, Barbara, 'The Astronomical Sundial in Saint Willibrord's Calendar and Its Early Medieval Context', *Archives d'histoire doctrinale et littéraire du Moyen Âge*, 67 (2000), 71–118.

Obtrist, Barbara, 'Démontrer, montrer et l'évidence visuelle. Les figures cosmologiques, de la fin de l'Antiquité à Guillaume de Conches et au début du XIIIe siècle', in E. C. Lutz, V. Jerjen, and C. Putzo (eds.), *Diagramm und Text. Diagrammatische Strukturen und die Dynamisierung von Wissen und Erfahrrung* (Wiesbaden: Reichert Verlag, 2014), 45–78.

O'Donnell, J. Reginald, 'Themestius's Paraphrasis of the *Posterior Analytics* in Gerard of Cremona's Translation', *Mediaeval Studies*, 20 (1958), 242–315.

404 BIBLIOGRAPHY

Orme, Nicholas, 'The Cathedral School before the Reformation', in Gerald Aylmer and John Tiller (eds.), *Hereford Cathedral: A History* (London: Hambledon Press, 2000), 546–78.

Paillard Didier, 'Climate and the Orbital Parameters of the Earth', *Comptes Rendus Geoscience*, 342 (2010), 273–85.

Pannekoek, Anton, 'The Planetary Theory of Ptolemy' *Popular Astronomy*, 55 (1947), 459–75.

Pannekock, Anton, 'Ptolemy's Precession', *Vistas in Astronomy*, 1 (1955), 60–6.

Panti, Cecilia, 'Robert Grosseteste and Adam of Exeter's Physics of Light: Remarks on the Transmission, Authenticity and Chronology of Grosseteste's Scientific Opuscula', in John Flood, James R. Ginther, and Joseph W. Goering (eds.), *Robert Grosseteste and his Intellectual Milieu* (Toronto: Pontifical Institute of Mediaeval Studies, 2013), 165–90.

Panti, Cecilia, *Moti, virtù e motori celesti nella cosmologia di Roberto Grossatesta. Studio ed edizione dei trattati 'De sphera', 'De cometis', 'De motu supercelestium'* (Florence: SISMEL—Edizioni del Galluzzo, 2001).

Paul, Suzanne, 'Catalogue Entry: Testaments of the Twelve Patriarchs (MS Ff.1.24)', University of Cambridge Digital Library: https://cudl.lib.cam.ac.uk/view/MS-FF-00001-00024/1 (accessed 16 July 2021).

Pedersen, Fritz Saaby, 'A Twelfth-Century Planetary Theorica in the Manner of the London Tables', *Cahiers de l'Institut du Moyen-Age Grec et Latin*, 60 (1990), 199–318.

Pedersen, Fritz Saaby (ed.), *The Toledan Tables: A Review of the Manuscripts and the Textual Versions with an Edition*, 4 vols. (Copenhagen: Reitzel, 2002).

Pedersen, Olaf, 'Anonymous: The Theory of the Planets', in Edward Grant (ed.), *A Source Book in Medieval Science* (Cambridge, MA: Harvard University Press, 1974), 451.

Pedersen, Olaf, 'In Quest of Sacrobosco', *Journal for the History of Astronomy*, 16 (1985), 175–220.

Pedersen, Olaf, 'Origins of the "Theorica Planetarum"', *Journal for the History of Astronomy*, 12 (1981), 113–23, at 118–22.

Pedersen, Olaf, *A Survey of the* Almagest, rev. ed. (New York: Springer, 2011), 236–60.

Pegge, Samuel, *The Life of Grosseteste* (London: John Nichols, 1743), 19–23.

Pelzer, Auguste, 'Les versions latines d'ouvrages de morale conservés sous le nom d'Aristote en usage au XIIIe siècle', *Rev. néo-scolastique de philosophie*, 23 (1921), 398.

Pereira, Michela, *Arcana sapienza. L'alchimia dalle origini a Jung* (Roma: Carocci, 2001).

Pereira, Michela, 'Cosmologie alchemiche', in Concetto Martello, Chiara Militello, and Andrea Vella (eds.), *Cosmogonie e cosmologie nel medioevo. Atti del convegno della Società Italiana per lo Studio del Pensiero Medievale* (S.I.S.P.M.), Catania, 22–24 September 2006 (Turnhout: Brepols, 2008), 363–410.

Perry, Guy, 'A King of Jerusalem in England: The Visit of John of Brienne in 1223', *History*, 100 (2015), 627–39.

BIBLIOGRAPHY   405

Petersen, Viggo M. and Schmidt, Olaf, 'The Determination of the Longitude of the Apogee of the Orbit of the Sun According to Hipparchus and Ptolemy', *Centaurus*, 12 (1968), 73–96.

Peyrière, Jacques, *Convolution, séries et intégrales de Fourier* (Paris: Ellipses, 2012).

Phillips, Jonathan, *The Crusades, 1095–1204*, 2nd ed. (London: Routledge, 2014).

Phillips, Jonathan, *The Life and Legend of the Sultan Saladin* (New Haven, CT: Yale Universiy Press, 2019).

Pingree, David, 'Masha'allah's Zoroastrian Historical Astrology', in G. Oestmann, H. D. Rutkin, and K. von Stuckrad (eds.), *Horoscopes and Public Spheres: Essays on the History of Astrology* (Berlin: De Gruyter, 2005), 95–100.

Pollard, Graham, 'The University and the Book Trade in Medieval Oxford', *Miscellanea Medievalia*, 3 (1964), 336–44.

Pollard, Graham, 'The Legatine Award to Oxford in 1214 and Robert Grosteste', *Oxoniensia*, 39 (1974), 62–72.

Poulle, Emmanuel, 'L'astrolabe médiéval d'après les manuscrits de la Bibliothèque Nationale', *Bibliothèque de l'École des Chartes*, 112 (1954), 81–103.

Poulle, Emmanuel, 'Les instruments astronomiques de l'Occident latin aux XIe et XIIe siècles', *Cahiers de civilisation médiévale*, 15 (1972), 27–40.

Poulle, Emmanuel, 'L'astronomie du Moyen Âge et ses instruments', *Annali dell'Istituto e Museo di Storia della Scienza di Firenze*, 6 (1981), 3–16.

Poulle, Emmanuel, *Les sources astronomiques (textes, tables, instruments)* (Turnhout: Brepols, 1981).

Poulle, Emmanuel, *Les instruments astronomiques du Moyen Âge* (Paris: Brieux, 1983).

Poulle, Emmanuel, 'Le traité de l'astrolabe d'Adélard de Bath', in Charles Burnett (ed.), *Adelard of Bath: An English Scientist and Arabist of the Early Twelfth Century* (London: Warburg Institute, 1987), 119–32.

Poulle, Emmanuel, 'L'instrumentation astronomique médiévale', in Bernard Ribémont (ed.), *Observer, lire, écrire, le ciel au Moyen Âge. Actes du colloque d'Orléans, 22–23 avril 1989* (Paris: Klincksieck, 1991), 253–81.

Powell, James M., *Anatomy of a Crusade, 1213–1221* (Philadelphia, PA: University of Pennsylvania Press, 1986).

Power, Dan, 'Who Went on the Albigensian Crusade?', *English Historical Review*, 128 (2013), 1047–85.

Price, Derek J., 'Precision Instruments: To 1500', in Charles Singer, Eric J. Holmyard, A. R. Hall, and Trevor I. Williams (eds.), *A History of Technology*: Vol. 3, *From the Renaissance to the Industrial Revolution c.1500–c.1700* (Oxford: Oxford University Press, 1957), 582–619.

Ragep, F. Jamil, 'Thābit's Astronomical Works', *Journal of the History of Astronomy*, 23 (1992), 61–3.

Ragep, F. Jamil, *Naṣīr al-Dīn al-Ṭūsī's Memoir on Astronomy (al-Tadhkira fī 'ilm al-hay'a)*, 2 vols. (New York: Springer, 1993), ii. 400–8.

Ragep, F. Jamil, 'Al-Battānī, Cosmology, and the Early History of Trepidation in Islam', in Josep Casulleras and Julio Samsó (eds.), *From Baghdad to Barcelona: Studies in the Islamic Exact Sciences in Honour of Prof. Juan Vernet*, Anuari de

406 BIBLIOGRAPHY

Filologia, XIX, 2 vols. (Barcelona: Instituto 'Millás Vallicrosa' de Historia de la Ciencia Arabe, 1996), i. 267–98.

Ragep, F. Jamil, 'Copernicus and His Islamic Predecessors: Some Historical Remarks', *History of Science*, 45 (2007), 65–81.

Ramírez-Weaver, Eric, '"So That You Can Understand This Better": Art, Science, and Cosmology for Courtiers in William of Conches' *Dragmaticon Philosophiae*', in E. C. Lutz, V. Jerjen, and C. Putzo (eds.), *Diagramm und Text. Diagrammatische Strukturen und die Dynamisierung von Wissen und Erfahrrung* (Wiesbaden: Reichert Verlag, 2014), 319–48.

Rashed, Roshdi, 'The Celestial Kinematics of Ibn al-Haytham', *Arabic Sciences and Philosophy*, 17 (2007), 7–55.

Rashed, Roshdi, *Ibn al-Haytham, New Astronomy and Spherical Geometry: A History of Arabic Sciences and Mathematics*, Vol. IV, trans. Judith Field (London: Routledge, 2014).

Raynaud, Dominique, 'Abū al-Wafā Latinus? A Study of Method', *Historia Mathematica*, 39 (2012), 34–83.

Richter-Bernburg, Lutz, 'Ṣāʿid, the *Toledan Tables*, and Andalusī Science', in David A. King and George Saliba (eds.), *From Deferent to Equant: A Volume of Studies in the History of Science in the Ancient and Medieval Near East in Honor of E. S. Kennedy* (New York: New York Academy of Sciences, 1987), 373–401.

Robson, Michael, 'Robert Grosseteste and the Franciscan School at Oxford (*c.*1229–1253)', *Antonianum*, XCV (2020), 345–82.

Rorem, Paul, *Hugh of St Victor* (Oxford: Oxford University Press, 2009).

Rosemann, Philipp, *The Story of a Great Medieval Book: Peter Lombard's 'Sentences'* (Toronto: Toronto University Press, 2007).

Rouse, Richard H., 'Boston Buriensis and the Author of the *Catalogus scriptorium ecclesiae*', *Speculum*, 41 (1966), 471–99.

Rudolph, Conrad, *The Mystic Ark: Hugh of Saint Victor, Art, and Thought in the Twelfth Century* (Cambridge: Cambridge University Press, 2014).

Russell, Josiah C., 'The Preferments and "Adiutores" of Robert Grosseteste', *The Harvard Theological Review*, 26 (1933), 161–72.

Russo, Luigi, *The Forgotten Revolution* (New York: Springer, 2003).

Saari, Donald G., 'A Visit to the Newtonian N-body Problem via Elementary Complex Variables', *The American Mathematical Monthly*, 97 (1990), 105–19.

Sabra, A. I., 'An Eleventh-Century Refutation of Ptolemy's Planetary Theory', in E. Hilftein, P. Czartoryski, and F. D. Grande (eds.), *Science and History* (Wrocław-Warszawa: *Studia Copernicana*, 1978), 117–31.

Sabra, A. I., 'Configuring the Universe: Aporetic, Problem Solving, and Kinematic Modeling as Themes of Arabic Astronomy', *Perspectives on Science*, 6 (1998), 288–330.

Saliba, George, 'The First Non-Ptolemaic Astronomy at the Maraghah School', *Isis*, 70 (1979), 571–6.

Saliba, George, 'Ibn Sīnā and Abū ʿUbayd al-Jūzjānī: The Problem of the Ptolemaic Equant', *Journal for the History of Arabic Science*, 4 (1980), 85–112.

BIBLIOGRAPHY 407

Saliba, George, 'The Role of the *Almagest* Commentaries in Medieval Arabic Astronomy: A Preliminary Survey of Ṭūsī's Redaction of Ptolemy's *Almagest*', *Archives Internationales d'Histoire des Sciences*, 37 (1987), 3–20.

Saliba, George, 'The Role of the Astrologer in Medieval Islamic Society', *Bulletin d'Études Orientales*, 44 (1992), 45–67.

Saliba, George, *Islamic Science and the Making of the European Renaissance* (Cambridge, MA: MIT Press, 2007).

Saller, Heinrich, *Operational Symmetries: Basic Operations in Physics* (Cham: Springer, 2017), 159–61.

Samsó, Julio, *Las ciencias de los antiguos en Al-Andalus* (Madrid: Mapfre, 1992).

Samsó, Julio, 'Sobre el modelo de Azarquiel para determinar la oblicuidad de la ecliptica', *Homenaje al Prof. Darío Cabanelas Rodríguez*, 2 vols. (Granada: Universidad de Granada, 1987), ii. 367–77, repr. in Julio Samsó, *Islamic Astronomy and Medieval Spain* (Aldershot: Variorum, 1994).

Samsó, Julio, 'Trepidation in Al-Andalus in the 11th Century', *Islamic Astronomy and Medieval Spain* (Aldershot: Variorum, 1994), 2–5.

Samsó, Julio, 'Astronomical Observations in the Maghrib in the Fourteenth and Fifteenth Centuries', *Science in Context*, 14 (2001), 165–78, at 169–74.

Samsó, Julio, 'Ibn al-Haytham and Jābir b. Aflaḥ's Criticism of Ptolemy's Determination of the Parameters of Mercury', *Suhayl*, 2 (2001), 199–225.

Samsó, Julio, *On Both Sides of the Strait of Gibraltar: Studies in the History of Medieval Astronomy in the Iberian Peninsula and the Maghrib* (Leiden: Brill, 2020).

Saxl, Fritz and Meier, Hans, *Catalogue of Astrological and Mythological Illuminated Manuscripts of the Latin Middle Ages*: Vol. 3, *Manuscripts in English Libraries* (London: Warburg Institute, University of London, 1953), 386–93.

Sayers, Jane E., *Papal Judges Delegate in the Province of Canterbury 1198–1254* (Oxford: Oxford University Press, 1971).

Schaldach, Karlheinz, 'Gli "schemi delle ombre" nel Medio Evo latino', *Gnomonica Italiana*, 16 (2008), 9–16.

Schröder, Stefan, 'Die Klimazonenkarte des Petrus Alfonsi', in Ingrid Baumgärtner, Paul-Gerhard Klumbies, and Franziska Sick (eds.), *Raumkonzepte. Disziplinäre Zugänge* (Göttingen: V&R Unipress, 2009), 257–77.

Schulman, Nicole M., 'Husband, Father, Bishop? Grosseteste in Paris', *Speculum*, 72 (1997), 330–46.

Schum, Wilhelm, *Beschreibendes Verzeichnis der Amplonianischen Handschriften-Sammlung zu Erfurt* (Berlin: Weidmann, 1887).

Skemer, Don C., *Medieval and Renaissance Manuscripts in the Princeton University Library*, 2 vols. (Princeton: Princeton University Press, 2013).

Smalley, Beryl, 'A Collection of Paris Lectures of the Later Twelfth Century in the MS. Pembroke College, Cambridge 7', *The Cambridge Historical Journal*, 6 (1938), 103–13.

Smalley, Beryl, *The Study of the Bible in the Middle Ages*, 3rd ed. (Oxford: Blackwell, 1983).

Smith, Lesley, *The Glossa Ordinaria: The Making of a Medieval Bible Commentary* (Leiden: Brill, 2009).

408 BIBLIOGRAPHY

Sønnesyn, Sigbjørn, 'Word, Example, and Practice: Learning and the Learner in Twelfth-Century Thought', *Journal of Medieval History*, 46 (2020), 513–35.

Southern, Richard, W., 'From Schools to University', in Jeremy I. Catto (ed.), *The History of the University of Oxford*: Vol. 1, *The Early Oxford Schools* (Oxford: Oxford University Press, 1984), 1–36.

Southern, Richard W., *Robert Grosseteste: The Growth of an English Mind in Medieval Europe*, 2nd ed. (Oxford: Oxford University Press, 1992).

Southern, Richard W., *Scholastic Humanism and the Unification of Europe*: Vol. I, *Foundations* (Oxford: Blackwell, 1995).

Southern, Richard W., *Scholastic Humanism and the Unification of Europe*: Vol. II, *The Heroic Age* (Oxford: Blackwell, 2001).

Stansbury, Ronald J. (ed.), *A Companion to Pastoral Care in the Late Middle Ages (1200–1500)* (Leiden: Brill, 2010).

Stella, Francesco, 'Poesie computistiche e meraviglie astronomiche. Sull'*horologium nocturnum* di Pacifico', in Francesco Mosetti Casaretto and Roberta Ciocca (eds.), *Mirabilia. Gli effetti speciali nelle letterature del Medioevo* (Alessandria: Edizioni dell'Orso, 2014), 181–206.

Stella, Francesco, 'The Sense of Time in Carolingian Poetry: Christianizing the Zodiac and Astronomical Observation in Pacificus of Verona', in Pascale Bourgain and Jean-Yves Tilliette (eds.), *Le sens du temps. Actes du VIIe Congrès du Comité International de Latin Médiéval (Lyon, 10–13.09.2014)* (Geneva: Droz, 2017), 193–219.

Stevenson, Francis, *Robert Grosseteste, Bishop of Lincoln* (London: MacMillan and Co. 1899).

Swerdlow, Noel M., 'Hipparchus's Determination of the Length of the Tropical Year and the Rate of Precession', *Archive for History of Exact Sciences*, 21 (1979–80), 291–309.

Thomas, Hugh, *The Secular Clergy in England, 1066–1216* (Oxford: Oxford University Press, 2014).

Thorndike, Lynn, 'Robertus Anglicus and the Introduction of Demons and Magic into the Commentaries upon the Sphere of Sacrobosco', *Speculum*, 21 (1946), 241–3.

Thorndike, Lynn, *The Sphere of Sacrobosco and Its Commentators* (Chicago: University of Chicago Press, 1949).

Thorndike, Lynn, 'Notes upon Some Medieval Astronomical, Astrological and Mathematical Manuscripts at Florence, Milan, Bologna and Venice', *Isis*, 50 (1959), 33–50.

Thorndike, Lynn, *Michael Scot* (London: Nelson, 1965).

Tolan, John, *Petrus Alfonsi and his Medieval Readers* (Gainsville, FL: University of Florida Press, 1993).

Tolan, John, *Saint Francis and the Sultan: The Curious History of a Christian-Muslim Encounter* (Oxford: Oxford University Press, 2009).

Toomer, Gerald J., 'A Survey of the Toledan Tables', *Osiris*, 15 (1968), 5–174.

BIBLIOGRAPHY 409

Tyerman, Christopher, *England and the Crusades* (Chicago: University of Chicago Press, 1988).

Vafea, Flora, 'From the Celestial Globe to the Astrolabe: Transferring Celestial Motion onto the Plane of the Astrolabe', *Medieval Encounters*, 23 (2017), 124–48.

Valleriani, Matteo (ed.), *De Sphaera of Johannes de Sacrobosco in the Early Modern Period: The Authors of the Commentaries* (London: Springer Nature 2020).

Van de Vyver, André, 'Les premières traductions latines (X^e–XIe s.) de traités arabes sur l'astrolabe', in F. Quicke, P. Bonenfant, Y. Barjon, and L. Jadin (eds.), *1er Congrès international de geógraphie historique sous le haut patronage de S. M. le Roi de Belges*, 2 vols. (Brussels: Secrétariat Général, 1931), ii. 266–90.

Van Dyke, Christina, 'The Truth, the Whole Truth, and Nothing but the Truth: Robert Grosseteste on Universals (and the *Posterior Analytics*)', *Journal of the History of Philosophy*, 48 (2010), 153–70.

Vernet, J. and Samsó, Julio, 'The Development of Arabic Science in Andalusia', in R. Rashed (ed.), *Encyclopedia of the History of Arabic Science*, Vol. 1 (London: Routledge, 1996), 243–75.

Vincent, Nicholas, *Peter des Roches: An Alien in English Politics, 1205–1238* (Cambridge: Cambridge University Press, 1996).

Vincent, Nicholas, 'England and the Albigensian Crusade', in Björn Weiler and I. W. Rowlands (eds.), *England and Europe in the Reign of Henry III (1216–1272)* (Ashgate: Aldershot, 2002), 67–97.

Vincent, Nicholas, *Magna Carta: A Very Short Introduction* (Oxford: Oxford University Press, 2012).

Wasserstein, David J., *The Caliphate in the West: An Islamic Political Institution in the Iberian Peninsula* (Oxford: Oxford University Press, 1993).

Wayno, Jeffrey M., 'Rethinking the Fourth Lateran Council of 1215', *Speculum*, 93 (2018), 611–37.

Weisheipl, James A., 'The Concept of Nature', in William E. Carroll (ed.), *Nature and Motion in the Middle Ages* (Washington, DC: Catholic University of America Press, 1985), 1–24.

Weisheipl, James A., *Nature and Motion in the Middle Ages* (Washington, DC: Catholic University of America Press, 1985).

Wenzel, Siegfried, 'Robert Grosseteste's Treatise on Confession, *Deus est*', *Franciscan Studies*, 30 (1970), 218–93.

West, Jeffrey J. and Palmer, Nicholas (eds.), *Haughmond Abbey: Excavation of a 12th-Century Cloister in its Historical and Landscape Context* (Swindon: English Heritage, 2011).

Wharton, Henry, *Anglia Sacra*, 2 vols. (London: Richard Chiswel, 1691).

White, Rebekah C. et al., 'Magnifying Grains of Sand, Seeds, and Blades of Grass: Optical Effects in Robert Grosseteste's *De Iride* (*On the Rainbow*) (circa 1228–1230)', *Isis*, 112 (2021), 93–107.

Wiesenbach, Joachim, 'Pacificus von Verona als Erfinder einer Sternenuhr', in Paul Leo Butzer and Dietrich Lohrmann (eds.), *Science in Western and Eastern Civilzation in Carolingian Times* (Basel: Birkhäuser, 1993), 229–50.

410 BIBLIOGRAPHY

Wilmart, D. André, 'Un opuscule sur la confession composé par Guy de Southwick vers la fin du XIIe siècle', *Recherches de théologie ancienne at médiévale*, 7 (1935), 337–52.

Włodarczyk, Jarosław, 'Observing with the Armillary Astrolabe', *Journal for the History of Astronomy*, 18 (1987), 173–95.

Wormald, Patrick, *The Making of English Law: King Alfred to the Twelfth Century*: Vol. 1, *Legislation and its Limits* (Oxford: Blackwell, 1999).

Zaitsev, Evgeny A., 'The Meaning of Early Medieval Geometry: From Euclid and Surveyors' Manuals to Christian Philosophy', *Isis*, 90 (1999), 522–53.

Index

Abbotsley 28–9, 66
Abelard, Peter *see* Peter Abelard 191
Abū Maʿshar, *Introduction to*
   Astronomy 59, 208
Achaia 18
Acre 18
Adam Marsh 27, 30–1, 58
Adelard of Bath 178, 187
affectus 80–1, 362
al-Battanī 178–9, 300
   Sabian Zīj 287–9, 291
al-Bīrūnī, *Qānun al-Masʿūdī* (Masudic
   Canon) 170
al-Biṭrūjī (Alpetragius), *On the Motions*
   of the Heavens 5, 11–12, 14,
   16, 212
al-Farghānī (Alfraganus) 4, 6, 87, 201,
   205, 243, 264, 266, 305, 363
   Liber de aggregationibus scientie
   stellarum (*Aggregatio*) 212–13,
   257, 259
al-Khwārizmī, *Zīj al-Sindhind* 176–9,
   188, 201
al-Malik al Kāmil, Ayyubid sultan 19
al-Ṣūfī 178, 292
al-Zarqālī, Abū Isḥāq Ibrāhīm
   (al-Zarqālluh, Arzachel,
   Azarquiel) 179, 212
   Treatise on the Motion of Fixed
   Stars 287–8
Alan of Lille
   Anti-Claudianus 191
   Plaint of Nature 191
Alberbury 77
Albinus, Master, chancellor of Hereford
   cathedral 74–5
Alexander III, pope 35–6

Alexander Neckham
   In Praise of Divine Wisdom 192
   On the Nature of Things 192–3
Alfonso I, king of Aragon 200
Alfraganus *see* al-Farghānī
Alhazen *see* Ibn al-Haytham
Almagesti minor 213–14, 243, 248–9,
   264, 270
Alpetragius *see* al-Biṭrūjī
Andrew, king of Hungary 19
anima mundi 191–3
Anselm, archbishop of Canterbury 69
ar-Raqqa 287
Arim (Udidjayn), India 113, 134–6,
   168, 176, 198–9, 201, 259, 296
Aristotle 12, 15, 18, 59, 92, 131,
   133–4, 137, 168–9, 173, 197,
   206, 209, 211–12, 240–1, 260,
   265, 362–3
   astronomy as Aristotelian
   science 214–25
   and demonstration 229–37
   and *episteme* 216–23, 236–7
   and hierarchical ordering of
   sciences 222–5
   and observation, role of 236–9
   On the Heavens 252, 254–9, 367
   On the Soul 214
   Physics 58, 214, 219, 233–4, 250, 253,
   256, 259; *see also* Grosseteste,
   Roberts, works, *Notes on Physics*
   Posterior Analytics 6, 11, 83, 214–18,
   222, 224–5, 227, 229, 232–3,
   244–5, 248, 253, 271, 274; *see also*
   Grosseteste, Roberts, works,
   Commentary of Posterior Analytics
   and subalternation 223–4, 226–9

412 INDEX

Artephius, *Clavis sapientiae* 193
Arzachel *see* al-Zarqālī, Abū Ishāq
    Ibrāhīm
aspectus 80–1, 362
astrology (general) 3, 16, 49–51, 57, 86,
    93–4, 166–7, 173, 179, 224, 303
  constituent parts 245–7
  horoscopes 5, 17, 49–50, 55, 57,
    180–2, 287
  judicial vs mundane 49, 180–3
  and MS Savile 21: 50 180–4
  and precession 286–7
  as pseudo-science 171–2, 366
  see also Tables of London; Tables of
    Pisa; Tables of Toledo; *ziges*
astronomy (general)
  as Aristotelian science 214–25
  constituent parts 245–7
  Islamicate (general) 166–73, 176–9,
    184, 186, 243–4, 249, 276, 363
  Latin (general) 166–7, 172, 178–9,
    187–90
  as liberal art 206–14, 246
  tools for observation 184–90,
    331, 352
  see also Tables of London; Tables of
    Pisa; Tables of Toledo; *ziges*
Athens 17–18
Avempace (Ibn Bājja) 171
Averroes 225
axes, fixed 260–1
Azarquiel *see* al-Zarqālī, Abū Ishāq
    Ibrāhīm

Bacon, Roger 18
Baghdad 183, 287, 295
Banū Mūsā 295
Baron's War, First 35
Basil of Caesarea, *Hexaemeron* 51–2
Bath 29
Becket, Thomas, archbishop of
    Canterbury 36
Bede 174, 295
  On the Nature of Things 198, 201

Bernardus Sylvestris,
    Cosmographia 191, 363–4
Berrow, Worcestershire 42
Boethius 217, 226, 241, 247
  *On the Consolation of
    Philosophy* 190–2
  On the Institutions of Arithmetic
    209, 211
Bologna 37, 62
Brahe, Tycho 300–2
Braose, Giles de 44
Braose, William de 44
Brecon 73
Burchard of Worms 67
Burford 72
Bury St Edmunds 51–2

Cahors 25
Cairo 19
Calcidus 197
Calne 27, 30
Cambridge 24–5
Canterbury 28, 37
Cassiodorus, *Institutions* 197
Chester 26
Chirbury 72
climate change 5, 174, 201–2, 281–3,
    294–9, 381
Clun 72
Coddington 41–2
Colcombe 42
Comestor, Peter 62
comets 168, 172, 295, 365; *see also*
    Grosseteste, Robert, works,
    On Comets
Constantinople, sacking of
    (1204) 17–18
Cook, Matthew 42
Cook, Roger 41–2, 45
Cordoba 176, 200
Cromwell, Thomas 58
crusades
  Albigensian 20–1
  Third 18

INDEX 413

Fourth 17–18
Fifth 19–20, 22
Culmington 35, 46, 48
Cunestabulus, Magister 296–8

Damietta 19
Daniel of Morley
 *Book on Inferior and Superior
  Nature* 193
 Philosophia 259
demonstration 233, 236
 doctrine of 229–37
 propter quid 69, 92, 219, 225, 229–32,
  235, 251–2, 254, 256, 258–9, 267,
  272–3, 277, 365
 quia 92, 229–31, 251, 258
Dover 22
Durham 28

eccentricity of celestial bodies 171, 212
 of earth 281–3, 370
 of moon 94, 127–8, 158–61, 269,
  306–7, 328
 of planets 12, 158, 197, 281, 307, 322
 of sun 93–4, 122–5, 127, 154–5,
  158–61, 195, 197, 201–2, 264,
  266–7, 294–5, 306–7, 314, 346,
  348, 370, 373, 380, 382
eclipses vi, 3–4, 57, 92, 94, 97, 109, 111,
  113, 128, 138–30, 135–6, 162–5,
  168, 187, 189, 194–6, 199–202,
  217, 237–9, 269–73, 307, 310,
  346–8, 371, 373, 380, 382
Edmund of Abingdon, *Speculum
  religiosorum* 69–70
epicycles 305–6, 346
 for earth-moon system 324–9
 and Fourier series 316–17, 319–22,
  327, 329
 modern 312, 315
 and Ptolemy 308–10
 for sun-earth system 318–24, 329–30
episteme 215–16, 236–7
 as science 221–2
 as understanding 216–21

Ethelbert, saint 73
Euclid 6, 232, 244–5, 248–9, 359, 363
 Elements 247, 307
Eustace, abbot of Ely 79
Exeter 24, 37

figura 89, 95–6, 106, 245–7, 249,
  252, 277, 337
Fitzwalter, Robert 19, 22
Foliot, Hugh, bishop (and archdeacon)
  of Hereford 7, 26, 34–5, 46–7,
  48, 71–2, 74, 76, 79, 241
 and justice 38–43
 pilgrimage to Santiago de
  Compostela 44–5
Foliot, Ralph, canon of Hereford 74
Fourier, Joseph 312, 315–17, 319–22,
  327, 329
Francis of Assisi, saint 19
Frederick II, Holy Roman
  Emperor 20

Gascoigne, Thomas 58
Geber *see* Ibn Ḥāyyān, Abū Muḥāmmad
  Jābir
Genoa 17
Gerald of Cremona 4, 17, 212, 225,
  266, 305
Gerald of Wales 25, 29, 37, 41,
  61, 211
 Gemma ecclesiastica 61–2
 Life of St Ethelbert 73
 Mirror of Two Men, letters of 73–5
Gerbert of Aurillac (Pope Silvester II) 186
Gervase of Canterbury 46
Gilbert, chaplain of Ledbury 41–2
Giles de Braose, bishop of Hereford
  35, 72
Gloucester 74, 202
Gratian, *Decretum* 37, 62
Gravesend, Richard, dean of
  Lincoln 58
Gregory I the Great, pope 63
Gregory Nazianen 63
Grosmont 43

414 INDEX

Grosseteste, John 27
Grosseteste, Richard, archdeacon of
Calne 27–30
Grosseteste, Robert, bishop of Lincoln,
biography 11
and Abbotsley, benefice of 28–9
as archdeacon 26–7
death of 58
family of 27–9
at Hereford 25–6, 29, 34, 48,
72, 241
legal activities 35–47, 48, 65
at Lincoln 18, 60–1, 78
at Oxford (possibly chancellor) 14,
23–4, 26, 29–34, 78, 375–8
at Paris 25–6, 33, 37, 45, 66–7
see also handwriting/manuscripts
copied by Grosseteste; pastoral
care of Grosseteste
Grosseteste, Robert, On the Sphere
(De sphera) (general)
astrology 166–7, 173, 180–4
astronomy
constituent parts 245–7
Islamicate 166–73, 176–9, 184,
186, 243–4, 249, 276, 363
Latin 166–7, 172, 178–9, 187–90
as liberal art 206–14, 246
audience for 1, 331–5, 352–61
context, historical 17–25, 47, 60
dating of 10–17, 24, 26, 49, 79
edition of 95–111
geometric content of vi–vii, 59, 89,
92–4, 131–65, 245–60, 261–3,
268–74, 362, 368
glossing of 2, 348
moon 268–76
planets, omission of details of
172–5, 180
purpose of 1–4, 82, 83, 92, 111, 205,
331–5
Sacrobosco, compared to 1–6, 13–17,
84, 167–8, 193–8, 204–5, 242–3,
249, 271, 276, 363
scope of 274–8

sources v, x, 2, 5, 11, 13, 57, 82, 89,
166–8, 196–9, 201, 213–14, 219,
222, 225–6, 242–9, 302, 363
stars, fixed 267–8
structure of 81–2, 172, 194–5, 205,
239–41, 242, 244–5, 249–50,
274, 363
synopsis of 92–4
and Tables of Toledo 175, 179–80,
182, 295, 300–2
technical commentary on 131–65
as textbook 3, 6–7, 166, 205, 240
translation of 111–30
translation, principles of 88–92
see also axes, fixed; climate change;
demonstration; eclipses;
epicycles; figura; horizon,
concept of; Grosseteste, Robert,
On the Sphere, manuscripts of;
illustrations/decorations of On
the Sphere; kinematics; machina
mundi; precession;
retrogradation; situs of universe;
sphericity and demonstration;
sun and zodiac; time, intervals of;
trepidation; Virtual Celestial
Model; world soul and elements
Grosseteste, Robert, On the Sphere
(De sphera), manuscripts
of v, 1–2, 83–8, 331–5,
379–82
Bologna, Bib. Uni. Lat. 1845
(Bu) 335n.12, 379
Cambridge, University Library MS
Ff.6.13 (Cu1) 84, 338, 340, 341
Cambridge, University Library MS
Gg.6.3 (Cu2) 335n.12, 338n.18,
342–3, 379
Cambridge, University Library MS
Ii.1.13 (Cu3) 333, 379
Cambridge, University Library MS
Mm.3.11 (Cu6) 335n.12,
342–3, 379
Dublin, Trinity College MS 441 (Dt)
333, 335n.12, 337, 342–3, 380

INDEX 415

Erfurt, Universitätsund
  Forschungsbibliothek, Bibliotheca
  Amploniana CA 4⁰ 351 (Ea2) 85,
  333, 335n.12, 338, 346–7, 349, 380
Erfurt, Universitätsund
  Forschungsbibliothek, Bibliotheca
  Amploniana CA 4⁰ 355 (Ea3) 85,
  87n.12, 335n.12, 342–3, 380
London, British Library, Add. MS
  27589 (La1) 85, 87n.12, 335n.12,
  342–3, 380
London, British Library, Egerton MS
  847 (Le2) 335n.12, 337, 380
London, British Library, Harley MS
  3735 (Lh2) 85, 87n.12, 335n.12,
  338, 339, 341, 343, 347, 354, 355,
  356–9, 357, 380
London, British Library, Harley MS
  4350 (Lh3) 85, 331, 332,
  335n.12, 352, 354n.43, 359, 381
Milan, Bib. Naz. Braidense AD XII 53
  (Mb) 381
New Haven, Yale University, Medical
  Library MS 11 (Ny) 335n.12,
  342, 381
New York, Columbia University,
  Smith Western Add. MS 1
  (Ns) 335n.12, 342, 347, 356, 381
Oxford, Bodleian Library MS Bodley
  676 (Ob2) 336n.12, 381
Oxford, Bodleian Library MS Digby
  98 (Od3) 336n.12, 381
Oxford, Bodleian Library MS Gough
  Linc. 13 (Og) 336n.12, 381
Oxford, Bodleian Library MS Laud
  Misc. 644 (Ol) 85, 87, 336n.12,
  343, 344, 359, 381
Paris, Bibliothèque nationale de
  France MS lat. 7195 (Pb1) 85,
  87–8, 336n.12, 381
Paris, Bibliothèque nationale de France
  MS lat. 7292 (Pb3) 336n.12,
  347, 382
Paris, Bibliothèque nationale de France
  MS lat. 7298 (Pb4) 336n.12, 338,
  341–3, 348n.35, 354n.44, 382

Princeton, University Library, Garrett
  MS 95 (Pg) 336n.12, 382
Salamanca, Biblioteca Universitaria
  111 (Su1) 336n.12, 382
Utrecht, Bibliotheek der
  Rijksuniversiteit 722
  (Ur) 336n.12, 347, 382
Vatican City, Vatican Library, Pal. Lat
  1414 (Vp) 84, 336n.12, 338,
  346–7, 382
Verdun, Bibliothèque municipale MS
  25 (Vmu) 85, 333n.4, 336n.12,
  343, 345, 346–7, 359, 382
Grosseteste, Robert, works
  Commentary on Posterior Analytics 1,
    33, 83, 206, 214, 216, 225–41,
    244–5, 250, 252–4, 262, 270–1, 274
  Compotus 1, 10, 13, 25, 33, 45, 57, 79,
    83, 87, 213–14, 248–9, 276–7
  Dicta 59, 89, 365
  God is 65
  God Known in Judea 65
  Hexaëmeron 1, 179–80, 182–3,
    362, 366
  Meditations 49, 65, 69–71, 80
  Mirror of Confession 65–9
  Notes on Physics 33, 233–5, 254, 256
  On Comets 26, 45, 295
  *On Corporeal Movement and On
    Light* 261
  On Light xii, 1, 84, 193
  On Supercelestial Movement 261, 268
  On the Generation of Sounds 1, 11,
    49, 67, 80
  On the Gifts 70
  *On the Impressions of the
    Elements* 193
  On the Intelligences 78
  On the Liberal Arts v, 1, 6, 11, 49, 59,
    67, 80, 93, 167, 179–80, 193, 206,
    209–11, 227, 236, 240, 245–7, 362
  On the Rainbow 225
  On the Six Differences 260
  On the Soul 80
  On the Sphere see Grosseteste, Robert,
    On the Sphere

416 INDEX

Grosseteste, Robert, works (*cont.*)
   On the Temple of God 33, 49, 65,
     70–1, 76–82, 241
   On the Way of Making of
     Confessions 49, 64, 71–2
   translations of other authors 18, 367
Guala, papal legate, cardinal 22, 38, 43
Guy of Southwick 62

Hallow 38–9
handwriting/manuscripts copied by
   Grosseteste 5, 49–60
   Aristotle, *Physics* 58
   Cambridge, Corpus Christi College,
     MS 480 51, 54
   Cambridge, Pembroke College, MS 7
     51–4
   Cambridge, University Library, MS
     Ff.1.24 51, 53–5
   Oxford, Bodleian Library, MS Savile
     21 16, 50–1, 54–7, 59, 179–84,
     213, 287
Haughmond Abbey 45–7
Henry I, king of England 200
Henry II, king of England 28, 36
Henry III, king of England 19, 21–3, 44
Henry de Bohun, earl of Hereford 19
Henry of Kirksted 84
Henry of Lexington, bishop of Lincoln 30
Herbert Poor, bishop of Salisbury 27–8
Hereford 24–6, 29, 34, 39, 43–5, 48, 61,
   65, 72, 334
   astrology and astronomy in 167,
     175, 179
   as ecclesiastical centre 73–6, 79
   Franciscans at 77
   Jewish community of 76
   St Guthlac's 202
Hermann of Carinthia, *On Essences* 193
Hipparchus 286, 301, 307–8
Honorius III, pope 19, 78
horizon, concept of 263
horologia 184–5
Hubert de Burgh, English justiciar 22–3,
   43–5

Huesca 200
Hugh Mapenor, bishop of Hereford 35,
   44, 72, 74, 76
Hugh of Avalon, bishop of Lincoln 61
Hugh of St Victor 62
Hugh of Santalla 188
Hugh of Wells, bishop of Lincoln 28,
   30–1, 34
Hypomnestikon biblion Ioseppou 53

Ibn al-Haytham (Alhazen) 189
   Shukūk 'alā Baṭlāmyūs (Aporias
     Concerning Ptolemy) 170–1
Ibn al-Muthannā 188
Ibn al-Zarqāllluh *see* al-Zarqālī, Abū
   Ishāq Ibrāhīm
Ibn Bājja *see* Avempace
Ibn Ezra, Abraham 178
   Book on the Principles behind
     Astronomical Tables 292
Ibn Ḥāyyān, Abū Muḥammad
   Jābir (Geber, Jābir ibn Aflaḥ)
     4, 270–1
   De astronomia 212–13, 243, 248–9,
     257–8
Ibn Hibinta 183
illustrations/decorations of *On the*
   Sphere v, 331–5, 347–52, *350,*
    357, 365
   dragons 346–7, 359–60
   initials, decorated *332, 333,* 354
   and instruments, links to 352–61, *355*
   rabbits 346, 359
   use by readers 352–61
   and World Machine diagram 335–46,
     338, 339, 344, 345, 351–2, 359,
     379–82
Innocent III, pope 18, 20–1, 25, 35–8,
   66, 73
Isidore of Seville
   Etymologies 198
   On the Nature of Things 198, 201,
     342, 348
Ivette (Juetta), sister of Robert
   Grosseteste 27–8

INDEX 417

Jābir ibn Aflaḥ *see* Ibn Ḥāyyān, Abū
    Muḥāmmad Jābir
James of Venice, translation of
    Aristotle 59, 215, 217, 219, 225
James of Vitry 79
Jerusalem 18, 20
Jews 292
  in Hereford 76
  in Iberia 199–200
  see also Petrus Alfonsi, *Dialogue*
    Against the Jews
John, king of England 21–5, 38, 44, 182
John of Basingstoke, Master 18
John of Brienne, king of Jerusalem 20
John of Damascus, saint 18
John of Monemuta, Master 376, 378
John of Sacrobosco *see* Sacrobosco,
    John of
John of Salisbury, bishop of Chartres 82
John of Seville, translation of
    Alfraganus 212, 266, 269
John of Worcester 178
John of Wroxeter, Master 46
John Philoponus 225
Jordanus 5, 57, 59
Juliana, wife of Roger Cook 41–2, 45
justice
  papal 35–41
  royal 41–5

Kepler, Johannes 312–14, 324–6
  Astronomia Nova 281
kinematics 169, 305–6
  modern 311–30
  orbits 312–17
  and Ptolemy 308–11
Kingston and Lambeth, peace of 22

Lacey, William de 31
Lacy, Geoffrey de, chancellor of
    Oxford 31
Lacy, Walter de, sheriff of
    Hereford 43–5
Langton, Stephen, archbishop of
    Canterbury 23, 38, 44, 79

Lateran Councils
  Third 64, 72, 75
  Fourth 18, 24, 60, 64, 66, 78, 241
law, canon 37, 48, 62
  papal judicial system 35–7
Leicester 26
Leland, John 58
Lincoln 23–5, 28–9, 32, 37, 59–62, 73
  battle of (1217) 22
Llanthony Secunda 74
Llewelyn ap Iorweth, Prince of Wales
    and of Gwynedd 43–4
Lombard, Peter, *Four Books of*
    Sentences 62
London 22
Louis VIII, king of France 21–2, 25
Lucan 196
  Pharsalia (*Civil War*) 195
Ludlow 72

machina mundi (world machine) 4,
    10, 82, 89–90, 92–3, 95, 111,
    246, 360, 362
  commentary on model of 131–65
  diagram of 335–46, 351–2, 359, 379
Macrobius 190–1, 196, 200–1
  Commentary on the Dream of
    Scipio 295
Magdeburg 186
Magna Carta 21–2, 43
Marsh, Adam 27, 30–1, 58
Martianus Capella 190
  The Marriage of Philology and
    Mercury 192, 197–8
Māshā'allāh, *Conjunctions, Religions*
    and Peoples 183
Maslama al-Majrīṭī 176–8
Matthew Paris 21, 25
Mecca 168
Menelaus of Alexandria 287
  Spherica 169
Michael Scot, translator of
    al-Biṭrūjī 5, 12–13, 15–16,
    212, 353
mnemonic verses 2, 14, 195, 204

418 INDEX

Nettleham 375, 377
Newton, Isaac 311
Nicholas, chancellor of Hereford
    cathedral 75
Nicholas of Tusculum, papal legate
    23, 38
Nicholas the Greek 18
Northampton 27–8

observation, role of 236–9
Oliver Sutton, bishop of Lincoln 23,
    30–4, 376–8
On the Motion of the Eighth Sphere
    see Pseudo-Thābit
On the Solar Year 298
Ordered Universe Project v, vii–viii, ix,
    84, 88, 364
Osbert, abbot of Haughmond 46
Otto II, Holy Roman Emperor 186
Ovid 14, 93, 115
  From Pontus 195
  Metamorphoses 195
Oxford 14, 23–6, 29–34, 58–9, 334,
    375–8

Pacificus of Verona 185
Pandulf, papal legate 22, 38, 43
Paris 28, 62, 200, 302
  St Victor 67, 77, 202
  University of 13–16, 25, 31, 33, 37,
    45, 60, 66, 78
Paschal II, pope 36
pastoral care 62–5, 241
  audience and intention of
    Grosseteste 71–6
  of Grosseteste 49, 60–2, 64, 82, 365
  works of *pastoralia* of
    Grosseteste 65–71
Pelagius, papal legate 19
Peter Abelard 191
Peter des Roches, bishop of
    Winchester 20, 22, 43–5
Peter of Abergavenny 75
Peter of Medbourne, Master 375–8
Peter the Chanter 62

Petrus Alfonsi 168, 259, 302
  Dialogue Against the Jews 199–202,
    296
  Letter to the Peripatetics 200
Petrus Collivachinus of Benevento,
    Compilatio tertia 37
Philip II Augustus, king of France 21
Philip of Aubigné, tutor of Henry III of
    England 19–20
Pisa 178
Plato 192
  Timaeus 190, 197, 349
Pliny the Elder 174, 196, 201
  Natural History 295
Pontesbury 72
precession 3, 94, 156, 159, 177–8, 188,
    279–94, 283, 297–9, 301–4,
    307–9, 364
Pseudo-Bede, *On the Constitution of
    the Celestial and Terrestrial
    Worlds* 194
Pseudo-Dionysius 18
Pseudo-Thābit, *On the Motion of the
    Eighth Sphere* (*De motu octave
    spere*) 4–5, 13, 16, 57, 205, 213,
    243, 268, 289–91, 293–4, 298,
    300–2, 363
Ptolemy 2–3, 92, 94, 167–70, 173–4,
    177, 248, 159–61, 300–2, 306,
    363–4, 366, 370
  Almagest 4–5, 12, 125, 156–7, 169,
    171, 175–6, 178, 196–7,
    201–2, 205, 212–14, 232, 243,
    249, 257–8, 268, 275, 283,
    286–7, 291, 293–4, 298, 305,
    307–11, 323
  astronomical tools of 188–9
  Handy Tables 287
  lunar theory of 310–11
  Optics 171
  Planetary Hypotheses 171
  planetary theory of 308–10

Ranulf, earl of Chester 19–20
Raymond of Marseilles 179, 187–8

INDEX 419

Raymond of Peñafort, *Decretals* 37
Reading 28
Reginald de Braose 44
retrogradation 12, 167–8, 173–4, 177, 183, 196–8, 307, 309, 322
Richard, Master, bishop of Hereford's official 46
Richard Ilchester, bishop of Winchester 28
Richard Poor, bishop of Salisbury 28
Robert, bishop of Lincoln 30, 58, 376, 378
Robert de Cotinton (Coddington), Master 41–2
Robert of Chester 178
Robert of Courson 79
Robert of Flamborough, *Liber poenitentialis* (*Book on Penitence*) 67, 71, 77, 82
Robert of Ketton 178
Robert of Northampton 190
Robertus Anglicus 192
Roger le Fevre 41–2
Roger of Hereford 175, 179, 189–90
Roger of Martival, chancellor of University of Oxford 375–6
Roger of Weaseham, Master, chancellor of University of Oxford 375–6
Rudolf of Bruges 187
Rufinus of Aquileia 63

Sacrobosco, John of 173, 271, 276
  Algorismus 13, 88
  biographical info 14–15
  Compotus 13–16
  On the Sphere 1–6, 13–17, 82, 84, 87, 167–8, 192–8, 204–5, 212, 242–3, 249, 258, 260–1, 264, 338, 341–2, 348, 353, 357, 363, 379
Saer, earl of Winchester 19
Ṣā'id al-Andalusī, qāḍi of Toledo 288
St David's 74
Salah al-Din, sultan of Egypt and Syria 18
Salisbury 27–30

Salop, Shropshire 38–9
Sandwich, Kent 22
Santiago de Compostela 44–5
Sapey, Shropshire 38
Seneca 196
Shobdon 77, 202
Silvester II, pope *see* Gerbert of Aurillac
Simon of Melun 75
Simon of Montfort 21
simony 80
situs of universe 89, 95, 245–7, 262, 265, 267–8, 277–8
Skenfirth 43
Soissons, Council of (1121) 191
sphericity and demonstration 249–60, 365–6
Stottesden 72
subalternation 223–4, 226–9
Suda 18
sun and zodiac 264–9, 275–6

Tables of London 292
Tables of Pisa 292
Tables of Toledo 175, 179–80, 182, 212, 288, 290, 294, 298, 300–2, 363
Thābit ibn Qurra (Thebit) 2–4, 6, 12, 16, 57, 59, 94, 157, 179, 290, 293–4, 298, 300; *see also* Pseudo-Thābit, *On the Motion* (*Movement*) *of the Eighth Sphere*
Thebes 17
Thebit *see* Thābit ibn Qurra
Themistus 225
Theodosius, *Spherica* 247, 249
Theon of Alexandria, *Little Commentary* 287
Theorica Planetarum 88, 212–13, 243, 266, 269, 357
Thessaloniki 17
Thierry of Chartres 191, 363
Thomas, Master, precentor of Hereford 46
Thomas Becket *see* Becket, Thomas, archbishop of Canterbury

420 INDEX

Thomas de Melsonby, prior of Durham 52
time, intervals of 264–7
Toledan Tables see Tables of Toledo
Toledo 179, 200
trepidation 4–5, 13, 17, 94, 157–8, 177, 179, 195, 279–82, 288, 290–2, 299, 300–4, 363, 370, 373, 380
  historical observations of 300–4
  and Virtual Celestial Model 370
Trevet, Nicholas 10, 33, 58, 83, 85, 192

Udidjayn see Arim
usury 79–80

Vacarius, Master 37
Venice 17
Virgil 14, 93, 115
  Georgics 195
Virtual Celestial Model vii, 6, 306, 365
  eccentricity 370
  as geometric construction 368
  instructions for use 371–4
  moon and eclipses 371
  seasons 369
  sun 368–9
  trepidation 370
  zodiac 370

Walcher, prior of Malvern 136, 187
  On the Dragon 200
Walter of St Edmund 77

Wells 28–9
Wenlock 72
Whitecastle 43
Wigmore 77
William, abbot of Haughmond 46
William, earl of Derby 19
William, son of Ralph of Hallow 38–9
William de Brailes, manuscript artist 334
William de Montibus 24, 60–2, 73, 82
William de Vere, bishop of Hereford 35, 37, 41, 61–2, 71, 202
William Marshal, earl of Pembroke 22, 43
William of Alnwick, Franciscan lector 59, 335
William of Auvergne 26
William of Conches 348, 363
  Dragmaticon 190, 192
  Glosses on Macrobius 190–1
  Glosses on the Consolation of Boethius 190–1
  Philosophy of the World 190
William of Kilpeck 74
William of Nottingham 58
Worcester 37–9, 178
world soul and elements 191–3

Yazdegird III, Sasanian ruler 176
York 24

ziges 176–8, 287